ROYAL COMMISSION ON ENVIRONMENTAL POLLUTION

CHAIRMAN: SIR TOM BLUNDELL FRS, FMedSci

Twenty-fifth Report

TURNING THE TIDE:

ADDRESSING THE IMPACT OF FISHERIES ON THE MARINE ENVIRONMENT

Presented to Parliament by Command of Her Majesty
December 2004

Cm 6392 £60.00

More information can be obtained from http://www.rcep.org.uk or from the Secretariat at Third Floor, 5-8 The Sanctuary, Westminster, London SW1P 3JS.

ROYAL COMMISSION ON ENVIRONMENTAL POLLUTION

TWENTY-FIFTH REPORT

To the Queen's Most Excellent Majesty

MAY IT PLEASE YOUR MAJESTY

We, the undersigned Commissioners, having been appointed 'to advise on matters, both national and international, concerning the pollution of the environment; on the adequacy of research in this field; and the future possibilities of danger to the environment';

And to enquire into any such matters referred to us by one of Your Majesty's Secretaries of State or by one on Your Majesty's Ministers, or any other such matters on which we ourselves shall deem it expedient to advise:

HUMBLY SUBMIT TO YOUR MAJESTY THE FOLLOWING REPORT.

'The light produced a thousand charming varieties, playing in the midst of the branches that were so vividly coloured. I seemed to see the membraneous and cylindrical tubes tremble beneath the undulation of the waters. I was tempted to gather their fresh petals, ornamented with delicate tentacles, some just blown, the others budding, while a small fish, swimming swiftly, touched them slightly, like flights of birds'.

Jules Verne, *20,000 Leagues Under the Sea*, 1870

'...the great and long iron of the wondrychoun [trawl] runs so heavily over the ground when fishing that it destroys the flowers of the land below the water there...'

Commons petition to the King of England, 1376
Auster *et al*, 1996

CONTENTS

Chapter 4

THE LEGAL FRAMEWORK FOR MARINE ENVIRONMENT AND FISHERIES

Chapter 5

A LEGACY OF OVERFISHING

Chapter 6

IS AQUACULTURE THE ANSWER?

Chapter 7

UNDERSTANDING THE IMPACTS OF FISHING ON THE MARINE ENVIRONMENT

Chapter 8
MARINE PROTECTED AREAS

Chapter 9

IMPROVING FISHERIES MANAGEMENT

Chapter 10

BRINGING ABOUT RADICAL CHANGE

Chapter 11

Recommendations

Note: Appendices can be found by referring to the CD-ROM of the full report.

xi

INFORMATION BOXES

Chapter 1

INTRODUCTION

CRISIS IN THE MARINE ENVIRONMENT

1.1   The marine environment is vast. Seas cover over 70% of the planet and they play an important role in determining conditions on the remaining 30%. They teem with myriad life forms in complex relationships. The unspoilt beauty of the ocean is largely hidden from us. We can only see it second hand through television programmes or reproduced in aquaria. These can never capture its scale, or even show some of its larger creatures (figure 1-I).

Figure 1-I
One of the larger species in UK waters: the basking shark[1]

Figure 1-II
Marine hydrothermal vents with thriving tubeworms[2]

1.2   The size of the ocean and the harshness of conditions in the deep sea mean that more is known about the surface of the moon than the bottom of the deep ocean. Many deep-sea species, possibly running into the millions, remain unstudied and even very large organisms like the great squid remain a mystery. It is only a few decades ago, with the discovery of hydrothermal vents and associated life forms, that we first discovered an ecosystem that is not ultimately dependent on sunlight (figure 1-II). We simply do not know what is there and know even less about how it all works in complex ecosystems.

1.3   Humankind has long interacted with the sea. Coastal communities have exploited marine life as a valuable source of food. The oceans have been a barrier to exploration which has been overcome. The sea can be a source of food and of beauty but also a fickle friend that can change to devastating destructiveness. The power of the sea helps to form our land and weather and may one day be harnessed to provide the energy we need for modern society.

1.4   The power and vast size of the ocean might give the impression that unlike the land it is invulnerable to any damage that humans do to it (figure 1-III). But that is not the case. Although at one time the human impact on the sea was limited largely to navigation and fishing in coastal waters, we now have the technology and the scale of activity to have significant effects.

1.5 The impacts take many forms. Land-based activities, including agriculture, can raise nutrient levels putting at risk whole ecosystems like coral reefs. Industrial pollution can render seafood in affected areas unfit to eat. Increasing levels of carbon dioxide are increasing the acidity of the sea with unknown long-term consequences. But the most direct effect is undoubtedly that of fishing. Our report focuses primarily on fishing in the seas around the United Kingdom.

1.6 OSPAR (the Convention for the Protection of the Marine Environment of the North East Atlantic) covers the area from 36°N to the North Pole, and from the 42°W meridian off the Atlantic Coast of Europe, to 51°E in the Arctic ocean (figure I-IV). It has ranked human activities in the North Sea in terms of the severity of their impact on the envrionment. Three of the top six relate directly to fisheries. The other three are addressed by controls elsewhere and are not the subject of this report (table 1.1).

Figure 1-III
Power of the Ocean[3]

Figure 1-IV
The OSPAR area[4]

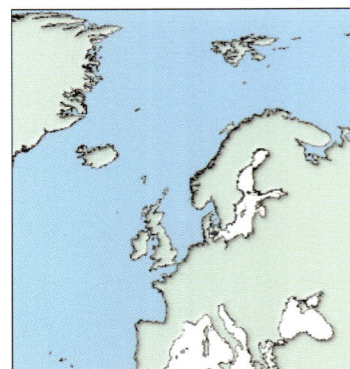

Table 1.1

Priority classes of human pressures based on OSPAR[5]

Activities attributable to fisheries are shown in italics

Human Pressure	Category
Removal of target species by fisheries	*Fisheries*
Inputs of trace organic contaminants (other than oil and PAHs from land)	Trace organic contaminants
Seabed disturbance by fisheries	*Fisheries*
Inputs of nutrients from land	Nutrients
Effects of discards and mortality of non-target species by fisheries	*Fisheries*
Input of TBT and other antifouling substances by shipping	Trace organic contaminants

1.7 A recent study reported that 90% of large predatory fish have been lost from the world's oceans with declines in entire communities across varying ecosystems.[6] The waters of northern Europe have been part of that global disaster. The fishing industry has contracted. But where it remains, fishing uses increasingly sophisticated technology. The propensity of many fish species to aggregate in schools has made it possible to catch those that are left, and has even given some fishers the impression that populations have not fallen as dramatically as more systematic scientific studies have shown.

1.8 Although the seas seem boundless and fishing vessels small by comparison, the nets can be large. In one recently reported case a net with a mouth of 40,000 square metres (50 football fields) has been introduced.[8] The relentless pressure of fishing has meant that the deep ocean and the seas around our coast are being depleted of fish and other living creatures at an alarming and unsustainable rate. Bottom trawling can plough furrows two metres wide and up to 30 cm deep, mile after mile across the seabed. Some areas are trawled this way five times a year.

Figure 1-V
Scallop dredging[7]

Those areas of the seabed are turned over much more often each year than arable fields. This has had a disastrous effect on the three dimensional habitat which is important in its own right and for the species that live there (figures I-V, 1-VIa and 1-VIb).

Figure 1-VIa
Fishing impacts in areas with low natural disturbance – before trawling[9]

Figure 1-VIb
After trawling

1.9 This report, which is concerned with protection of the marine environment, focuses on the impact of marine fishing as the greatest individual threat to that environment in the seas around the UK. We consider the impact of the fishing industry on the fish themselves, and more importantly, on the wider ecosystem. Marine ecosystems in the area with which we are concerned are described in chapter 2.

1.10 Chapter 3 of our report looks in some depth at the fishing industry, why we eat fish and the methods the industry uses. Fishing is a form of hunting. That conjures up a misleading image from land based hunting of a well-targeted activity that impinges little on other species. Some forms of fishing are like that, but most are not; collateral damage to other caught fish and the environment can be severe. Fishing can be likened to unsustainable tropical logging where a tree is cut down and the surrounding area devastated in the effort to remove the timber. We need to care as much for the seas around us as we do for the rainforests.

1.11 Fish is a traditional part of the diet in many communities. There is a growing appreciation that long-chain n-3 polyunsaturated fatty acids, found in oily fish, are needed for optimal health and should be part of a balanced diet. We recognise the health benefits of these fatty acids, but simply encouraging increased consumption of fish ignores the problem that existing global consumption levels are unsustainable. Some of these fatty acids can be produced from other sources, for example, marine algae and taken as a supplement. In our view, more effort should be put into developing sustainable sources of long-chain n-3 fatty acids and into assessing their health benefits.

1.12   Chapter 4 of the report looks at the current legal position in relation to managing the marine environment and fisheries. It describes rather ad-hoc arrangements at the global, European and national level. On the conservation side, there are measures to protect specific species and a raft of more general measures, often largely aspirational. There is a separate legal framework dealing with fisheries and little linkage between the two. At EU level, fisheries effectively take precedence over environmental legislation. The end result is that the environmental declarations have not had much impact other than, for example, some restrictions on sandeel fisheries to protect food supplies for sea birds. Despite recent attempts to promote a more holistic and ecosystem based approach, fishery regulations are largely devoted to trying, often unsuccessfully, to manage the maximum sustainable yield of particular species of fish.

1.13   In 2002 the UK government published its first marine stewardship report *Safeguarding our Seas.*[10] This endorsed the use of an ecosystem approach, which it defined as 'the integrated management of human activities based on knowledge of ecosystem dynamics to achieve sustainable use of ecosystem goods and services and maintenance and ecosystem integrity." Little has actually happened so far except for the protection (under the Common Fisheries Policy) of a very small area, the Darwin Mounds (*ca* 0.2% of UK territorial waters), off the west coast of Scotland to conserve a particularly sensitive area of cold-water coral. These were only discovered in 1998 (figure 1-VII). Protecting the Mounds set a useful precedent but protecting even such a small area was controversial and was implemented only just in time to save them from damage by bottom trawling.

1.14   Chapter 5 describes how fishery policies have failed even in terms of their own narrow objectives. The worst case undoubtedly has been the, perhaps irreversible, collapse of the Grand Banks cod fishery, once the world's most prolific. In many areas around the world, quotas on the amount of fish which can be caught have been set too high, and there is often widespread evasion. Catches are significantly greater than quotas. More intense enforcement of the existing system is not the answer; a different approach which is more easily enforced is necessary.

Figure 1-VII
Deepwater corals (*Lophelia* and *Madrepora*) at Darwin Mounds[11]

1.15   UK waters and EU waters outside the Mediterranean fall into an area covered by the International Council for the Exploration of the Seas (ICES). This body co-ordinates and promotes marine research in the North Atlantic including adjacent seas such as the Baltic and the North Sea. The fact that marine science and fisheries has been operating on this international basis since 1902 illustrates the international nature of the problem. ICES have assessed that from 1996 to 2001 the proportion of fish stocks in their area that are within safe biological limits, fell from 26% to 16%. The absolute numbers are shocking with less than one species in six deemed to be safe. The rate of decline shows that the situation is not only disastrous, but rapidly moving to catastrophic. Many populations are now at less than 10% of the previous levels while some have completely collapsed. Once abundant fisheries such as cod are now at such low levels that in many areas there is no commercial fishing at all. The fish are just no longer there.

1.16 This is a global problem affecting not just the North Atlantic. A few countries have woken up to this and have begun to act. South Africa and New Zealand have targets for marine protected areas covering between 10 and 20% of their territorial waters. Australia has also taken a conscious decision not just to conserve 33% of a hot spot like the Great Barrier Reef, but other areas as well. In the US, a similar realisation of the depth of the crisis has led to two recent reports (one commissioned by Presidential order) that recommend radical action.

1.17 The very significant changes that people have inflicted on the marine ecosystem have led not only to lower catches of fish. Compared with fifty years ago, the fish we eat are smaller than they were and there has been a change in the type of fish we eat (figure 1-VIII). This is not just a consequence of an increasingly cosmopolitan diet; it also reflects the fact that there simply is not so much of the traditional fish available. Cod, which used to be commonplace and cheap, is now expensive. Traditionally, the fish we ate tended to be the top predators in the food chain, but as these are fished out, fishing has moved further down the food chain to exploit different species (for example, from cod to prawns). Some fish populations increased temporarily with the removal of predator pressure, but are now themselves being heavily fished.

1.18 Chapter 6 of our report explains why fish farming cannot easily expand to meet any potential shortfall in the supply of wild fish. The types of fish farmed in temperate climates are carnivorous and depend on fishmeal for a significant proportion of their diet. This is not

Figure 1-VIII
The average size of cod has fallen over the last fifty years[12]

true of shellfish where production could increase. The difficulty here is to find new sites given the need for water of a certain quality and the other priorities for the use of potential sites such as leisure and landscape.

WHAT NEEDS TO CHANGE

1.19   Chapter 7 and subsequent chapters look at the changes that are urgently needed to deliver a significant improvement in the way in which the marine environment and fisheries are regulated. Marine fishing is close to collapse in many areas, and behind the figures on the fish populations lies severe degradation of the marine environment. This is evident from pictures of the damage that is caused by fishing to fragile marine structures, some of which can take many years to re-establish.

1.20   We recognise that there is a considerable degree of scientific uncertainty in our understanding of marine ecosystems. Our report recommends changes in the emphasis of research away from management of fish populations towards a wider focus on the marine environment. Although more work needs to be done to improve our understanding of the marine environment, it is clear that significant damage is being done now. We cannot wait for the results of the research. **We need to adopt a precautionary approach and not use scientific uncertainty as a reason for delay.**

1.21   Chapter 8 therefore looks at the protection of marine areas; the obvious analogy being with the regulation of development on land. We see such protection as a logical consequence of the intensifying exploitation of the seas. Without it, we will be accepting degradation of the marine environment to a degree that we would not tolerate on land. Protection for the marine environment needs to catch up.

1.22   The impact on the marine environment of the changes we propose needs to be studied and the measures adopted need to be adjusted in the light of what it discovered. But the evidence we present on the degree of damage means that national and EU regulators need to act. Otherwise there will be so little left to preserve and to study that we will have lost the opportunity to act. Once environments are lost they may be lost for generations; some may vanish forever.

1.23   The changes we propose in fishing management and technologies are set out in chapter 9, while chapter 10 concludes with a more detailed look at the overall new framework for fisheries in the marine environment recommended in the report. Our recommendations are radical – but are in the same direction as the recent report by the Prime Minister's Strategy Unit that focused on fishing.[13] Our recommendations go further however: in our view the situation is so serious that we need a radical solution; incrementalism will deliver too little too late. Such change will be painful for the industry but a continued policy of overly timid measures will not serve the best interests of fishers in the longer term. Not only will we have done irreparable damage to the marine environment but we will also have a smaller fishing industry longer term than we could have if firm action is taken now.

1.24 Our recommendations are summarised in chapter 11. We recommend measures that will need to be introduced across the EU. We recognise that this will not be easy. Fisheries policy like any other will continue to be decided at a political level. In our first report,[15] we said "Public opinion must be mobilised in such a way that elected representatives regard themselves as trustees for the quality of air, water and the landscape." Much progress has been made since that was written in 1971, but protection of the marine environment has lagged far behind. The fact that we cannot easily see what damage is being done means that damage to the ocean is a less obvious public concern than damage on land except for special issues like marine mammals (figure I-IX). This needs to change so that the public understand what is happening and what is at stake. There is a role for greater public engagement and education. Without this, the political process will continue to be dominated by pressure from particular interest groups and will not deliver the radical changes necessary to save the marine environment.

Figure 1-IX

Dolphins are at risk of bycatch[14]

1.25 This report comes after a series of reports and books (e.g. *In a perfect ocean* by Daniel Pauly[16] and *The end of the line* by Charles Clover[17]) both in the UK and worldwide, that have called for action. We make no apology for adding to the list. The situation is worse than many thought, and the need for change is urgent. It has been said that "The oceans are the planet's last great living wilderness, man's only remaining frontier on Earth, and perhaps his last chance to prove himself a rational species."[18] Our report is intended to inform that process and recommends a package of measures to discharge humankind's stewardship of the oceans to enable future generations to enjoy a diverse and healthy marine environment and to continue to have fish to eat.

Chapter 2

THE MARINE ENVIRONMENT

The marine environment is a complex interconnected system. It comprises the physical aspects of the coastline, seabed, water column, water surface and the overlying atmosphere, as well as a variety of marine life, habitats and ecosystems. What do we know about marine ecosystems and their vulnerability?

INTRODUCTION

2.1   Oceans and seas cover over 70% of the Earth's surface and contain 90% of the biosphere; the regions of the Earth's surface and atmosphere where living organisms exist. They are thus a major source of natural resources and biodiversity. The marine environment is of enormous scale and importance with a physical, chemical and biological complexity that we have barely begun to understand. It has been often said that we know more about the surface of the moon than the bottom of the deep oceans. In this chapter we examine the extent of our knowledge of these complex ecosystems and their relation to the broader marine environment.

DESCRIPTION OF OSPAR AREA

2.2   This report covers the five regions of the OSPAR Convention area (including most of the north-east Atlantic Ocean), with a specific focus on the area around the UK (figure 2-I). The Atlantic Ocean began to develop as Earth's last supercontinent, Pangea, broke apart. A fissure formed during the Jurassic period (206-144 million years ago), dividing the American continent from what became Europe and Africa. The Atlantic Ocean continued to expand from this early rifting, and the seismically active zone, where new oceanic crust is formed, now constitutes the mid-ocean ridge. The Atlantic Ocean is still expanding in an east-west direction by a couple of centimetres a year. The mid-ocean ridge system is a huge volcanic chain, with Iceland thought to represent the highest point over an underlying mantle plume. The ocean can be divided into three distinct regions: the coastal continental shelf, the continental margin and the deep oceanic basin (figure 2-II).

Figure 2-I
Three-dimensional bathymetric Map of OSPAR area showing the mid-Atlantic ridge[1]

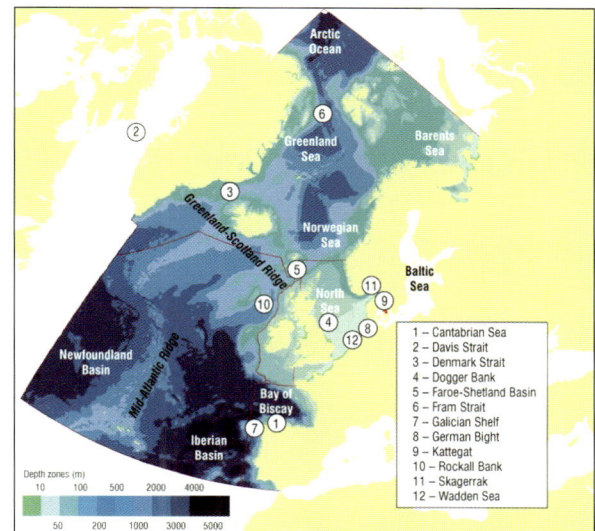

1 – Cantabrian Sea
2 – Davis Strait
3 – Denmark Strait
4 – Dogger Bank
5 – Faroe-Shetland Basin
6 – Fram Strait
7 – Galician Shelf
8 – German Bight
9 – Kattegat
10 – Rockall Bank
11 – Skagerrak
12 – Wadden Sea

2.3   The continental margins consist of large wedges of sediments ranging from sand to mud up to 10 km thick that extend down into deeper sea basins. The continental shelf stretches from the shoreline to a depth of about 200 m. It is covered with sediments; in the North and Irish Seas much of it is of glacial terrestrial origin.[2] The most extensive continental shelf areas in the OSPAR area are in the North and Celtic Seas (see appendix E for a detailed description of the seas around the UK).

2.4   The depth of the light (or euphotic) zone (box 2A) depends largely on the concentration of organic and inorganic materials dissolved or suspended in the water column. In coastal zones, light will penetrate only a short distance compared with the open ocean or tropical waters where there is little suspended material and concentrations of plankton are sparse. The maximum visible depth from above the sea surface is typically a few tens of metres. Surface or satellite measurement in most other wavebands can sense only the first millimetre or so. Other types of measurement are made difficult by the fact that the pressure increases with depth by 1 atmosphere for every 10 metres, as well as by the corrosive nature of salt water and often hostile wave and wind conditions. Importantly, this means that the impact of human activity on the marine environment often goes unseen and unrecognised.

2.5   The deepest part of the ocean, ranging in depth from 2,000 m to 5,000 m, consists of an abyssal plain. This is a flat area on the sea floor having a very gentle slope of less than one metre per kilometre, extending from either side of the mid-ocean Atlantic ridge. The deep-sea floor consists of bare rocks, which outcrop in places, and muddy sediments up to 2 km thick, made up principally of the remains of microscopic organisms and clay-sized particles. In some places

Figure 2-II

Different regions of the ocean[3]

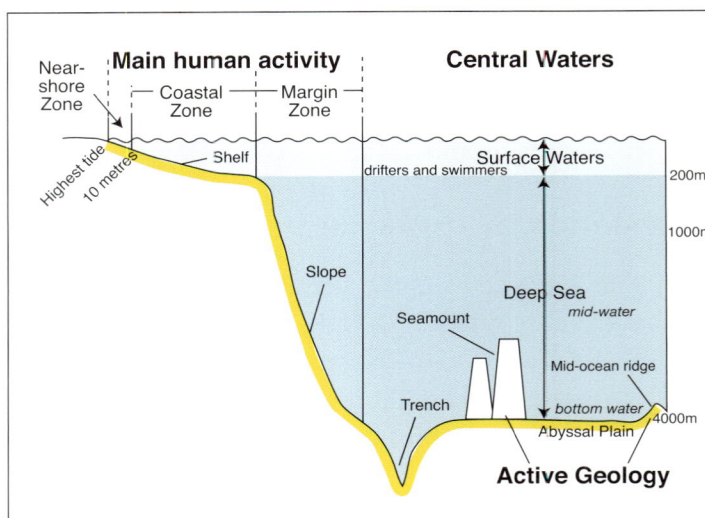

seamounts occur as submerged single mountains or chains of mountains on the ocean floor.

MAJOR OCEAN CURRENT SYSTEMS IN THE OSPAR AREA

2.6   Currents are influenced by prevailing winds, the rotation of the earth, the moon's gravity and temperature. The major currents in the ocean are driven from its surface by solar heating and by interaction with the atmosphere which lead to temperature contrasts. The difference between the evaporation and precipitation of water, and also the melting and freezing of sea ice, leads to salinity contrasts. The combined effect of these thermal and salinity contrasts is to drive the overturning circulation of the ocean known as the Thermohaline Circulation. This circulation brings relatively warm water into the high latitudes of the north-east Atlantic where, particularly in the Norwegian and Greenland Seas, the water is dense enough to sink about two to three kilometres and then return south as North Atlantic Deep Water.

2.7   The winds blowing over the ocean lead to an exchange of momentum that also drives major currents in the ocean. The subtropical, easterly Trade Winds, along with the mid-latitude westerlies, drive a clockwise circulation in the ocean, The Subtropical Gyre, with southward drift in the open ocean and a compensating strong northward current on the western side. In the North Atlantic this is the Gulf Stream. The Gulf Stream transports warm, salty, nutrient-rich near-surface waters northward near the East Coast of the US at a flow rate (50 million cubic metres per second) more than a hundred

Figure 2-III

Major surface currents in the North Atlantic, the mean flow rates of the currents are shown on the diagram[4]

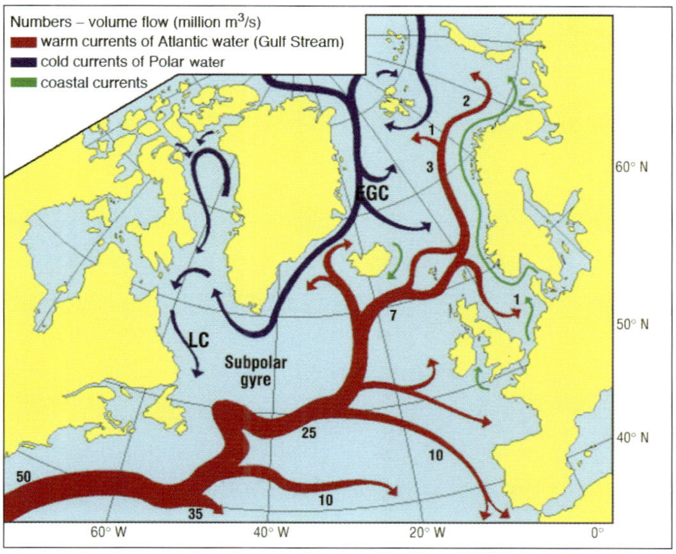

times that of the Amazon River. The Gulf Stream separates from the coast near Cape Hatteras. Much of its waters recirculate southwards. However, associated also with the Thermohaline Circulation, some waters continue north-eastwards towards north-west Europe as the North Atlantic Drift.[5] The mid-latitude westerly winds and higher latitude easterlies drive a weaker anti-clockwise Subpolar Gyre including a cold current down the east coast of Greenland and the coast of Labrador (figure 2-III).

2.8   The topography of the ocean bed is extremely important in determining the details of the flow. On the continental shelf there are important currents driven by local processes and by interaction with the deep ocean. The deep circulation of the north-east Atlantic is influenced by Thermohaline Circulation and the bathymetry of the ocean basins. Typically the net flow is southwards via North Atlantic Deep Water, but closer to the continental margin, and in the vicinity of important fishing grounds, substantial deeper currents such as the north-west Scotland Slope Current flow in a north-easterly direction. It is the interaction between the circulation of waters of the shelf seas with that of the deep ocean that is reflected in everything from the supply of deep-water nutrients, to the distribution of larval fish, to the migration pathways of important pelagic species and cetaceans.[6]

2.9   Minor currents such as those in the seas west of the UK have local effects on marine ecosystems, through the movement of nutrients, pollution and plankton, and so also influence fish behaviour. Tides driven by the gravitational pull of the moon are also important in producing mixing of the water column and the nutrients within it.

CLIMATE SYSTEMS

2.10 The ocean plays an important role in determining the composition of the atmosphere and in the climate system. Algal photosynthesis in the ocean is a major source of atmospheric oxygen. A significant portion of the carbon dioxide produced by human activities is removed from the atmosphere by exchange with the ocean. The heat, water and momentum exchanges between the atmosphere and the ocean play a vital role in climate with relation to wind speed, evaporation rates and rainfall levels.

2.11 Two and a half metres of seawater contains as much heat as the whole atmosphere above it. This large heat content reduces the difference between winter and summer temperatures in maritime regions such as north-west Europe. The warm water brought to the north-east Atlantic in the Thermohaline Circulation ensures that north-west Europe is warmer than north-east America or indeed Alaska, since the Pacific does not have a comparable Thermohaline Circulation. There is some concern that the Atlantic Thermohaline Circulation may weaken with global warming, thereby removing the special warmth of north-west Europe.

2.12 The general westerly winds over the north-east Atlantic are associated with low pressure in the Iceland region and high pressure in the Azores region. One of the major modes of atmospheric variability in the Atlantic region is a fluctuation in the strength of these westerlies and of the related pressure field. This North Atlantic Oscillation (NAO) has associated changes in the strength and region of storms, precipitation and temperature anomalies.

Figure 2-IV

A schematic diagram showing the components of the North Atlantic Oscillation (NAO) in high or positive mode[8]

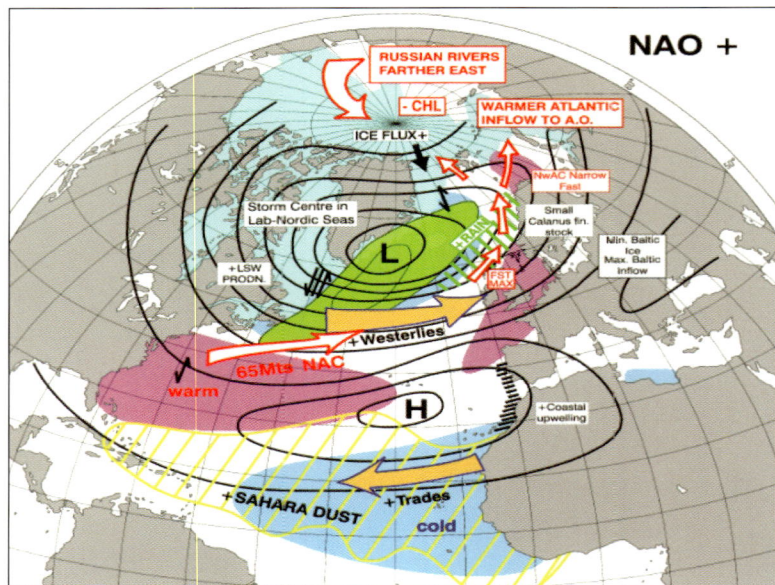

All these changes lead to changes in the marine environment in the north-west Atlantic region (figure 2-IV).

2.13 The NAO has been mostly very positive since 1990 and there is current scientific debate over whether this is an indication of anthropogenic climate change or an extreme period of natural fluctuation. Associated with this extreme NAO behaviour have been changes in the physical and biological state of the northern Atlantic marine environment.[7]

PHYSICAL, CHEMICAL AND BIOLOGICAL INTERACTIONS

2.14 Biological activity in the sea is driven by the interaction of chemical processes and physical forces (global ocean circulation, currents, tides, day/night cycle and weather). The Convention on Biological Diversity defines 'ecosystem' as a dynamic complex of plant, animal and micro-organism communities and their non-living environment interacting as a functional unit.[9] The term marine ecosystem is used to refer to all the marine organisms in an area together with the physical, chemical, and structural features of their environment.

2.15 Marine ecosystems are affected by long-term environmental processes such as climate change over thousands of years, medium-term weather and climate variability, such as the North Atlantic Oscillation, that can last from a few weeks to a decade, and short-term processes, such as tides and daylight, that occur on daily, seasonal, or annual cycles.[10,11] Ecosystems are generally adapted to cope with short-term environmental variability and fluctuation, but when there are strong longer-term trends in the environment, organisms may change their geographical distribution and so become locally extinct or abundant.

2.16 Most marine life derives energy from sunlight and/or nutrients (the exception being various organisms that are part of marine ecosystems driven by chemical energy sources available at deep-sea volcanic vents). Living organisms in the sea therefore respond to local processes that determine delivery of nutrients to the sunlit surface waters that in turn determine the amount of primary production (2.17).

TROPHIC LEVELS AND THE FOOD WEB

2.17 In some regions, deep oceanic waters carrying nutrients come to the surface and there is abundant marine life. Such 'upwellings' can result from several different processes. They can occur for example where cold, deep ocean currents on the ocean-floor encounter a physical underwater obstruction, such as a seamount, continental shelf, or offshore bank, as a result of surface wind stress, or as the result of the divergence of major surface currents in the open ocean. On the continental shelf, stirring by tides and by winds can also distribute nutrients throughout the water column fostering the development of rich marine ecosystems. The boundary between major current systems is often marked by a 'front', which acts as a zone for accumulating particulate material including food and larvae of marine organisms.

Figure 2-V

A simplified marine foodweb, indicating the stage or trophic level in the food web occupied by various types of organism, which is determined by the number of energy-transfer steps to that level, leading from primary producers (lowest trophic level) through herbivores to primary and secondary carnivores[12]

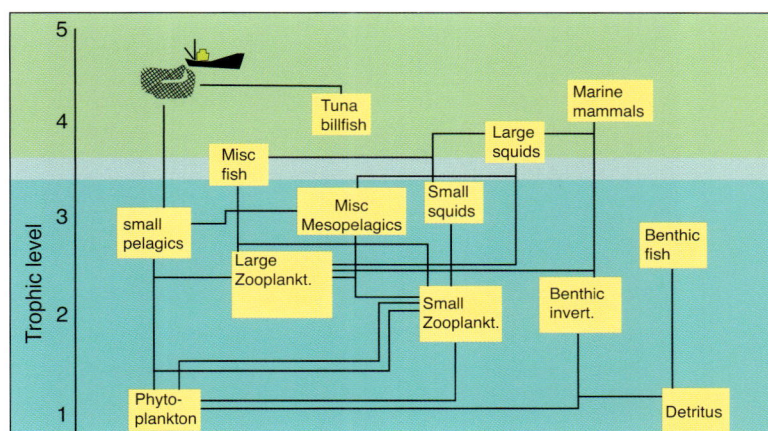

BOX 2A   **PHYTOPLANKTON: PLANTS AT THE BASE**
OF THE FOOD WEB

Most plankton occur in the top 20 m of the sea, within the euphotic zone (where light penetrates). This zone varies in depth, depending on turbidity and light intensity. It generally extends to around 100 m in the tropics and 50 m or so in temperate zones. The factors that initiate phytoplankton growth in spring are vertical mixing and stratification of the water column, along with increasing day length. In the north Atlantic, the water becomes warmer and stratified in the spring, and causes a bloom of diatoms (single-cell algae, figure 2–VI). As summer comes, surface waters warm and a more permanent thermocline develops. The thermocline is the boundary where the greatest vertical change in temperature occurs. It is the transition zone between the layer of warm water near the surface that is mixed by tides and wind and the cold deep-water layer. Colder, nutrient-rich waters sink away from the euphotic zone; primary production slows and tends to be confined to deeper layers near this boundary. The resulting phytoplankton community is one that can cope with reduced nutrient levels. When silica becomes limited, other groups of phytoplankton, such as flagellates, bloom. With autumn, and the increase in wind strength, the sea becomes mixed once again, and at this time a secondary bloom, of dinoflagellates (a unicellular microscopic plant, figure 2-VII) may occur. As the light levels diminish later in the year, phytoplankton growth is inhibited and primary production once again decreases. The water mixes over winter aiding the distribution of nutrients throughout the water column.

Figure 2-VI
A colony of the diatom *Eucampia zodiacus,* which can form long chains of cells in helices several millimetres in length. It is one of the component species of the phytoplankton spring bloom in the southern North Sea (scale 1mm=7.14µm)[13]

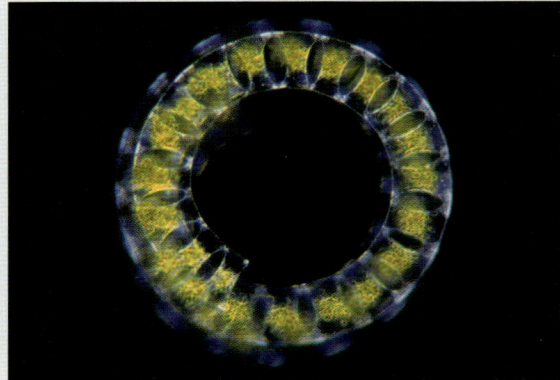

Since the 1970s, there has been a change in the phytoplankton community of the North Sea. The extent of the spring bloom of diatoms has declined in recent decades, but a longer growing season for other species, such as flagellates, now extends from March to early winter. If this reflects a shift in the ecosystem of the North Sea, then many other dynamics of the food web may also have changed.[14] There are major implications for the food web because the newly dominant flagellates may be of less food value and in

Figure 2-VII
The dinoflagellate *Ceratium fusus*, the most abundant species of phytoplankton in the North Sea (scale 1mm=6.15µm)[17]

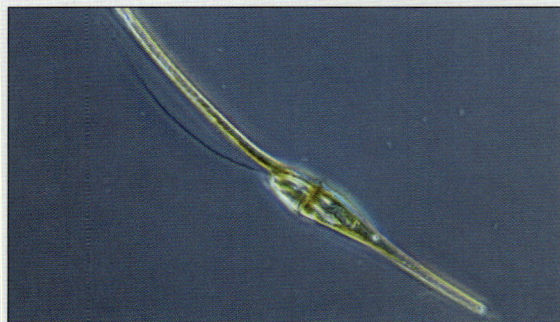

some cases may not be eaten by zooplankton at all. Phytoplankton biomass has increased in the last four decades over most of the North Sea, but this increase in biomass is not being passed up the food web because of the change in species composition between diatoms and dinoflagellates. [15,16]

2.18 All marine ecosystem components link together in a food web (figure 2-V). Primary productivity in the ocean is high at upwellings, because nutrients are available and are used by micro-organisms (plants, animals, fungi and bacteria) known as plankton. Most large animals in the sea are carnivores, with microscopic free-floating plants (phytoplankton) and herbivorous animals (zooplankton) at the base of the food web. Each type of marine organism has a position in the marine food web, with phytoplankton at the lowest level (trophic level 1 in figure 2-V); they are eaten by herbivorous zooplankton that are eaten by larger animals that themselves may be eaten by even larger carnivores. Typically, the marine food web has five trophic levels in contrast to the terrestrial food web, which have only three. Humans consume protein primarily available from large terrestrial herbivores (level 2), in contrast when humans consume marine protein it is usually from carnivorous fish at the third or fourth trophic level (e.g. salmon and tuna). This contrast in the exploitation of the basic food chains between the marine and terrestrial biosphere is of great importance in understanding the future of marine ecosystems.

2.19 The planktonic microbial community plays a major role in the transfer of energy and nutrients to other marine organisms. Phytoplankton are responsible for most marine primary production (and for 50% of primary production taking place on the planet). They form the base of the food web in surface waters, providing food that can be passed upwards to other parts of the web in all but the winter months (box 2A).[18] The biomass of phytoplankton is greater than that of all other marine organisms. The primary consumers, the herbivorous zooplankton, have only 10% of the biomass of the phytoplankton. The consequent loss of 90% of the energy between trophic levels is standard for oceanic food webs. In surface waters, phytoplankton are consumed by unicellular protozoa (protozooplankton) and larger invertebrate zooplankton (box 2B). In addition, phytoplankton excrete dissolved organic compounds, which are consumed by planktonic bacteria (bacterioplankton). These bacteria may then themselves be consumed by protozoans, which in turn may be grazed by the larger zooplankton. Planktonic micro-organisms therefore comprise a complex and productive community.

2.20 Plankton are an important food source for fish and other marine animals such as shellfish. For example, planktonic organisms, primarily copepods (box 2B), constitute a major food resource for many commercial fish species, such as cod and herring. Dead phytoplankton and other dead organisms, faecal pellets and waste products, collectively known as particulate organic matter, fulfil the same role at the base of deep-sea food webs in deep-sea parts of the water column and seabed. Seabed (benthic) organisms (2.28) display greater diversity than their mid-water (or pelagic) counterparts with the highest biodiversity being found in the coral reefs and the dark, deep-sea benthic realm (2.38).[19]

2.21 Plankton are also central to a variety of important biogeochemical processes. They play an important role in the climate system, as they grow they absorb carbon dioxide or bicarbonate from the water. They are eaten by zooplankton and are in turn consumed by larger species. In the process some carbon falls to the bottom of the ocean in the form of shells and faecal pellets. This draw-down of carbon dioxide from the atmosphere to the ocean and then to the seabed is important in moderating the rate of increase in carbon dioxide due to anthropogenic emissions, and is known as the 'biological pump'. Changes in plankton populations are therefore of considerable importance, whether natural or as a result of human influences on the environment.

BOX 2B   ZOOPLANKTON AND BACTERIOPLANKTON: ANIMALS THAT ARE A VITAL LINK IN THE FOOD WEB

Small shrimp-like crustaceans, such as krill, copepods and amphipods, constitute the largest proportion of zooplankton biomass. Many spend their entire life cycle floating at the surface of the ocean. As the biomass of phytoplankton increases, the herbivorous zooplankton, primarily copepods (figure 2-VIII), begin to grow, as do the organisms that feed on them, such as the larvae of other marine organisms. Zooplankton include the surface-dwelling eggs and larval stages of fish and organisms that live on the seabed such as starfish (figure 2-IX), worms, crustaceans and molluscs, which release millions of eggs to the surface water so that a few offspring may survive to disperse over a wide area before returning to the seabed as adults (broadcast spawning, box 5A).

As the level of nutrients declines and the number of copepods increases, phytoplankton productivity slows with a knock-on effect on the zooplankton until the entire system finds a steady state controlled by the rate of nutrient re-supply. As a result of inefficient feeding habits of herbivorous zooplanktonand the leakage of dissolved organic matter from phytoplankton cells, organic matter is released into the water column. The bacterioplankton utilise this food source, recycling these nutrients back into the food web. The microbial loop is critical for maintaining primary production when the availability of nutrients is low. As summer stratification sets in and surface waters warm, nutrients become less and less available and the oceanic food web switches from the classical food web to the microbial food web. Protozoans (unicellular animals such as ciliates and phagotrophic flagellates), gelatinous zooplankton (e.g. comb jellies) and filter-feeding organisms may be able to proliferate in the summer months while other organisms reliant on large-sized phytoplankton, such as copepods, decrease.[22]

Figure 2-VIII

The calanoid copepod species *Calanus finmarchicus* is an important component of the food web in the north Atlantic (scale 1mm=25µm)[20]

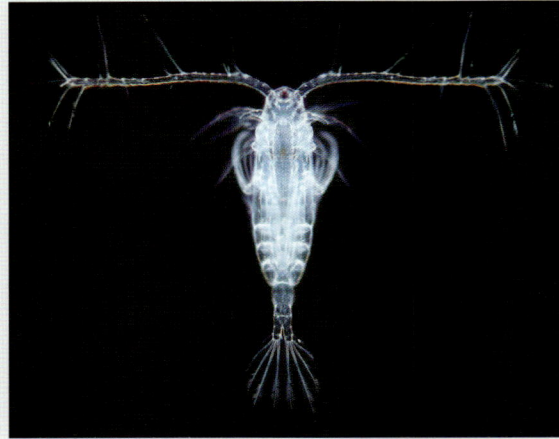

Figure 2-IX

Brittlestar larvae (*Ophiura* species) have an internal skeleton of silica to support their soft body. These species spawn in the spring and larvae settle as juveniles on the seabed in the autumn (scale 1mm=14.03µm)[21]

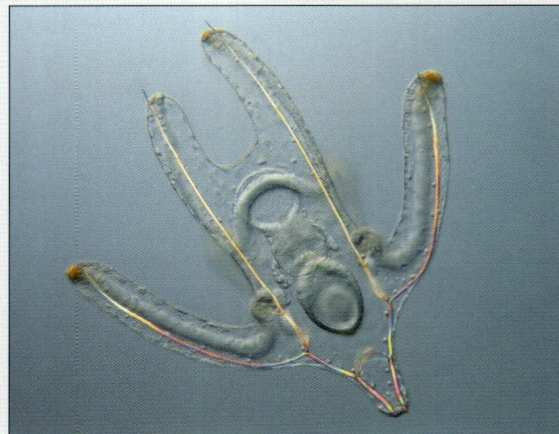

RECENT CHANGES IN PLANKTON COMMUNITIES AND THEIR EFFECTS ON MARINE ECOSYSTEMS

2.22 The phytoplankton and zooplankton communities in the north-east Atlantic have changed in abundance and composition in the last 40 years (boxes 2A and 2B). In the case of zooplankton, cold-water species have retreated northwards and warm-water species have moved north by as much as 10° of latitude. Of these zooplankton species, the dominant copepod genus is *Calanus*. These copepods are major predators of diatoms and an important food resource for higher trophic levels including commercial fish species.

2.23 In the North Sea, the dominant copepod species *Calanus finmarchicus* (figure 2-VIII) and *Calanus helgolandicus* have been studied for many years. Analysis of data has shown that east of 20°W in the north Atlantic and European waters there has been a significant poleward movement of warm water copepod species and an associated clear decrease in subarctic species. In this time the abundance of *C. finmarchicus* (a subarctic species) has decreased dramatically and *C. helgolandicus* (a temperate water species) has increased (figure 2-X). The major bloom period within the year differs for these plankton species, the *C. helgolandicus* bloom taking place in the autumn months.

2.24 As fish larvae rely on these plankton species for food during the spring months of the year, this species shift is likely to have had a detrimental effect on fish population recruitment levels and will, if these changes continue, lead to modifications in the abundance of various fish species in UK waters. In particular, the spring/summer bloom of *C. finmarchicus* is an important component of the cod diet in the early stages of its life cycle, whereas the *C. helgolandicus* population peaks in the autumn and is not available to the cod larvae at the appropriate point of the life cycle.

2.25 The northward shift in plankton species has been linked to the general rise in temperature in the northern hemisphere, along with the additional effect of the positive phase of the North Atlantic Oscillation that in recent decades has brought warmer conditions to the region.[23] This shift has led to marked changes in the dynamic regime of North Sea ecosystems and is likely to have profound consequences for exploited resources and biogeochemical cycles.[24] There are complex links between species in food webs across trophic levels, indicating a high degree of ecological interdependence. The effects of these changes in plankton levels and species on marine ecosystems are substantial.

2.26 It should be stressed that while changes in copepod distribution have exacerbated the population decline of cod,[25] the main cause of the collapse is overfishing.[26] Recent work has indicated that when cod populations are at a low spawning population biomass they are more vulnerable to the effects of environmental change. A positive North Atlantic Oscillation has no effect on recruitment when cod populations are high, but significantly reduces the recruitment rate when populations are low, thereby increasing the likelihood of a population collapse.[27] This means that the cod populations in North Sea are unlikely to recover to the levels seen in the 1970s and 1980s while the North Atlantic Oscillation remains positive, the plankton communities remain in an altered state and the populations continue to be fished at their present very low levels.[28]

Figure 2-X

The intensity of the plankton blooms of the two copepod species from the 1960s to the 1990s (shown on the horizontal axis) for January to December (denoted 1 to 12 on the vertical axis); the areas of red colouration denote high numbers of copepods, blue colouration low. The dashed lines indicate the period of cod larvae occurrence in the North Sea (March-September). Graphs (c) and (d) indicate that cod recruitment covaries positively with changes in plankton abundance. The high levels of recruitment in the Mid 1960s to 1980s (the 'gadoid outburst') was a consequence of a plankton ecosystem that was favourable for cod larvae/juveniles.[29]

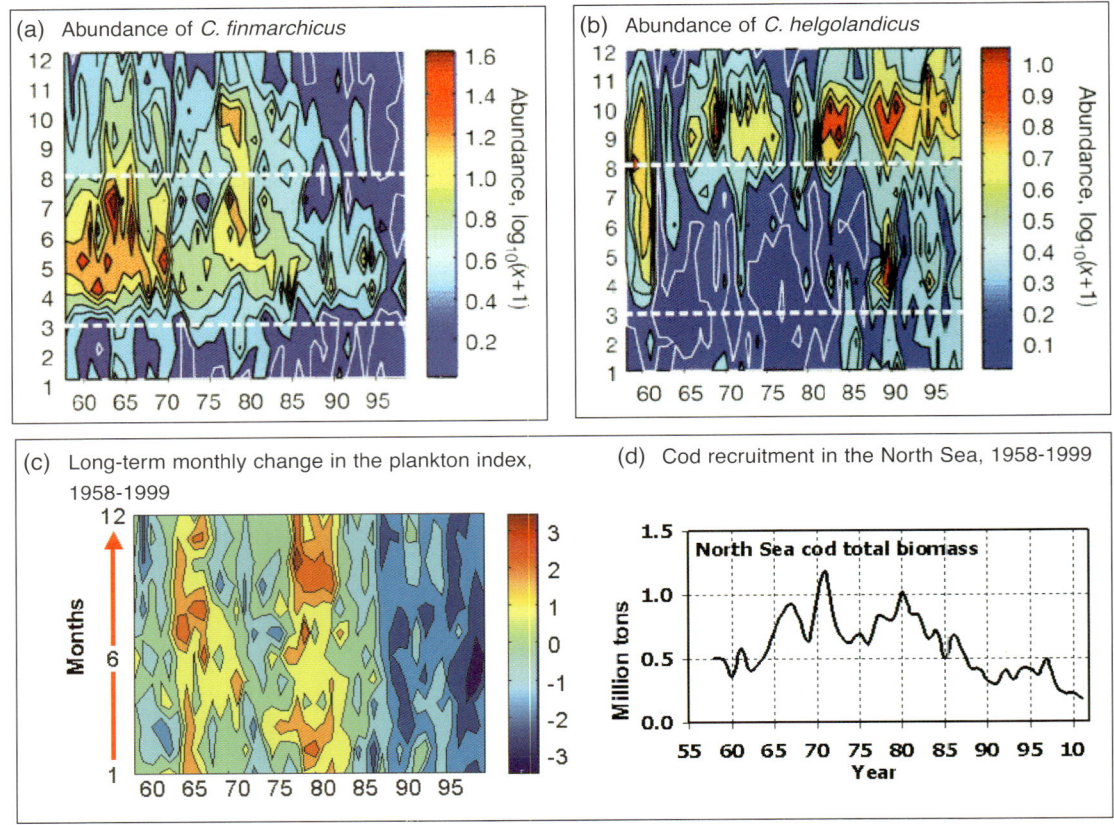

2.27 Careful consideration needs to be given to the exploitation of all marine resources within this changed environment. The complexity of marine food webs and the dependence of species on one another are poorly understood. However, temperate marine environments may be particularly vulnerable to climate change because the recruitment success at higher trophic levels, including commercially important fish species, is highly dependent on synchronisation with pulsed planktonic production. Such changes will affect trophic interactions and lead to eventual ecosystem level changes.[30] It is unlikely that we will fully understand the implications of climate change until better parameterised ecosystem models encompassing the full range of trophic levels are developed (chapter 7).

THE SEABED ENVIRONMENT — THE BENTHOS

2.28 The seabed is a mosaic of different habitats that contributes to the large diversity of marine life. The biota living near, on or in the seabed are collectively known as the benthos (figure 2-XI). The benthos can be further divided into plants known as phytobenthos (box 2C) and animals known as zoobenthos (box 2D). The habitats occupied by the benthos range from submerged tide-swept rocky reefs to relatively undisturbed muddy sediments.

2.29 The diversity and biomass of the seabed are dependent on several factors including substrate (sediment or rock), water depth, salinity and hydrodynamics.[32] The faunal communities of the seabed reflect the range of bottom sediments, and the degree of exposure to benthic currents and food supply. Characteristic benthic communities occur in different habitats as a result of adaptation to differences in the nature and stability of various sediments, as well as to depth, wave action and tidal streams. In general, shallow areas on the continental shelf experience more frequent disturbances from wave action and strong currents than deeper ones and the organisms that occupy these environments reflect these influences. Species diversity is generally higher inshore as there are more varied sediment types and spatial niches,[33] although wave exposure may reduce variety under some circumstances.

Figure 2-XI

A number of benthic species in an undisturbed benthic environment[31]

BOX 2C	**PHYTOBENTHOS: PLANTS THAT LIVE ON THE SEABED**

Algae are simple plants that do not have all the more complex tissue structures of higher plants. Since they are dependent on light for photosynthesis, plants that grow on the seabed are restricted to shallow waters. Microscopic algae (less than 0.1 mm in length) may thrive on any substrate, often contributing along with other micro-organisms such as fungi and bacteria to the stabilisation of loose sediments. Seaweeds, or macroalgae, that are between 1 mm and many metres long, attach to a hard substrate, such as rock, although some green seaweeds may thrive on more diverse substrates such as mussel beds or solid sediments. In shallow, well-lit and constantly churned coastal water, seaweeds grow rapidly, absorbing nutrients directly from the surrounding seawater. The mass of organic detrital material produced by them forms the basis of the food web in shallow waters. Higher plants, such as beds of seagrass (five species of which are present in the UK, Dwarf eelgrass *(Zostera noltii)*, narrow-leaved eelgrass *(Zostera angustifolia)*,

eelgrass *(Zostera marina)*, beaked tassel weed *(Ruppia maritima)* and spiral tasselweed *(Ruppia cirrhosa))*, may also be found on sandy and muddy sediments in shallower, calmer waters sheltered from significant wave action, such as marine inlets, lagoons and bays (figure 2-XII). They stabilise the substratum as well as providing important nursery grounds for young fish. Besides providing habitat and shelter for other plants that grow on them, the phytobenthos provides food for numerous grazing species such as periwinkles and limpets.[35]

Figure 2-XII

Eelgrass *(Zostera marina)* is found below the lower intertidal zone[34]

The number of macroalgal species decreases from south to north within the Arctic and northern temperate areas, but their total biomass can be considerable. Rocky shores have the most developed macroalgal communities, with vegetation down to approximately 15 m in the southern part and 30 m in the northern part of the North Sea.[36] The dominant seaweeds of northern areas are large, brown algae (Laminarians or kelp),[37] which form an important habitat below the low-water mark (figure 2-XIII). Together with their under-storey of red seaweeds, such as sea beech *(Delesseria sanguinea)*, dulse *(Rhodymenia palmata)* and sea oak *(Phycodrys rubens)* that extends into deeper water beyond the kelps' limits, kelp forests support a diverse and important faunal community that includes bryozoans, hydroids, sponges and colonial sea squirts. Excessive grazing of macroalgae by herbivores, due to low levels of predation, leads to loss of the associated ecosystem and is indicative of overfishing.

Maerl is a collective term for several species of calcified red algae that grow as unattached nodules or branching structures on the seabed in clear, warm and shallow bay waters to form extensive beds on the sea floor (figure 2-XIV). Live maerl has been found at depths of 40 m but typically occurs from 20 m to the low-tide level. Maerl beds, like tropical coral reefs, provide shelter for many other marine animals that live amongst or are attached to the nodules or burrow in the dead maerl beneath the living top layer. Molluscs, sea cucumbers and sea urchins, and a number of commercially important species of fish and shellfish, e.g. scallops, use maerl beds as their nursery grounds.

Figure 2-XIII
Kelp grows in 'forests' below the low water mark *(Laminaria hyperborea)*. Two other species, sugar kelp *(Laminaria saccharina)* and tangle kelp *(Laminaria digitata)*, can be seen on rocky shorelines above the low tide mark[38]

Figure 2-XIV
Maerl *(Phymatolithon calcareum)* forming a maerl bed, Loch Carron, West Scotland[39]

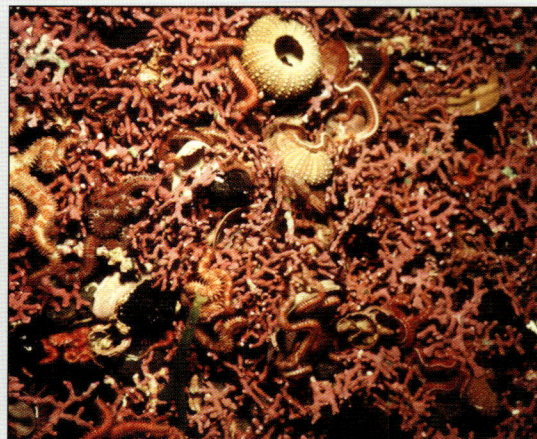

2.30   Much of the seabed around the UK consists of sediments. Most are derived from either glacial deposits, reworked by wave or tidal action, or inputs from rivers which are redistributed by waves, tides and currents. The sediments vary from coarse gravel and sand in areas of high current speed or wave action, through to soft mud where there is little natural disturbance. Extensive rocky substrates are also present in the English Channel, central Irish Sea, west of the Outer Hebrides, along sections of the continental slope and on the banks and mounts beyond the shelf break.[40] No systematic compilation or collation of broad habitat types has yet been undertaken, but the Irish Sea Pilot project (2.58-2.60) calculated that, for the Irish Sea, sands and coarse sediments (including gravel) accounted for over 60% of the area, muddy sands and sandy muds 22%, muds 10%, and rocky habitats about 3%.[41] Other than sediment type, factors that play a role in determining the biota present in a particular habitat include water temperature, salinity, depth and nutrient input. For example, the Greenland-Scotland Ridge is a major biogeographical boundary for benthos within the OSPAR area (appendix E). This ridge forms a barrier between warm- and cold-water areas, and hence cold and warm water species.

2.31   In shallow areas in the North Sea, benthic and pelagic (the seabed and mid-water) processes are often strongly coupled and work in concert to make the region highly productive. Frontal regions (2.17) normally have a high pelagic primary production resulting in productive benthic communities. Such frontal regions occur throughout the OSPAR area, including the North Sea and the Irish shelf front to the west of Ireland. There are estimated to be more than 15,000 benthic species (not including bacteria and viruses) in the shelf seas around Britain reflecting the wide range of environmental conditions. A 1986 survey covering the whole of the main North Sea basin[42] showed clear north-south differences in diversity, abundance, biomass and average individual weight of the soft-bottom organisms. Those from the deeper northern regions had higher diversity, lower biomass and lower individual weights than those from the shallow southern regions. The main causes of variation were thought to be differences in the grain size of the sediment and the supply to the bottom of organic matter from pelagic primary production. It appears that the benthic-pelagic coupling is stronger in the more shallow southern areas.

2.32   There can be considerable short-term temporal changes in the diversity and structure of the benthic community in the central part of the North Sea.[43] This variability may be driven by climate-induced fluctuations in the overlying pelagic communities. Most of the seabed in the North Sea hosts soft-bottom communities, with the exception of the land margins of Norway and the United Kingdom where rocky shores dominate. Some benthic habitats supporting high levels of biodiversity are protected under the categories defined in the EC Habitats Directive Annex I classification (chapter 4).

2.33   Highly productive benthic communities can also be found in tidal areas, for example in the Wadden Sea along the south-eastern border of the North Sea and in several estuaries along the western European coast. Some substrates, such as rocky littoral habitats (the inter-tidal zone of the shoreline), support species assemblages of particularly high biodiversity. Other rare, unique but diverse communities of marine organisms are found only in association with habitats created by other organisms such as mussel beds, various structures built by invertebrate worms (box 2D), seagrass beds and maerl beds (box 2C). Some species perform other vital ecosystem roles, for example shellfish that remove organic matter and microbes from the water column by filter feeding. In some coastal marine habitats, such as estuaries, their contribution prevents eutrophication.

Box 2D   ZOOBENTHOS: ANIMALS THAT LIVE ON THE SEABED

Sandy and muddy seabeds mostly lack the larger material that seaweeds and hydroids (polyp-forming organisms such as sea anemones) need in order to settle. Life is therefore limited on the surface of sandy seabeds, although a range of animals live in the sand, such as burrowing anemones and worms, shells and sea urchins. A range of creatures burrow into muddy seabeds, including the Dublin Bay prawn *(Nephrops norvegicus)*, other shrimps and a wide range of worms, molluscs and other invertebrates (figure 2-XV). The burrowing activity of these organisms creates a complex habitat, enlarging the surface area and oxygenating the deep sediment. Many of the species are exploited for human consumption or are the dominant food for demersal fisheries species.

Extensive areas of gravel are also found around the Irish Sea, south-eastern coasts and in the English Channel. These areas are generally stable, with tidal currents preventing the deposition of fine sediments and keeping the gravel surface clear. Because of the lack of surface sedimentation and the presence of larger pebbles, a range of animals can gain a foothold on the seabed, including deposit- and filter-feeding animals such as scallops, hydroids and anemones, while between the stones worms, such as the peacock fan worm, and the sand mason worm can find a suitable habitat. Scallops provide a particularly economically valuable shellfish fishery (figure 2-XVI).

There is also a range of rock communities which vary with depth, wave exposure, currents and rock type. In deeper water, there are fewer seaweeds, and rock surfaces are covered by encrusting animals such as anemones, sponges and hydroids (figure 2-XVII). These communities are less affected by wave action but are strongly influenced

Figure 2-XV
A Dublin Bay Prawn[44]

Figure 2-XVI
A great scallop *(Pecten maximus)*, which is usually found in a shallow depression in the seabed. This species is associated with areas of clean firm sand, fine or sandy gravel, but is occasionally found on muddy sand[45]

Figure 2-XVII
A yellow cluster sea anemone *(Parazoanthus axinellae)*. This species is usually found attached to organic substrata including sponges shells and worm tubes or on rocks from the shallow sublittoral offshore to about 100m[48]

by the strength of tidal currents. Most animals are fixed to the rock surface and rely on the movement of the sea to bring them a constant supply of food, mostly small particles of organic debris, and microscopic plants and animals. Tidal currents represent a predictable conveyor belt, into which animals can extend tentacles or funnel-like filters to collect passing food. If current speed increases beyond a certain level, fewer organisms are able to keep a foothold on the rock, and communities become characterised by fewer more resilient species.[46,47]

It is likely that the seafloor looked quite different prior to the advent of widespread trawling (chapter 5), with many more sponges, hydroids, oyster, mussel beds and other biogenic structures. There are a number of invertebrate worms that can form substantial and often quite solid aggregations in areas which would otherwise be composed predominantly of sediments, thereby increasing the stability of otherwise unstable habitats. Some species may form reefs several centimetres thick raised above the seabed which persist for many years (figure 2-XVIII). They are of particular importance where they occur on sediment or mixed sediment areas as they provide a habitat for a range of species that would otherwise not occur in these environments.[49,50] Mussel reefs are composed of layers of living and dead mussels at high densities, bound together by the threads secreted by the mussels and sometimes overlaying a great deal of accumulated sediment. They have an associated assemblage of fauna and flora. Mussel beds also provide shelter for large numbers of organisms and form a rare hard substrate in a soft-bottomed environment. The horse mussel *(Modiolus modiolus)* forms dense beds at depths of 5-70 m in fully saline, often moderately tide-swept areas (figure 2-XIX). True beds, forming a distinctive habitat, support one of the most diverse sub-littoral communities in north-western Europe, that includes tubeworms, sea squirts, variegated scallops and sea cucumbers. Such beds, however, are rare although scattered individuals of the species are common.

Figure 2-XVIII
Sandy reefs at Heysham, Morecambe Bay, created by the honeycomb worm *(Sabellaria alveolata)* exposed at low tide[51]

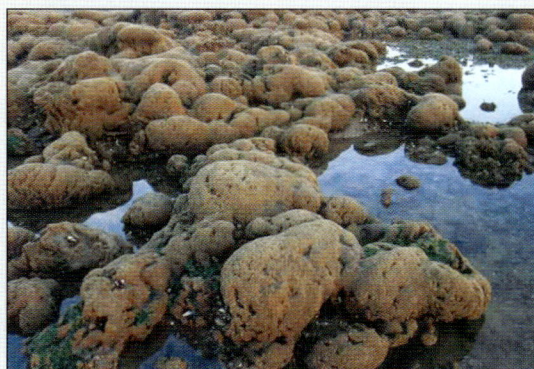

Figure 2-XIX
A horse mussel *(Modiolus modiolus)* bed. This species is found part buried in soft sediments or coarse ground or attached to hard substrata down to about 280m[52]

Continental margin and deep-water habitats

2.34 The ocean depth and sea floor topography (bathymetry) of the OSPAR area ranges from the continental shelf margin at 200 m to the abyssal plains at around 5,000 m depth (figure 2-I). Deep seabed habitats and communities are not isolated from the water mass above and there is ecological continuity throughout the water column.[53] Seabed currents, or benthic storms, have been detected during research undertaken at 2,400 m depth on the North Feni Ridge west of Scotland. These submarine storms are capable of remobilising sediments and dispersing them vertically in the water column as well as horizontally across the seabed.[54]

2.35 A number of important habitat types occur in the deep-water areas of the north-east Atlantic. For example, there are an estimated 50,000 seamounts over 1,000 m high in the world's oceans usually surrounded by abyssal plains. The majority of seamounts in the OSPAR area lie along the mid-Atlantic ridge as they are of volcanic origin. There are also major seamounts closer to the continental margin such as the Anton Dorhn seamount and Rosemary banks within UK territorial waters. Seamounts have unique current and sediment conditions and act as islands for marine life. What scientific information is available suggests that they are associated with high levels of biodiversity, with the number of species unique to each particular seamount averaging 15% or higher. They are also often used as reproduction and feeding grounds by migratory species and as 'stepping stones' for the trans-oceanic dispersal of continental shelf species (figure 2-XX).

Figure 2-XX
The benthos on the granitic seamount of Haig Fras, characterised by cup sponges and erect breaching sponges, located approximately 110 km north west of the Isles of Scilly within the Celtic Sea. This outcrop, which measures approximately 45 km by 15 km, rises from from a surrounding seabed depth of 100-110 m to within 38 m of the surface[56]

Figure 2-XXIa
Carbonate mound[55]

Figure 2-XXIb
Hydrothermal vents

2.36 Another specific habitat found in the deep sea are carbonate mounds and other types of hydrothermal vent, which occur in small localised clusters mainly on the eastern sea margin of the North Atlantic (figures 2-XXIa and b). These mounds are usually dominated by filter-feeding communities and can support rich deep-sea coral communities, which form a secondary hard substrate for an abundant and diverse range of marine organisms (figure 2-XXII).[57] Deep-sea sponge aggregations are also known to support a rich, diverse epibenthic fauna (seabed organisms that live on other seabed organisms). Dense aggregations are known to occur in various places in the north-east Atlantic, from a depth of 250-500 m close to the shelf break. It has been reported that one study off the coast of northern Norway took samples from an area of less than 3m², yielding 4,000 sponge specimens belonging to 206 species, 26 of which had not previously been described.[59]

Figure 2-XXII
Benthic fauna on and around carbonate mounds and hydrothermal vents including tubeworms, molluscs and crustaceans[58]

2.37   Deep cold-water coral reefs (*Lophelia pertusa*) have a wide geographical distribution, from 55°S to 70°N. Large areas of deep-water coral reefs occur in the Atlantic Ocean near the continental shelf break off Ireland, Scotland, the Faeroe Islands, Norway, and off the south coast of Iceland, as well as on the continental shelf off Scotland and Norway. These cold-water coral reefs are likely to be thousands of years old (figure 1-VII). These reefs create biodiversity by providing a complex three-dimensional habitat that is attractive to many other species. These reefs include an undersea feature that was discovered during surveying of the deep seabed north-west of the Hebrides in 1998, the 'Darwin Mounds' (box 5H). Hundreds of seabed mounds, about 5 m high and 100 m across with tails several hundred metres long, were found. Deep-water corals grow on the mounds as well as giant single cellular organisms twenty centimetres across, called xenophyophores. Similarly, a large cold-water reef complex has been recorded in Scottish waters off the southern Outer Hebrides.

2.38   Diversity among open ocean species has three peaks depending on habitat: the euphotic zone, benthic environments at 2,000-3,000 m on the margins of the continental slope, and the abyssal plains.[60] It had been thought that species diversity increases with depth in the continental shelf regions to a maximum just seaward of the continental rise, and then decreases with increasing distance towards the abyssal plain.[61,62] Recent surveys, however, have shown that deep-sea ecosystems beyond the continental shelf can possess unexpectedly high species abundance and diversity. The deep seabed off the Atlantic seaboard supports an abundance and variety of life including cold-water corals, sponges, sea slugs, sea urchins, starfish, deep-water fish and many other benthic organisms (figure 2-XXIII). Deep-ocean species tend to be much smaller in size than their shallow-water counterparts.

2.39   The deep sea is the habitat for many species of fish, including bottom-living species caught by fishers such as roundnose grenadier (*Coryphaenoides rupestris*), Atlantic orange roughy (*Hoplostetus atlanticus*) (figure 2-XXIV) and blue whiting (*Micromesistius poutassou*);

pelagic oceanic species (living in water above the continental shelf edge and deep ocean) such as albacore tuna (*Thunnus alalunga*) and oceanic Atlantic redfishes (*Sebastes* species); and rays and sharks. Little if anything is known about the behaviour, populations and ecology of deep-living fish species (chapter 5). They are adapted to a cold, dark habitat that provides low levels of nutrients, where energy must be conserved as much as possible, so that as a result the fish grow extremely slowly and reproduce at a slow rate. Many species become mature only between the ages of 10 to 30 years and can live to over one hundred years old. Some deep-water species have an extensive geographical distribution owing to the small environmental variation between deep-sea habitats in different parts of the world. We will examine the important consequences of these factors for the sustainability of fishing in these areas in later chapters.

Figure 2-XXIII
A giant sea spider (*colossendeis* species) on the sea floor of the Faroe-Shetland Channel[63]

Figure 2-XXIV
An orange roughy[64]

COMMERCIALLY EXPLOITED SPECIES

2.40 In the seas around the UK, 330 fish species have been recorded[65] and over 1,000 have been recorded in the OSPAR area. Of these only about 5% are commercially exploited. About 2% of species make up 95% of the total biomass and commercial species make up a substantial proportion of this. Exploited species can be divided into two types: pelagic (species that live and feed in the open sea, associating with the surface or middle depths of a body of water feeding on plankton or on species that are planktivores) and demersal (living on or near the bottom and feeding mainly on benthic organisms). Although fish species can be pelagic and demersal at different parts of their life cycle, these terms are usually used to refer to where they are located at harvest.

2.41 Commercially important demersal fish species include cod (*Gadus morhua*), haddock (*Melanogrammus aeglefinus*) and hake (*Merluccius merluccius*); flatfish such as halibut (*Hippoglossus hippoglossus*), plaice (*Pleuronectes platessa*) and sole (*Solea solea*) and others such as monkfish (*Lophius piscatorius*) that are caught as part of a mixed fishery by trawling. As a result, a large number of other organisms and fish species associated with the benthic environment will inevitably be caught as well. Industrial trawling, using very fine mesh sizes of less than 16 mm takes place in the North Sea for sandeels (*Ammodytes marinus*), which are used to produce fishmeal and fish oil.

2.42 By comparison, pelagic fish are caught in single-species shoals. Commercially important oceanic pelagic species include whiting (*Merlangius merlangus*), herring (*Clupea harengus*), Atlantic mackerel (*Scomber scombrus*) and sea bass (*Dicentrarchus labrax*). There are also some commercially exploited pelagic species that live in shallower water above the continental shelf such as Atlantic salmon, rays and sharks. Large commercially important pelagic predators (tuna and marlin) are highly migratory, ranging far beyond the boundaries of the OSPAR region.

2.43 In addition to finfish, many other species of marine organisms are exploited such as cephalopods (octopus and squids). The biology of squid is poorly known despite these species being very abundant especially in the Atlantic. Only a few species are exploited commercially, but squid are of considerable ecological importance as predators and as the food of some whales, fish and seabirds.[66] In the north-east Atlantic, populations of shrimps (*Pandalus borealis*) and other crustaceans such as Dublin Bay prawn, lobster (*Homarus gammarus*) and various crab species, as well as molluscs (shellfish) such as scallops (*Pecten maximus*), mussels (*Mytilus edulis*) and oysters (*Ostrea edulis*) are also of commercial importance.

2.44 In the North Sea and areas around the British Isles the main commercial species in terms of monetary value are demersal (such as cod, haddock, whiting, plaice, sole, shrimps and Dublin Bay Prawn), although there are also some important pelagic fisheries for mackerel and herring. Most species show annual or seasonal migrations related to feeding and spawning. Variability occurs naturally in the size of fish populations due to variations in egg and larval survival and the resulting number of juveniles joining the adult population from year to year. Most variability in the size of populations, however, is now caused by the levels of mortality rate induced by fishing (chapter 5). Further information on some commercially exploited fish species is given in appendix E.

LARGER MARINE ANIMALS

2.45 Large marine vertebrates – such as sea turtles, cetaceans, sharks and large fish species – often play a major role in shaping and maintaining the stability of the ecosystems of which they are a part. In the wider Atlantic, top predators such as sharks play an important role in maintaining the structure and diversity of fish assemblages. As they feed near the top of the food web, their status may reflect conditions of the wider ecosystem in terms of food availability. Many are the subject of conservation initiatives. These are often focused on species of marine mammals that occur in the north-east Atlantic, namely dolphins, porpoises, whales and seals.

2.46 About 16 of the 80 or so known species of cetaceans in the world can be seen off the British coast. These include the baleen whales such as fin (*Balaenoptera physalus*), sei (*Balaenoptera borealis*) and minke (*Balaenoptera bonaerensis*) whales, but also blue whales (*Balaenoptera musculus*), the largest of all marine mammals, and humpback whales (*Megaptera novaeangliae*). The largest toothed whale, the sperm whale (*Physeter macrocephallus*), also occurs around Britain, although only adult males are seen. Medium-sized whales are represented by the pilot (*Globicephala melaena*) and killer whales (*Orcinus orca*), while small species seen are Risso's dolphin (*Grampus griseus*), white-sided (*Lagenorhynchus acutus*), white-beaked (*Lagenorhynchus albirostris*), common (*Delphinus delphis*) and striped dolphins (*Stenella coeruleoalba*), as well as the more familiar common porpoise (*Phocoena phocoena*) and bottlenosed dolphin (*Tursiops truncatus*) (figure 2-XXV). Less frequent visitors include the beluga (*Delphinapterus leucas*) and narwhal (*Monodon monoceros*), which are Arctic species, and bottlenosed whales (*Hyperoodon ampullatus*) and various beaked whale species.[67]

2.47 Only two species of seal breed in UK territorial waters, the common (*Phoca vitulina*) and grey seals (*Halichoerus grypus*) (figure 2-XXVI), although others such as the hooded seal, bearded seal, ringed seal, harp seal and walrus may be occasional visitors from Arctic waters. Around 40% of the world's population of grey seals breed in the OSPAR area and Britain hosts 95% of the EU population. In 2000 it was estimated that there were about 120,000 grey seals in British waters. Common seals are more difficult to count; the present minimum estimate is 36,000, only a small proportion of the world population of about 500,000.[69] The status of some larger marine vertebrate species is discussed in appendix E.

SEABIRDS

2.48 Almost all parts of the OSPAR area support breeding and migratory birds that are dependent on the marine ecosystem. A total of around 30 million pairs of seabirds nest in the north-east Atlantic; of these the greatest numbers nest on the coasts of Arctic waters and the North Sea. Total numbers of individuals in these northern areas are several orders of magnitude greater than those in the southern regions of the OSPAR area (figure 2-XXVII). Twenty-five seabird species breed in the UK, with a total coastal population of just under eight million in 1998-2002. Only the great skua (*Catharacta skua*) is endemic to the OSPAR

Figure 2-XXV
A bottlenosed dolphin[68]

Figure 2-XXVI
Grey seal[70]

Figure 2-XXVII
Seabird distribution in the north-east Atlantic. Distribution of large seabird colonies in the OSPAR area[71]

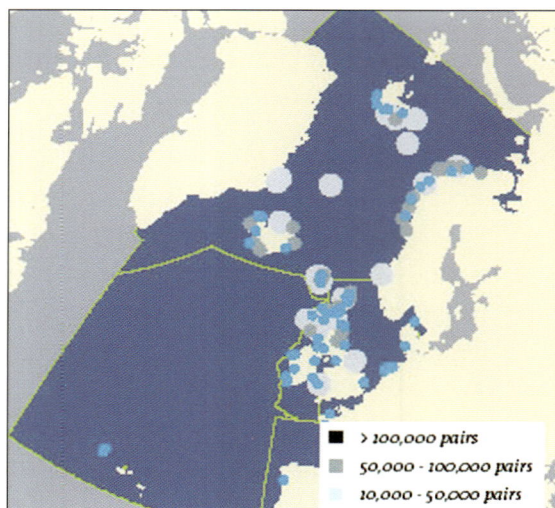

> 100,000 pairs
50,000 - 100,000 pairs
10,000 - 50,000 pairs

area, although others are near endemic (the Manx shearwater, *Puffinus puffinus*) or have endemic sub-species (the shag, *Phalacrocorax aristotelis*). Surveys of distribution at sea have not been carried out in all parts of the OSPAR area, but in those areas that have been studied, surveys show shelf seas to have substantially higher density of populations than oceanic waters.

2.49   Bird populations of the North Sea area are of global importance. Thirty-one species of seabirds breed along the coasts and major seabird colonies have been established along the rocky coasts of the northern part of the North Sea. Some 10 million seabirds are present at most times of the year, but migrations and seasonal shifts are pronounced. The North Sea coasts support more than 50% of the world's common terns (*Sterna hirundo*) and great skuas (figure 2-XXVIII). A further twelve species, such as the common scooter (*Melanitta nigra*) around the Flemish Banks, are

Figure 2-XXVIII
Great skua[73]

present in numbers exceeding 10% of their total estimated global populations. The UK holds about 90% of the global population of Manx shearwaters, about 68% of Northern Gannets and some 60% of great skuas.[72] Inter-tidal and inshore areas are also important for passage and wintering water birds.

2.50   In general, seabird populations have been increasing in the past fifty years, and it may be that some species have increased due to the availability of discarded fish as a food source. Numbers of seabirds breeding in Britain and Ireland have risen steadily over the last 30 years from around 5 million in 1969-70, to over 6 million in 1985-88, to almost 8 million in 1998-2002. The coastal populations of 13 species have increased in size by more than 10%, three have decreased by more than 10% and five have changed by less than 10%. However, some species are declining, especially in recent

Figure 2-XXIX
Puffin[75]

years. The recent poor breeding record of some seabird species (e.g. kittiwakes, puffins (*Fraticula artica*) (figure 2-XXIX)) and razorbills may have been caused by an unprecedented slump in the number of sandeels available as prey; sandeels are reliant on plankton species that have shifted northwards.[74] The status of some seabird species is discussed in more detailed in appendix E.

DISTRIBUTION OF MARINE ECOSYSTEMS

2.51 The United Kingdom's exclusive economic zone extends over 867,000 km² – three and half times its land area (figure 2-XXX). Marine ecosystems in the waters around the UK are diverse and productive. They are influenced by colder arctic waters in the north, temperate waters more usual at this latitude and warmer influences such as the Gulf Stream. Ecosystems are distributed throughout the region according to a set of diverse physical influences such as temperature, salinity, water depth and the mosaic of different benthic habitats present. In this section we describe how the distribution of ecosystems can be used to classify marine regions on an environmental basis.

Figure 2-XXX
This figure shows the extent of UK territorial waters or exclusive economic zone (EEZ) and the inshore 12nm limit. The UK 200 nm fisheries zone covers a smaller area than the EEZ, as the 200 nm limit relates to the British Isles rather than Rockall[76]

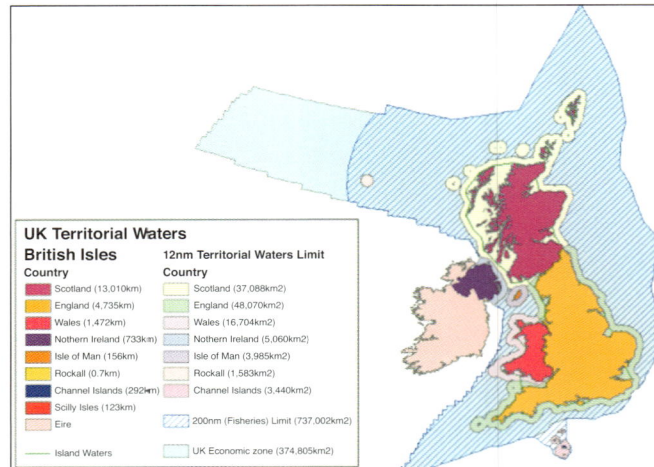

BIOGEOCHEMICAL PROVINCES AND LARGE MARINE ECOSYSTEMS

2.52 Four marine biomes (a biogeographical region characterised by distinctive plant and animal species) have been defined by the dominant oceanographic processes that determine vertical density structure of the water column, based on archived chlorophyll profiles and satellite remote sensing. These chlorophyll profiles can be seen as an indicator of vertical nutrient flux. The Polar biome contains polar and subpolar oceans; the Westerlies biome contains temperate and subtropical areas of the oceans; the Tradewinds biome contains the tropical sea areas; and the Coastal Boundary biome comprises all continental shelf waters adjacent to land. The highest level of primary production occurs on the continental shelf areas (the Coastal Boundary biome), which are areas of strong water mixing. Of the four biomes defined, three – Polar, Westerlies and Coastal Boundary – occur in the OSPAR area.[77]

2.53 Biomes can be further subdivided into ecologically meaningful units based on the global system of 57 biogeochemical provinces (BGCPs).[78] These provinces contain living organisms that respond to local environmental processes that determine nutrient delivery. The BGCP system is based on the recognition of distinct natural regions of the ocean characterised by pelagic production patterns which respond to a characteristic pattern of physical forcing, as for example the seasonal development of the thermocline. The coastal areas of these provinces can be further classified into distinct regions. The term 'Large Marine Ecosystems' (LMEs) is used for these regions of ocean space encompassing coastal areas of the seaward boundary of continental shelves and the outer margins of coastal current systems.[79] These areas are the source of 95% of the world's annual fishery yield at the present.

2.54 Large Marine Ecosystems cover around 200,000 km² of the world's oceans; they have distinct bathymetry, hydrography and productivity patterns and are all associated with unique ecological communities. Worldwide, 64 such areas have been defined.[80,81] Fifteen of these occur in the North Atlantic, and eight of these are covered by the OSPAR area; two occur in UK territorial waters, the North Sea and the Celtic-Biscay shelf. Large Marine Ecosystems can be incorporated into the framework as subunits of the biogeochemical provinces in order to define ecoregions.[82] The Large Marine Ecosystems around the UK roughly equate to the regional seas that have been traditionally referred to as the North Sea, the Celtic Seas and the English Channel.

CLASSIFICATION OF UK REGIONAL SEAS

2.55 The LMEs described above are on too coarse a scale for the purposes of environmental management. The traditional or legal dermarcation of regional seas has no scientific basis and cannot be used for environmental management either. For this purpose, areas within UK territorial waters have been classified by the Joint Nature Conservancy Council (JNCC) into nine 'regional seas', using a range of geophysical and hydrographical parameters such as temperature, depth and currents (figure 2-XXXI). The regional seas are the northern North sea; the southern North Sea; the Mid- and Eastern English Channel; the South-Western Peninsula, the Western Approaches; the Irish Sea; the Western Isles; the Atlantic waters off Scotland; and the area around Shetland and the Orkneys. These are further described in appendix E.

2.56 This draft framework model of regional seas will be further subdivided by the use of a system for classifying marine seabed landforms and identifying ecological units using survey data and biogeographical analysis. The aim is to use all available data and information to identify features and landforms of national importance requiring conservation action.[83] Geophysical information can be used as a proxy for biological information in order to develop a classification of medium-scale marine habitats. The justification for this assumption is the very strong relationship that exists between geophysical and hydrographic factors and the character of biological communities present.[84]

2.57 On the basis of this relationship, the marine habitats present in the water column (using parameters such as water temperature, depth/light, and stratification/mixing regime), and on the seabed (using parameters such as water temperature, depth/light, substrate type and exposure/slope) can be inferred. The marine habitats identified by this type of classification are termed marine landscapes.[85] This terminology has now been adopted by the recent government-sponsored Review of Marine Nature Conservation. Having defined regional seas around the UK, and ecological units within each regional sea, it should be possible to identify the location of habitats, sites and species which are nationally important and require conservation action across the geographical area of UK waters.

2.58 This approach has been used to identify all the marine landscapes present in the Irish Sea as part of a conservation initiative by the JNCC.[86] These have been further validated by biological surveys to check that the landscapes reflect the biological characteristics expected and to estimate the nature conservation value of the various landscapes and their susceptibility to harm. Biological characterisation was achieved by linking the available biological data to the relevant marine landscapes by joining the data spatially within Geographic Information System. This involves aggregating data to the smallest geographical unit of a habitat characterised by its biota according to national classification criteria, so that it can be delimited by convenient boundaries, although these data are sparse or unavailable for many offshore areas.

2.59 Three main groups of marine landscape have been identified in UK waters.[89] These are:

- coastal marine landscapes, such as estuaries and rias (a type of estuary often known as a drowned river valley) where the seabed and water body are closely inter-linked;

- seabed marine landscapes which occur away from the coast, i.e. the seabed of open sea areas. In this group, the marine landscapes comprise the seabed and water at the substrate/water interface;

- water column marine landscapes of open sea areas, such as mixed and stratified water bodies and frontal systems. These are the marine landscapes above the substrate/water interface.

2.60 Eighteen coastal and seabed marine landscape types were identified for the Irish Sea (figure 2-XXXII). Surveys found a good correlation between the survey data and the marine landscapes identified on the basis of geophysical and hydrographic data. In general, the predictions of biotope complexes (the smallest geographical unit of the biosphere that can be delimited by convenient boundaries and identified by its biota) were validated by the surveys, with a good level of confidence that the marine landscape types are ecologically relevant. Four water column landscape types were also identified, although with less confidence. Nonetheless, the work carried out shows the basic concept is sound

Figure 2-XXXI
Regional Seas around the UK, as classified by the JNCC[87]

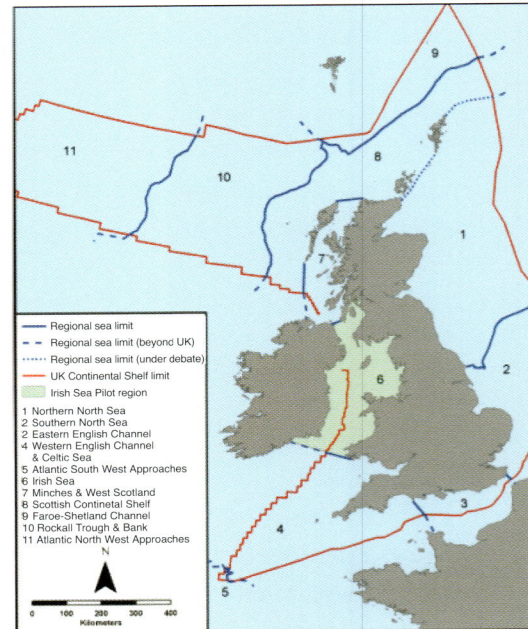

Figure 2-XXXII
Seabed and coastal marine landscapes in the Irish Sea defined by the JNCC[88]

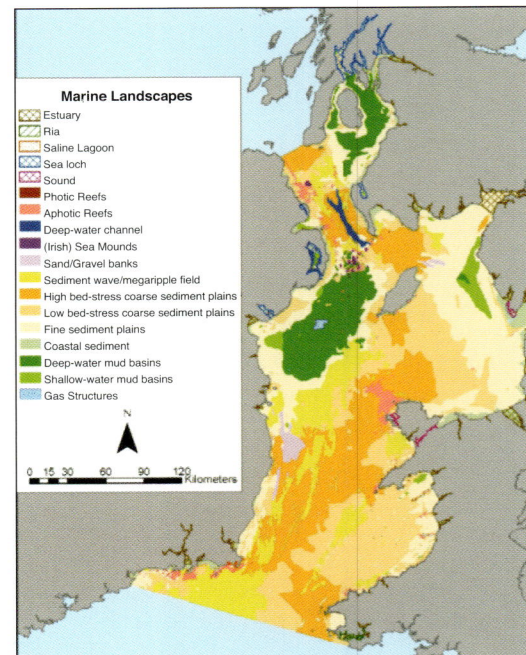

and could be extended to all areas of UK waters. Classification of marine landscapes can be used to predict the susceptibility of their biological communities to human impacts and also to identify specific areas of national importance, rare habitats and areas necessary to sustain nationally important or rare species. The use of the regional sea framework and marine landscapes for the development and implementation of strategic and spatial planning in the marine environment is further discussed in chapters 7 and 10.

CONCLUSIONS

2.61 Up to half of the UK's biodiversity – over 44,000 species – may be found in our seas,[90] but only within the last few years have attempts been made to map and locate the habitats in the marine environment and understand our impacts on them. Vast numbers of undescribed species exist in familiar oceanic habitats, ranging from plankton and worms of shelf muds to tiny nematodes and highly colourful sea slugs in tropical lagoons. New species are still being discovered in relatively unexplored deep-sea and polar habitats. Research on deep-sea ecosystems indicates that far more species may exist there than were previously thought, with estimates ranging from 500,000 to 10 million species.[91] By comparison, estimates of total global species diversity have varied from 2 million to 100 million species, with a best estimate of somewhere near 10 million; only 1.4 million have actually been named.[92]

2.62 Thirty-two out of a total of 33 animal phyla (a rank within the hierarchy of classification of organisms) are found in marine habitats and of these, 14 have no representatives in freshwater or terrestrial habitats.[93] The ecologically important benthic and planktonic protists (algae, fungi and protozoans) alone may comprise 34 phyla and 83 classes (another rank within the hierarchy of classification of organisms), and there is a vast complexity of undescribed parasites (plant, animal and microbial) that live on and in other marine organisms.[94] Newly recognised biological habitats that contain novel species assemblages – such as hydrothermal vents, whale carcasses, brine seeps and wood debris – continue to emerge, especially in deep water.[95]

2.63 This rate of discovery is likely to increase with the ongoing Census of Marine Life Project, which is a growing global network of researchers in more than 50 nations engaged in a ten-year initiative to assess and explain the diversity, distribution and abundance of marine life in the world's oceans, past, present and future. The emphasis of the programme is on field studies, which are to be conducted in poorly known habitats as well as those assumed to be well known. In both coastal and deep waters, projects will identify new organisms and collect new information on ocean life. The field studies aim to gather the information necessary to define the 'Known, Unknown and Unknowable' about marine populations and ecosystems, this information being key to managing global ocean resources wisely, both now and in the future.[96]

2.64 Understanding of the mechanisms responsible for the creation, maintenance, and regulation of habitat-specific marine biodiversity is incomplete, fragmentary or in some cases entirely lacking. Inadequate knowledge of the species present in a given marine community or ecosystem limits understanding of ecosystem function and the prediction of how human activities impact that function.[97] In UK waters new species and habitats, such as the Darwin Mounds cold-water coral reefs off the north-west of Scotland identified in 1998, are still being discovered.

2.65 The assemblages of marine organisms are extremely diverse throughout the OSPAR area, with the north-east Atlantic supporting some of the potentially most productive fisheries in the world. Although commercially important fish species make up a significant proportion of the fish biomass, they represent only a small part of the larger marine environment. Individual marine ecosystems are very different, with few intermediate prey species in the north, more mixed feeding ecologies in the North Sea, and very specialised ecosystems in the deep sea. Although there is comparatively poor understanding of the connections between species in the marine food web, it is known that changes in the abundance of one species, such as a commercial fish species, can have dramatic effects on the rest of the ecosystem. Marine biological diversity is changing, human impacts are the primary cause of these changes and the biggest of these impacts is fishing, an issue we return to in chapter 5.

Chapter 3

THE ROLE OF THE FISHING INDUSTRY

Fishing has become an increasingly intensive activity over the past fifty years. How has this affected global fishing patterns and the industry's role in our national life and economy? What are the prospects for the sector's future? How could they affect our diet, health and the quality of the environment?

INTRODUCTION

3.1   Fishing has long played an important role in the life of our nation and that of many of our European neighbours. It has been at the heart of many coastal communities and provided us with much-loved staples of our national diet. While some traditional fishing methods survive, the face of the industry has changed over the past half century with the emergence and growth of a large-scale fishing industry.

3.2   Global levels of fishing effort have increased dramatically as technological advances have allowed boats to fish further out to sea and for longer periods of time. This trend is now tailing off as fishing grounds become over-exploited. This has resulted in ever smaller, but more powerful fishing fleets, employing fewer people and representing a more minor economic activity.

3.3   In contrast to the trend in capture fisheries, aquaculture is expanding rapidly. As global consumption of fish continues to rise, aquaculture provides an increasing share of the fish and seafood that is consumed. This does not however imply a straightforward substitution of wild fish for farmed ones. A large part of the aquaculture industry in north-west Europe is heavily reliant on capture fisheries to supply the huge volume of fishmeal and fish oil required for aquafeed.

3.4   Demand for fish and seafood products remains buoyant with new products coming onto the market and consumers recognising the health benefits of eating fish. Indeed, this aspect is receiving increasing attention, as the Food Standards Agency urges the nation to eat more fish to reduce the prevalence of one of the UK's largest killers – coronary heart disease. We examine the current interest in the health aspects of fish consumption, including the benefits linked to long-chain n-3 polyunsaturated fatty acids (also known as omega-3 fatty acids) and concerns over the levels of environmental contaminants such as dioxins and mercury.

3.5   The continued drive to improve the technology of vessels and gears can have effects for both good and ill by offering the possibility of improving the selectivity of gears and so reducing environmental impacts but also by increasing the industry's ability to catch fish from declining populations. We examine the present and future state of these technologies; in later chapters we examine their environmental impacts and some possible solutions.

THE GROWTH OF CAPTURE FISHERIES OVER THE LAST FIFTY YEARS

3.6   The fishing industry is very diverse. It ranges from factory ships on the high seas to inshore craft with a small capacity and crew. Some of the terms that help distinguish these various sectors are described in box 3A.

BOX 3A	THE FISHING INDUSTRY: SOME DEFINITIONS[1]

Capture fisheries: catching wild fish in inland waters or at sea.

Large-scale or intensive fishing: carried out by large commercial boats with strong financial backing.

Industrial fishing sector:[2] concerned with catching fish that will be processed into fishmeal and oil for aquaculture and livestock feed. It targets pelagic species, such as mackerel and herring, and smaller fish such as sprats, collectively known as **forage fish**.

Pelagic fishing: mid-water fishing of species such as mackerel and herring.

Demersal fishing: catching fish associated with the seabed such as flatfish and cod.

Artisanal fishing or small-scale fishing: involves commercial boats that are usually concentrated within a few miles of the coast and are dependent on local resources and closely linked to the community.

Recreational fishing: fishing whose main purpose is enjoyment. It may be carried out from the shore or from small boats.

Aquaculture: farming fish, shrimps, lobsters, plants and other products. It can take place in fresh or salt water, and may be carried out in ponds or rivers, off the coast or out at sea.

EXPANSION OF GLOBAL FISHERIES

3.7   After the second world war, technological advances allowed large boats with powerful engines and advanced gear to move across most of the world's continental shelves. According to official figures,[3] in the two decades following 1950, world capture fisheries production (marine and inland) increased by an average of nearly 6% per year, trebling from 18 million tonnes in 1950 to 56 million tonnes in 1969 (figure 3-I). During the

Figure 3-I
Total world capture fisheries production, 1950–2001[4]

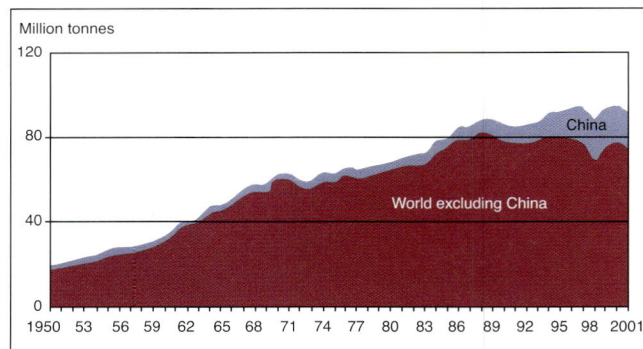

1970s and 1980s, the average rate of increase was 2% per year. This fell to almost zero in the 1990s. In 2000, total capture fisheries reached their highest ever level of nearly 95 million tonnes, followed by a fall to 91 million tonnes in 2001.[5]

3.8 The stagnating trend in the total catch, echoed in many regional fisheries, indicates that the maximum catch levels may have been reached. This is not a reflection of the long-term stability of established fisheries, but of over-exploitation and of declines in many fisheries being offset by exploiting new species, including deep-sea fish. This strategy is unlikely to be sustainable beyond the short term. Moreover, while the proportion of fish destined for livestock feed has decreased in recent years, an increasing proportion is being used to feed carnivorous farmed fish such as salmon (figure 3-II).

Figure 3-II

World use and supply of fish from capture fisheries, excluding China, 1950-2000[7]

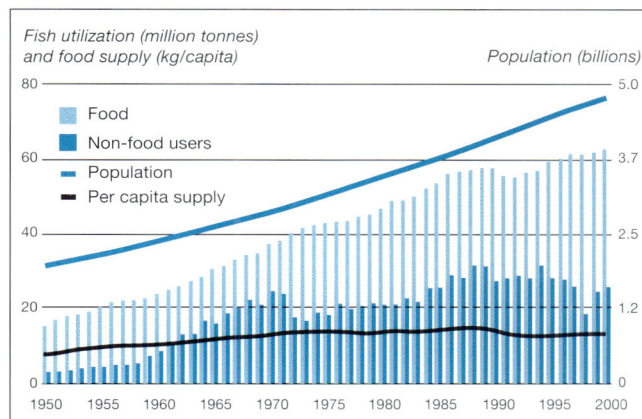

3.9 In 2001, the main world producers in terms of capture fisheries were China, Peru, the (then) 15 countries of the European Union, the US, Japan, Indonesia, Chile, India and the Russian Federation.[6] The European Union is the world's third biggest producer of wild caught fish, accounting for 5% of the global catch. This is equivalent to 6-7 million tonnes fish per year, some of which is caught in the waters of developing countries.

3.10 Chinese fish capture has been rising rapidly and now is said to account for nearly a third of total production from capture fisheries and aquaculture. Recent studies have, however, indicated that the data from China may be unreliable. Excluding the activities of China, global catches have reached a plateau of around 78 million tonnes[8] (figure 3-I).

The emergence and growth of aquaculture

3.11 Production from aquaculture is now growing by around 5% a year and supplies 30% of total global production of fish and seafood. In 2001, global aquaculture supplied around 38 million tonnes of fish, crustacean and mollusc products (figure 3-III). Only about two million tonnes of this was marine fish, representing 2.7% of total marine fish supplies but 4.8% of marine fish for human consumption.[9]

Figure 3-III

World aquaculture production, 1950-2001[10]

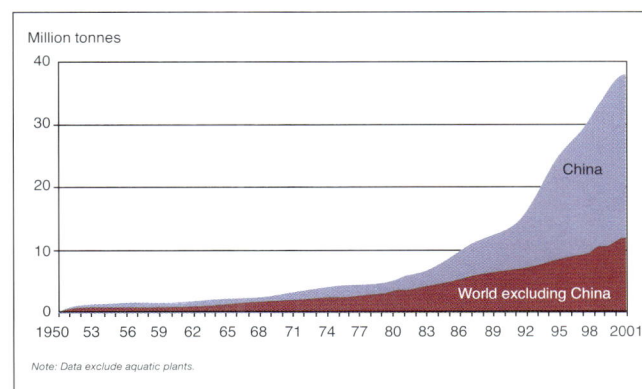

Note: Data exclude aquatic plants.

GLOBAL CONSUMPTION PATTERNS

Figure 3-IV
Projected decline in per capita seafood availability[14]

3.12 Global fish consumption has doubled over the last fifty years, driven in part by population growth. It has been predicted that seafood provision would need to double by 2020 to keep up with demand driven by rising population and income levels in developing countries.[11] Marine capture fisheries are unlikely to be able to meet this demand. This could lead to a significant food gap (figure 3-IV) unless the aquaculture industry

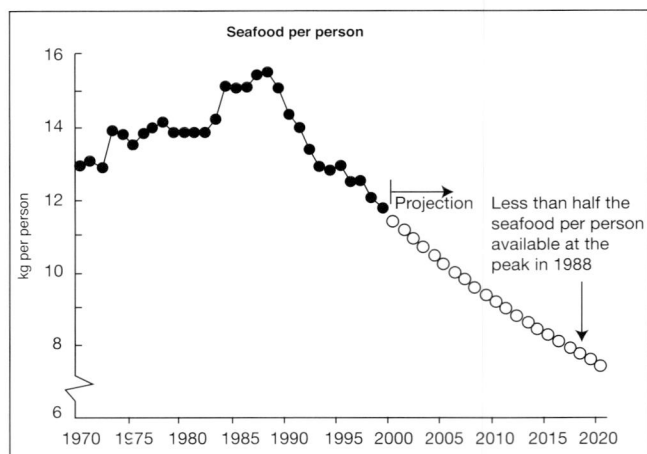

expands substantially. Aquaculture is in fact predicted to grow to the point where it provides 40% of fish for human consumption.[12] Thus, at the global level, freshwater aquaculture, especially in Asia, is replacing the traditional image of trawlers operating in northern seas as the source of fish and seafood.[13] In chapter 6, we examine the extent to which the further expansion of aquaculture may be possible and the potential environmental consequences.

THE EUROPEAN FISHING INDUSTRY

3.13 The North Atlantic is one of the most productive fishing grounds in the world.[15] The waters off northern Europe are exploited by fishing vessels from the EU, Norway, the Faeroe Islands, Iceland and Russia. Only boats from EU Member States however are allowed to fish within European Community waters: Denmark, Spain and the UK land the largest volume of fish.[16] Fishing employs less than 1% of the population of any EU country; however, in some isolated coastal areas with small populations this can rise to 10%.[17] The European Commission estimates that one job at sea generates one and a half on land in associated sectors such as fish processing.

3.14 The main division in the industry is between demersal and pelagic fisheries. In 2003, the European Community's total allowable catch (TAC) of demersal fish species was 3.3 million tonnes, of which nearly one-third (0.9 million tonnes) was industrial sandeel capture.[18] The European pelagic fisheries catch around 4 million tonnes annually.[19] Within this sector, industrial fishing in the North Sea has accounted for catches of around 0.6 to 1 million tonnes per year over the past three decades.[20] There is also a substantial industrial blue whiting fishery of up to 1 million tonnes per year.

3.15 Many more fish are caught than are landed because fish are often discarded at sea. Discarding occurs because fish are of an unmarketable species or size, or because a vessel does not have a quota for that species. A crew may also discard smaller fish once larger ones are caught in order to maximise the landing value of the quota. Discards rarely survive even if thrown back soon after capture. Many more fish escape from nets and other gears only to die of their injuries, as do marine mammals and birds. Discarding therefore represents a large waste of resources and can have severe ecological impacts. It also makes it more difficult to manage fishing sustainably because the true level of fishing activity is hard to estimate.

3.16 Unofficial estimates are that on average a quarter of catches are discarded. In the southern North Sea, beam trawlers discard over half the fish caught, and the discard rate can rise to 90% in some fisheries. The quantity of fish discarded in the North Sea in 1990 has been estimated at around 600,000 tonnes.[21] Official estimates are, however, likely to under-represent the true picture because fishermen have little incentive to report discards accurately.

THE UK FISHING INDUSTRY

3.17 In 2001, the UK was responsible for around 12% of the European catch.[22] Its commercial fleet consisted, in 2002, of some 7,590 registered fishing vessels, the majority of which are less than 10 m long and thus largely outside of the rules on fishing quotas.[23] The number of vessels and fishers engaged in the UK capture sector has steadily declined over the past decade while the fleet registered tonnage has risen, indicating a steady trend towards fewer, larger vessels. Few new vessels have been built in the past decade; the larger boats in the UK fleet (those over 10 m) are on average 23 years old.

Figure 3-V

Trends in the main activities of the UK fishing fleet over 10 m 1994-2002[25]

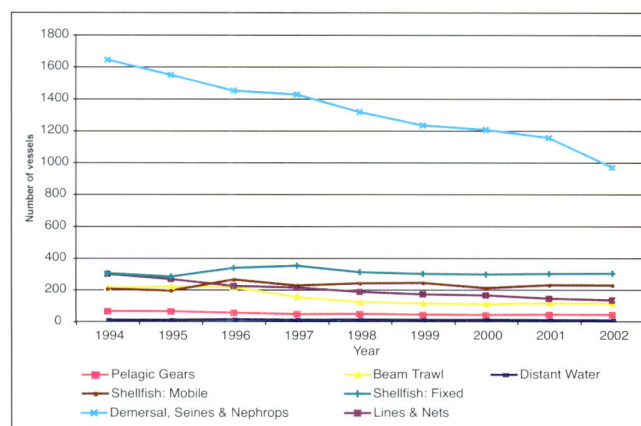

Types of fishing

3.18 Demersal trawling is one of the most commonly practised fishing techniques in the UK, but profitability and investment in demersal trawlers has fallen dramatically due to quotas and fuel costs, with the number of UK beam trawlers over 10 m declining by 50% over the past decade. In contrast, the UK pelagic fleet is profitable and reinvestment in new vessels in this sector is high.[24] Fishing techniques also vary geographically across the UK, for example, over 60% of the UK's pelagic fishing fleet is concentrated in Aberdeenshire.

Figure 3-VI

Decline in the number of UK fishing vessels under 10 m, 1994-2002[26]

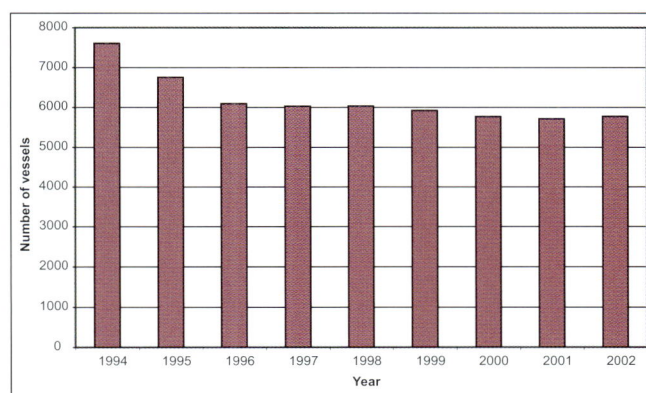

3.19 Trends in the types of commercial fishing practised over the past ten years are shown in figure 3-V. These data show a marked reduction in the number of beam trawlers, which now stands at less than 120 vessels, equivalent to 1.6% of the fleet. Figure 3-VI indicates the reduction in the number of UK vessels under 10 m.

Economics of fishing

3.20 According to a 2004 report by the Prime Minister's Strategy Unit,[27] the UK fishing industry had a turnover of nearly £550 million worth of fish in 2002, which is linked to around £800-1,200 million of associated economic activity. Around £90-100 million of public money was spent on managing the industry in 2002; compared to the industry's gross operating profits of £130 million. Occasional rounds of decommissioning have cost over £100 million in public funds.

3.21 In 2002, exports of fish and fish products were worth around £750 million and imports about £1,400 million. The trade gap has been gradually increasing since the early 1990s and the trend is not expected to change.[28]

3.22 Recreational fishing is also an economically important sector in the UK. In 2002, members of over 1 million households went sea angling at least once in England and Wales, and it is estimated that their direct spending is over £500 million. Comparisons are difficult, but it broad terms it has been suggested that the economic scale of sea angling is similar to that of commercial fishing.[29]

3.23 Fishing comprises only a relatively small part of the wide variety of goods, services and other benefits that the oceans and seas provide. Major contributions come from oil and gas, renewable energy generation potential, transportation corridors and recreational opportunities. The full range of marine-related activities in the UK – including fishing, tourism and offshore oil and gas extraction from the seabed – has been estimated to contribute £69 billion to the UK economy (some 3 to 4% of gross domestic product) and to directly employ around 423,000 people.[30]

3.24 The Prime Minister's Strategy Unit report set out the goods and services provided by the marine environment and allocated monetary values to some of them.[31] Generally, we are more sceptical of monetary valuations, the further they get from real markets, especially when the hypothetical valuations are seeking to express complex environmental or social values. Table F.1 in appendix F reproduces the Strategy Unit's figures, re-arranged into different categories of valuations, including categories for which no monetary valuations are available.

3.25 Whatever significance is given to the more speculative monetary valuations in table F.1, it is clear that the marine environment is a source of great wealth for humans that is worth protecting, quite apart from any moral responsibility for such protection which many people will feel to be important.

Employment in fishing

3.26 The total number of vessels and employees in the UK fleet has decreased every year since 1994 (figure 3-VII). The figures for 2003 indicate that almost 12,000 people were directly employed in fisheries. The drop since 1995 has been dramatic – with full-time employment falling by a third and part-time employment by nearly 40%. These trends have been driven by quota reductions and decommissioning schemes, as well as technology gains.[32]

3.27 The UK fleet is dominated by Scottish fisheries, which account for about 60%. In 2001 (when total fishing employment was higher), there were 6,640 fishermen accounting for 0.2% of total employment in Scotland. However, rates of fishery-based employment levels are considerably higher in some local communities.[34]

3.28 Total UK fisheries-dependent employment, estimated on travel to work areas is shown in figure 3-VIII. The figure for a community's dependency on fishing-related jobs does not however fully capture its vulnerability to the effects of declining employment in this sector. For example, a medium-sized town may have a relatively high level of dependency on fishing, but alternative employment opportunities and access to other job markets may also be good. The most vulnerable communities are those in remote areas cut off from other sources of employment.

3.29 The modernisation of the fleet, together with decommissioning schemes, means that the size of the fishing industry is diminishing. This affects those directly involved in catching fish and the associated processing and support industries. The latter may have some room for diversification since they can also process imported fish and other types of food. A decline in the fishing industry may also have other knock-on effects on the community including on tourist revenue at fishing ports.

Figure 3-VII

Direct fisheries employment and fleet size in the UK[33] Total number of vessels as at 31 December of each year.

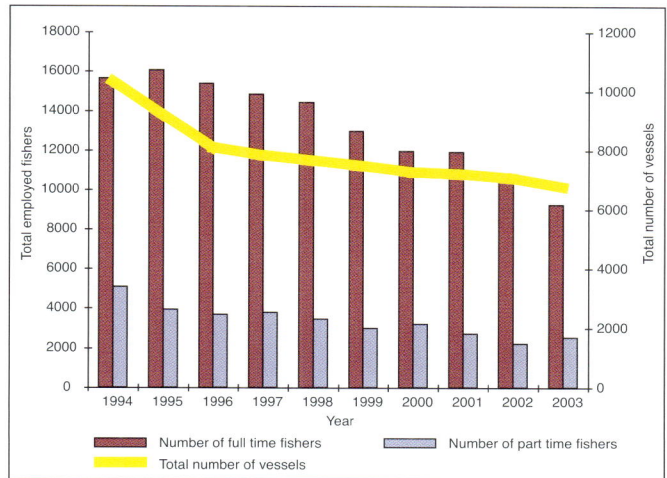

Figure 3-VIII

Dependence on fish catching employment by travel to work area in UK communities.[35] Percentages indicate fish catching employment in travel to work area

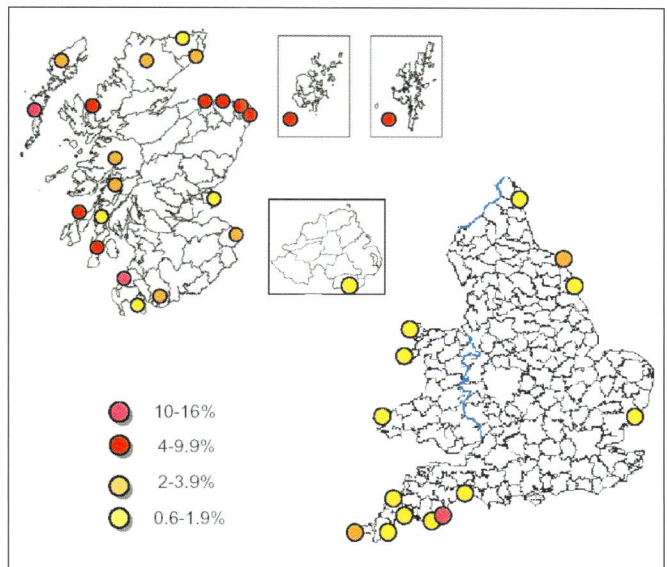

UK DEMAND FOR FISH

3.30 Fish has long been a staple part of the diet, but tastes and markets have changed over time. Consumption of oily fish such as herring and mackerel was far higher in the mid-twentieth century than at present. Cod was a staple British food for many hundreds of years, and fish and chips remains the number one take-away food in the UK. There are over 8,000 fish and chip shops in the UK which employ more than 61,500 people.[36]

3.31 Over the last thirty years, the total consumption of fish in the UK has been fairly stable (figure 3-IX). Similarly, household expenditure on fish has remained broadly the same over the last decade, allowing for the effect of inflation.[37] But the type of fish and the form in which it is purchased has changed considerably, with a move towards chilled and frozen convenience foods. This change is driven by social factors and reinforced by the dominance of supermarkets which now account for 80% of the retail market.[38]

Figure 3-IX
Fish consumption, 1970-2002[39]

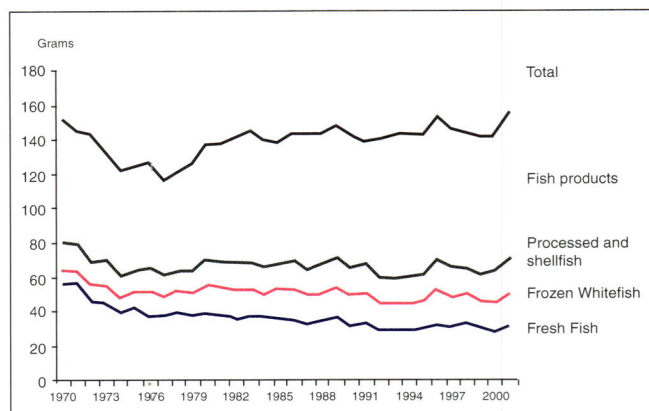

3.32 Appearance and taste are among the major influences affecting consumers' choice of seafood products. Only 4% of those interviewed considered healthy eating as the most important factor affecting purchase, but a much larger proportion – over 40% – ranked it as a significant influence.[40]

HEALTH BENEFITS FROM EATING FISH

3.33 It is clear that eating fish can contribute to a healthy diet. Fish are low in cholesterol and high in protein. They are also a source of all the essential amino acids, vitamins A, D and B complex, and valuable minerals such as iodine, calcium, iron, zinc and selenium.

3.34 Of real significance for Western diets is the fact that, unlike most other foods, fish, especially oily ones such as mackerel and herring, contain high levels of compounds known as long-chain n-3 polyunsaturated fatty acids (n-3 PUFAs) which are associated with numerous health benefits.

3.35 Long-chain n-3 polyunsaturated fatty acids are sometimes known as omega-3 fatty acids and are a family of naturally occurring compounds. It is the position of the double bonds within their hydrocarbon chain that gives n-3 fatty acids their name and their special properties. In these respects, they differ structurally from the more common n-6 family of PUFAs (figure 3-X). The simplest members of the n-6 and n-3 fatty acid families are linoleic acid and alpha-linolenic acid, respectively. Each of these fatty acids has 18 carbon atoms, and they are termed 'essential' fatty acids because only plants, not animals, have the enzymes to insert either n-6 or n-3 double bonds.[41]

Figure 3-X
Structures of n-6 linoleic and n-3 alpha-linolenic acids[42]

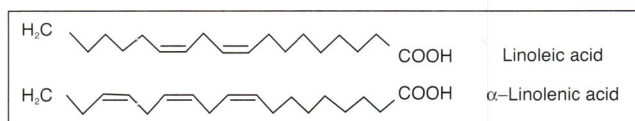

3.36 In 1929, a syndrome of 'essential fatty acid deficiency' was identified in humans, which is similar to other essential nutrient deficiencies.[43] It is estimated that the minimum human requirements for these n-6 and n-3 fatty acids are 1% and 0.2% of daily energy respectively.[44]

3.37 The two most important long-chain n-3 PUFAs necessary for human health are the 20 carbon eicosapentaenoic acid (EPA) and the 22 carbon docosahexaenoic acid (DHA), (box 3B and appendix G). These are abundant in marine life as they are synthesised by marine algae and are passed up the food chain into fish and shellfish. In contrast, EPA and DHA are scarce or absent in terrestrial plants and animals.[45]

3.38 Oily fish have a relatively high proportion of fat in their flesh to facilitate rapid movement up and down the water column. Since the long-chain n-3 PUFAs are fat-soluble, oily fish such as herring, mackerel, pilchards, sardines and fresh tuna are particularly good sources of these beneficial fatty acids. Whitefish (such as cod) have lower levels of fat in their flesh, and hence less fatty acid content, which tends to accumulate in the liver of the fish instead.

3.39 Farmed fish such as salmon are also a good source of long-chain n-3 PUFAs, when fed a diet high in fish; the concentration of long-chain n-3 PUFAs declines if the fish are fed a predominantly vegetable diet (chapter 6). Long-chain fatty acid content is also low in naturally herbivorous farmed fish. Some freshwater fish can produce PUFAs if they are fed vegetable oils, but these do not match the levels produced by oily marine fish.

BOX 3B   **WHERE DO LONG-CHAIN n-3 POLYUNSATURATED FATTY ACIDS COME FROM?**

Mammalian cells cannot synthesise linoleic acid and alpha-linolenic acid (ALA), but they can metabolise them (albeit slowly) by desaturation and elongation of the carbon chain. In the case of ALA this leads to the formation of eicosapentaenoic acid (EPA) and with addition of two further carbon atoms, to docosahexaenoic acid (DHA) (see figure below).

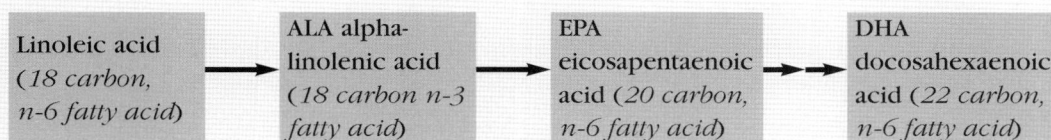

Linoleic acid (*18 carbon, n-6 fatty acid*)	→	ALA alpha-linolenic acid (*18 carbon n-3 fatty acid*)	→	EPA eicosapentaenoic acid (*20 carbon, n-6 fatty acid*)	→	DHA docosahexaenoic acid (*22 carbon, n-6 fatty acid*)

Tissues and oils in some plants (such as linseed, rapeseed, leeks and green leafy vegetables) are good sources of linoleic acid and alpha-linolenic acid (ALA). However, while up to 50% of fatty acids in plants may consist of ALA, their low fat content limits the bioavailablity of this to humans. In contrast, plant oils used in cooking are rich sources of linoleic acid, and, to a lesser extent ALA. Several, but not all, epidemiological studies have shown an association between ALA consumption and reduced cardiovascular risk. However, this is not proof of a causal relationship and dietary intervention studies do not support this either. The reported effects may also be due, not to ALA itself, but to EPA to which it can be converted.

The conversion of ALA to EPA and DHA in the human liver is slow and inefficient (less than 10% of ALA is converted to EPA, and much less to DHA).[46] The process can also be inhibited by high levels of n-6 fatty acids in the diet. Thus, while terrestrial plants, and to a lesser extent, animals, are a source of ALA-rich oils for direct human use, they only provide a limited source of EPA or DHA. In contrast, these compounds are present in high amounts in oily marine fish (see table below).

Typical content of EPA, DHA and total long-chain n-3 PUFAs in selected foods[47]			
Food	EPA g/100g	DHA g/100g	Total long-chain n-3 PUFA/portion g
Cod	0.08	0.16	0.30
Herring	0.51	0.69	1.56
Mackerel	0.71	1.10	3.09
Sardines (tinned)	0.89	0.68	1.67
Salmon (wild)	0.50	1.30	2.20
Prawns	0.06	0.04	0.06
Mussels	0.41	0.16	0.24
Roast beef	0.02	0.00	0.04
Roast lamb	0.03	0.02	0.08

Fish oils and the flesh of oil-rich fish remain the most important sources of EPA and DHA. However, recent studies have shown that ruminant products such as milk, lean beef and lamb are also a source of ALA and long-chain fatty acids, and the quantities of these compounds can be increased by altering the animals' diet.[48,49,50] For example, the meat and milk from cattle grazed on grass and clover have been reported to contain higher levels of ALA, and to a lesser extent EPA, than those from cattle fed intensively on grain.[51,52,53]

It may also be possible to artificially produce or enhance the levels of long-chain n-3 PUFAs in plants and animals through conventional breeding or genetic manipulation.[54] Although such techniques are in an early stage of development, recent research has succeeded in producing genetically modified linseed plants that accumulate significant levels of n-3 and n-6 PUFAs in their seeds.[55]

It is possible that healthier, more nutritious plant oils might be produced for human consumption. However, it is not clear whether increasing the amount of EPA and DHA in alternative dietary sources could ultimately produce the equivalent health effects as fish oils.

Another way to increase intake of long-chain n-3 PUFAs is to take supplements. For example, cod liver oil is taken by 13% of the UK population. Cultures of marine algae can also be used to produce DHA. This product is already being used as a supplement and as a component of baby formula. It is less clear whether taking this product confers the same health benefits as eating oily fish, which contain both EPA and DHA.[56]

WHY ARE LONG-CHAIN *n-3* POLYUNSATURATED FATTY ACIDS IMPORTANT?

3.40 The UK's biggest killer is heart and circulatory disease. In 2003, over 117,000 people died of coronary heart disease and 2.7 million people continue to live with the illness.[57] Research strongly links the intake of long-chain n-3 PUFAs with the prevention or amelioration of coronary heart disease, along with a number of others listed below, in descending order of the strength of the available evidence:[58]

- coronary heart disease and stroke;

- essential fatty acid deficiency in infancy;

- autoimmune disorders;

- Crohn's disease;

- cancers of the breast and prostate;

- mild hypertension; and

- rheumatoid arthritis.

3.41   Studies across 36 countries have shown links between fish consumption and reduced levels of mortality from all these diseases. The evidence base is discussed in more detail in appendix G.

NUTRITIONAL ADVICE

3.42   Given the benefits of eating fish, the Food Standards Agency (FSA) advises that "people should consume at least two portions of fish a week, of which one should be oily". In the UK, the figure for total fish consumption averaged across the adult population is around 1.55 portions a week, of which only a third of a portion is oily.[59,60] There is also a marked variation in intake between individuals, related partly to age and gender. In a recent dietary survey, seven out of ten people interviewed did not eat any oily fish at all during the study period.[61]

3.43   How much extra fish would be required if people were to adhere to the FSA's recommendations? Meeting the suggested intake level for oily fish, i.e. increasing overall consumption by 0.67 portions of oily fish, would also mean that the recommendation to eat two portions of fish a week would be met. To increase fish consumption by this amount for 49 million adults in the UK, would require an extra 33 million portions of oily fish per week. This implies an increase in present levels of total fish consumption of over 40%, and of oily fish by 200%.

3.44   The FSA's advice equates to an intake of 0.45 g/day long-chain n-3 PUFAs. This represents a minimal average population goal, not the level that would be required for maximum nutritional benefit.[62] In intervention trials with people who have already suffered some form of cardiovascular disease, intakes of around 1 to 1.5 g/day of long-chain n-3 PUFAs have been needed in order to show demonstrable benefits.[63] The incentive to improve the intake of PUFAs could thus conceivably drive demand for fish even higher.

3.45   There is therefore an important environmental dimension to the consumption of fish, especially if this reflected in higher demand in other countries as well. While some consider populations of most, but not all, pelagic stocks to be relatively sustainable, fishing on this scale has effects on the wider ecosystem. For example, some species of tuna are associated with high levels of by-catch of birds and mammals. Moreover, farming carnivorous fish such as salmon relies heavily aquafeed derived from supplies of wild fish.

CONTAMINANTS IN FISH AND FISH OIL

3.46   Although there is strong evidence for the health benefits of eating seafood, there are also concerns about contaminants. The most significant of these are dioxins, dioxin-like polychlorinated biphenyls and methylmercury, which are examined in the following sections and in appendix G.

Dioxins and dioxin-like polychlorinated biphenyls

3.47   These are persistent compounds that accumulate in lipids and are likely to be present in oily fish as a result of environmental contamination. The Food Standards Agency advises that men, and women not intending to have children, can eat up to four portions of oily fish a week before the possible risks might start to outweigh the known health benefits. Girls and women who may become pregnant at some point in their lives are advised to consume between one and two portions of oily fish per week, with extra restrictions on some types of fish at particular times (3.50). These contaminants also occur in supplements derived from fish products such as cod liver oil (see appendix G).

3.48   Risk-benefit analysis of eating salmon has shown that there would be adverse health effects from not eating salmon in order to avoid contaminants, unless the long-chain n-3 polyunsaturated fatty acids are supplied from other sources.[64, 65]

Mercury

3.49   Another important contaminant found in fish is mercury. Its toxicity depends on whether its form is elemental, inorganic or organic (e.g. methylmercury). Methylmercury affects the kidneys and nervous system. The fetal central nervous system is particularly at risk because methylmercury can cross the placenta and the immature blood-brain barrier.

3.50   Methylmercury accumulates within the marine food web and is at its highest levels in large, carnivorous fish. Current Food Standards Agency advice is that women who are pregnant, or are expecting to become pregnant, should eat between one and two portions of oily fish weekly, but avoid eating shark, swordfish or marlin, or large amounts of tuna.[66] The Food Standards Agency also advises that other adults should eat no more than one portion of shark, swordfish or marlin a week, and children should avoid eating them.

3.51   There are also concerns about other heavy metals, high levels of fat-soluble vitamins, the use of colourants, toxins and seafood allergies which are described in appendix G.

CONCLUSIONS ON HEALTH BENEFITS OF EATING FISH

3.52   Recognising the health benefits of the long chain n-3 polyunsaturated fatty acids in fish, the Food Standards Agency has recommended that people should eat two portions of fish a week, one of which should be oily. Higher levels of fish consumption will increase pressure on already depleted fish populations. As discussed, there are also concerns over the contamination of fish and fish oil with toxic products such as dioxins and polychlorinated biphenyl compounds, and over consuming excessive levels of fat-soluble vitamins. **We therefore recommend that:**

- **studies are undertaken to examine the full environmental implications of the Food Standards Agency's advice on eating fish;**

- **every effort is made to introduce alternative sources of long chain polyunsaturated fatty acids (n-3 PUFAs) from biological sources other than fish;**

- **an urgent effort is made to discover efficient chemical synthetic pathways to generate the fatty acids, EPA and DHA;**

- **further consideration is given to providing advice to the public about adding long-chain n-3 PUFAs as dietary supplements rather than relying solely upon an increase in oily fish consumption; and**

- **further research is undertaken to discover the mechanisms by which long-chain n-3 PUFAs benefit human development and health.**

TECHNOLOGICAL DEVELOPMENT IN THE MARINE FISHING INDUSTRY

3.53 Future levels of fishing activity will be affected by levels of consumer demand, but also by economics, legislation and environmental change. Technology is another important driver. We know that technological advances over the last fifty years have led to a dramatic increase in the amount of food produced from marine fisheries and aquaculture,[67] but this is now tailing off in the marine capture sector. The extent to which newer technologies are developed will affect future levels of production and, crucially, the impact of fishing on the environment.

TECHNOLOGY TO INCREASE PROFIT

3.54 Technology can increase the fishing industry's profit by increasing revenue, or reducing costs, or by doing both. Revenue can be increased by using technologies that enhance the catching power of a fishing gear, increase the encounter rate with target species, or that add value to the catch once it is onboard. Income can also be increased by opening up new markets for species previously ignored or discarded, or by gaining access to new fishing grounds. Recently, most new fishing opportunities have arisen in deep water, because the bulk of the fishing grounds and fish populations on continental shelves are already fully exploited.

3.55 Technological advances have been made in fishing gear, fishing methods, bridge electronics to 'look' under the sea to find fish, vessel design, propulsion systems, deck machinery and catch preservation. In most cases, there is little quantifiable information on the environmental impact of such advances. But environmental impacts are likely to have increased where technological advances have led to boats expanding their range or time at sea or their ability to find fish.

3.56 The biggest advances in propulsion technology were experienced during the first 50 years of the twentieth century, when vessels changed from sail to engine power allowing heavier gear to be towed. Since then the real advances have been in electronics.

3.57 Modern fishing vessels are equipped with the most sophisticated electronic aids available to mariners. This equipment increases the encounter rates with target species by reducing uncertainty in the fish capture process and provides information and tools that aid the skipper to choose when and where to fish most profitably. Advances have been made in the technologies of accurate vessel position location (Global Positioning Satellite, GPS), radar, sonar, fish finding, echo sounders, seabed mapping, electronic chart plotting, fishing gear sensors and autopilots.

3.58 Of these, the GPS system has been particularly important and has enabled fishers to record their vessel location at sea to an accuracy of 5 m. The availability and growth of powerful microprocessors has facilitated the development of electronic charts and plotters, and has allowed once discrete electronic aids to be networked, giving rise to new integrated functionality. Technological advances in bridge electronics and instrumentation have precipitated changes in fishing patterns in recent decades that are likely to have had both positive and negative environmental impacts. Improved technology has at least in part offset declining stocks, allowing fishers to seek out what fish remain.

FISHING GEAR DEVELOPMENTS

3.59 In the past 100 years, northern European fishing fleets have embraced new fishing techniques and abandoned less effective and profitable ones. For instance, in European demersal fisheries, seine-netting was the preferred fishing technique for significant numbers of UK and Scandinavian fishing vessels but during the past few decades, seine-netting has given way to demersal trawling and beam trawling and only a few seine-netters remain in the UK. Similarly, in the past 50 years, side trawling has been largely replaced by stern and by beam trawling. In the pelagic fisheries, small numbers of efficient large pelagic trawlers and purse-seiners now dominate where once the fishery was prosecuted by many hundreds of small driftnet and ringnet fishing vessels. Many new vessels are constructed as multi-purpose vessels and are able to switch between fishing techniques, choosing the most profitable options at any particular time. Thus, once stocks or quotas are exhausted or prices fall, they can switch to new species or areas. It is probable that the evolution of new, more effective and more profitable fishing techniques will be accompanied by increases in environmental impact. Appendix H gives details of the basic fishing gears and techniques used by the fishing industry.

3.60 There has also been a major shift from use of biodegradable natural fibres such as hemp and sisal in the construction of fishing gear to stronger, more durable and hard-wearing synthetic polymer-based fibres. Synthetic polymers allow bigger nets to be used but are resistant to degradation, so they persist in the environment much longer than natural fibres when gear is lost or abandoned. This can result in protracted periods of 'ghost fishing' where a net continues to destroy sea life after being abandoned or lost. This is a more serious problem in deeper water where the nets persist longer than in shallow water.

3.61 Technological advances in deck machinery have had some effects upon the environmental impacts of capture fishing. The advent of hydraulic systems has enabled much larger and heavier gears to be hauled aboard than was previously possible. Powerful net haulers and net flaking machines have permitted the gillnet and tangle net sectors to increase the amount of netting they can shoot and handle each day. The introduction of power blocks and net drums on trawlers has allowed vessels of all sizes to increase the size of trawls used and to carry out twin- and multiple-rig trawling. Auto-winch control systems keep an even load on the trawl winches and maintain optimal fish catching net geometry, so improving the efficiency of capture.

3.62 The ability to preserve the catch, or to add further value to the catch through onboard processing, are practices that have been adopted by virtually every modern fishing vessel. The onboard use of ice, liquid ice, vivier tanks (which hold live crustaceans in circulating seawater tanks for extended periods), refrigeration and freezing plants has made longer voyages economically viable and helped pave the way for the expansion of fishing into more distant fishing grounds. This technology has served to help fishing expand its range and increase the associated environmental impacts.

3.63 Tactics and technologies have developed in response to fisheries regulations targeting larger vessels, in order to dilute or circumvent proposed management measures. For example, many vessels are now designed and constructed to fall just below the cut-off point (in terms of length or engine power) at which rules apply. These vessels are designed and built to have the same fishing capacity as larger vessels, but are exempt from the restrictions. This has contributed to the evolution of 'short and stumpy' fishing vessel designs and to new classes of vessel.

3.64 Significant advances have been made through the automation of longlining which has greatly increased the catching power of longliners by permitting more hooks to be baited and deployed per unit of time. Advances in hook technology and in the use of swivels and stronger backing lines have also served to make this fishing technique much more effective. Pot and trap design improvements have resulted in the parlour pot and collapsible/stackable pots. These static gear technical innovations are likely to have increased environmental impact by increasing fishing effort.

3.65 There can be little doubt that much of this technological development has contributed to an overall increase in the environmental impact of the capture fisheries by increasing temporal and spatial fishing effort.

TECHNOLOGY TO REDUCE ENVIRONMENTAL IMPACT

3.66 Considerable work has gone into improving the size selectivity of fishing gears by manipulating mesh size, and this has been used as a measure to protect young fish in over-exploited stocks. Widespread use of this technology has however failed to halt the decline of many fish stocks in many European fisheries. More recent developments in the field involve the use of escape panels and modified codends, which are often constructed from alternative mesh shapes. Most European demersal fisheries are mixed-species fisheries, and developments have taken place to make fishing gears more species-selective, particularly against the background of quota restrictions. This research has so far resulted in grids, sieve nets, separator trawls and cut away trawls. Much research and development work continues in this field.

3.67 Successful technologies have been developed for use in the longline fisheries to reduce incidental catches of birds. These include streamers, sinker weights and setting tubes. Pingers, grids and modified sieve nets are being developed to reduce the incidental capture of mammals in the pelagic and static gear fisheries, although these have yet to solve the problems completely.

3.68 Until recently, gillnets were set out in lengths up to 50 km: international legislation now limits the length to about 2 km. Reliable figures on the extent of ghost fishing are difficult to find but one study reported that 7,000 km of drift nets were being lost each year in the North Pacific fishery.[68] Fairly simple technologies can reduce the impact of lost and abandoned fishing gears. These include retrieval programmes, addressing the causes of loss/abandonment and biodegradable release mechanisms on fish traps/pots. A return to traditional biodegradable net materials would reduce the length of time a net would 'ghost fish' if lost (chapter 9).

3.69   The advent of the Vessel Monitoring System (VMS, onboard satellite tracking) is facilitating the enforcement of fisheries regulations aimed at reducing environmental impacts. Enforcement could be further aided through the use of electronic logbooks and the tagging/marking of fishing gears. The strategic placing of sensors onboard a vessel and on the fishing gears which are linked to the VMS system has significant potential to provide extremely valuable data for scientists and managers.

3.70   Some technologies aim to reduce the damaging impact of gears on the seabed, but these are mostly at the developmental stage and are not used commercially to any extent. This area is one of the least advanced.

3.71   Further information about each of these techniques is given in the consultants report on technological innovations in the capture fishing industry prepared in support of this study. The report is available on the Royal Commission's website.[69]

CONCLUSIONS

3.72   The 'industrialisation' of fishing has seen global catches at sea increase fourfold over the last 50 years. This has inevitably had an impact on fish populations and the ecosystems to which they are intimately linked. Fishing vessels have increased in range and power and they also have access to a wide variety of ever more sophisticated gear. Much of it is heavy and capable of damaging the seabed. These operations may also be very large – some nets are capable of catching so many fish that it is not safe to pull them on board and the fish have to be pumped onto the boat instead.

3.73   The rise in wild caught fisheries has begun to level off in recent years while aquaculture has been growing rapidly, and there is no indication of a reversal of this trend. We are likely to see rising demand for fish products over coming years as global populations and incomes rise. This will undoubtedly put more pressure on the environment but it may also lead to restricted access to fish protein, which is an important component of many people's diets.

3.74   This is of particular concern as fish appear to be the best source of long-chain n-3 polyunsaturated fatty acids that are linked to decreased incidence of coronary heart disease and are thought to have a range of other important but perhaps less well-understood health benefits. There is an urgent need to understand more fully the environmental consequences of nutritional strategies and to speed up the search for alternative sources of these compounds.

3.75   The environmental impacts of capture fisheries could be reduced through the introduction of better-targeted technologies. These are being developed by virtually every fishing nation. In addition there is good international collaboration between workers in this field, facilitated by the International Council for the Exploration of the Seas. On the other hand, fishermen are normally only willing to adopt mitigating technologies if they also reduce costs or increase income through larger catches. If such technologies hit profits, fishers may try to find ways to avoid using them.

3.76 Aquaculture also faces many technological challenges. In Northern Europe, there are efforts underway to both increase the size of units, but also to address their environmental impacts (which we discuss in chapter 6). In developing countries, there is the opportunity to develop the industry in a sustainable way that does not cause unacceptable environmental impacts, by improving efficiency and designing aquaculture diets to reduce the reliance on industrially caught forage fish.

3.77 We examine in the next chapter the attempts to regulate capture fisheries and aquaculture.

Chapter 4

THE LEGAL FRAMEWORK FOR THE MARINE ENVIRONMENT AND FISHERIES

INTRODUCTION

4.1 Fish do not respect political boundaries and cannot be fenced off and allocated to individuals. Each fish that is caught means one less fish for someone else. This lack of property rights has led to a sharp difference between short-term interests of individual fishers on the one hand, and societal and long-term interests on the other. This has motivated fishers to race each other for fish, leading to overcapitalisation in the industry and depleted fish populations.

4.2 Agriculture has long been carried out within a framework of land ownership, that often provides incentives to exercise stewardship to ensure the future fruitfulness of the land. This has not been true of fisheries. Capture fisheries are a form of marine hunting over territory that, with the exception of some narrow coastal margins, was until recently outside the constraints of national and international law.

4.3 From the 17th century[1] onwards, the prevailing doctrine was formalised in the doctrine of *mare liberum*, 'freedom of the seas', under which most uses of the world's oceans remained unregulated except for limited constraints that nations placed on their own citizens within coastal zones. This approach emerged for a very practical reason: no single nation or group of nations could effectively either monitor or control activities on the oceans except within fairly close proximity to land, and thus the 'freedom' doctrine emerged as a negotiated compromise.

4.4 For many years, fish were generally so abundant there was no reason to establish national jurisdiction over the seas or oceans. After the industrial revolution, however, the increasing scale of fishing caused conflicts, resulting in treaties and other agreements from the 1870s onwards. Negotiations leading to the 1982 United Nations Convention on the Law of the Sea (UNCLOS) gave rise to the emergence of a nation's right to claim an exclusive economic zone (EEZ). The EEZ extends out to 200 nautical miles (nm) and generally covers the continental shelf – the most productive area.

4.5 Most of the world's oceans, however, remain outside national EEZs and so remain in a state of 'freedom' as an unregulated common resource. As this chapter shows, conflict over high seas fisheries in these areas is still a problem.

4.6 There is already a wide range of measures to control fishing's impact on commercial stocks. These tend to view fish as stocks to be managed to deliver a maximum sustainable yield rather than as populations to be protected as part of a complex ecosystem. The international framework is gradually evolving to regulate the widespread environmental damage caused by fishing and to promote sustainable development. Increasingly, environmental measures are being introduced into traditional fisheries management territory, and we examine the overlap between these two areas in this chapter. The situation is complex and needs to move much further towards providing environmental protection and to do so as a matter of urgency.

4.7   We begin by looking at a selected range of treaties, soft law guidance and codes that address the management of fisheries and the marine environment. We also cover other international obligations that play an important role in constraining national measures in these areas, such as the agreements overseen by the World Trade Organisation. We then examine the EU Common Fisheries Policy in detail and outline some of the main regulations applying to aquaculture. We also look briefly at the alternative management approaches taken by other countries.

MULTILATERAL AGREEMENTS FOCUSING ON FISHING AND TRADE

UN CONVENTION ON THE LAW OF THE SEA (UNCLOS)

4.8   The UN convened the first conference on the Law of the Sea in 1958. This conference produced four conventions, dealing respectively with the territorial sea and the contiguous zone, the high seas, fishing and conservation of the living resources of the high seas and the continental shelf. The legal framework continued to develop, culminating in the adoption of the UN Convention on the Law of the Sea in 1982,[2] and its entry into force in 1994.

4.9   The scope of the Convention is vast: it covers all ocean space, with all its uses, including navigation and over-flight; all uses of all its resources, living and non-living, on the high seas, on the ocean floor and beneath, on the continental shelf and in the territorial seas; the protection of the marine environment; and basic law and order.[3] The European Community is a party to the Convention, along with 145 individual states. The US has not ratified the Convention. The key provisions of the Convention are covered in box 4A, including the establishment of the current geographical limits to state fisheries and the EEZ limit of 200 nm.

4.10   The Convention establishes the international framework for conservation and management of marine living resources. It makes two types of distinction: the first between fisheries taking place on the high seas beyond national jurisdiction and those fisheries subject to coastal state sovereign rights (within the EEZ and on the continental shelf); and the second based on species behaviour when a species' migratory path or life cycle takes it outside the boundaries of a single state. Species behaviour distinctions arising from UNCLOS are given in box 4B.

4.11   All states are under a general obligation, arising from customary international law and contained in Article 192 of the Convention 'to protect and preserve the marine environment'. This obligation applies everywhere in the sea, including the high seas. Under another customary obligation reflected in the Convention, the measures taken 'shall include those necessary to protect and preserve rare or fragile ecosystems as well as the habitat of depleted, threatened or endangered species and other forms of marine life'.[6] This obligation covers any kind of vulnerable marine ecosystems or species, wherever they are located.[7]

BOX 4A   KEY FEATURES ARISING FROM UNCLOS[4]

Territorial seas

- Coastal states have sovereignty over their territorial seas, which they can establish up to a maximum limit of 12 nm. Foreign vessels are allowed 'innocent passage' through those waters. The latter term does not include fishing activity.

Exclusive economic zones

- Coastal states have sovereign rights in a 188 nm EEZ beyond the 12nm territorial sea to explore and exploit the living and non-living natural resources (Ninety percent of the world's fisheries fall within coastal state jurisdictions).

- Coastal states are responsible for conserving and managing living resources and for protecting the marine environment within their EEZ.

Continental shelf

- Coastal states have sovereign rights over their continental shelf, their national area of the seabed, for exploring and exploiting its natural resources (non-living and sedentary living organisms). The shelf often extends at least 200 nm from the shore. States may claim more under certain circumstances.

- Where the shelf extends beyond 200 nm, coastal states are to share with the international community part of the revenue they may derive from those resources.

BOX 4B   SPECIES BEHAVIOUR DISTINCTIONS MADE BY UNCLOS[5]

- Fish stocks that occur entirely within a single EEZ;

- Fish stocks or stocks of associated species that occur in more than one EEZ (transboundary straddling stocks);

- Fish stocks or stocks of associated species that occur both within an EEZ(s) and in the adjacent high seas (commonly called straddling stocks);

- Highly migratory species like tuna, listed in Annex I of the Convention, which migrate long distances, usually through several nations' EEZs and the high seas;

- Marine mammals like whales, which range throughout the oceans, and other cetaceans whose range is more regional;

- Anadromous species like salmon, which spawn in freshwater rivers and streams but spend most of their life cycle at sea;

- Catadromous species like eel, which spawn at sea but spend most of their life cycle in freshwater; and

- Sedentary species of the continental shelf, such as crab, lobster, and coral, defined as living organisms 'which, at the harvestable stage, either are immobile on or under the seabed or are unable to move except in constant physical contact with the seabed or the subsoil'.

4.12 In addition, coastal states are allowed, through the competent international organisation, to adopt more stringent measures for the prevention, reduction and control of pollution from vessels in an area 'for recognized technical reasons in relation to its oceanographic and ecological conditions' – a limited form of special area where general international rules and standards are inadequate to meet such special circumstances.[8]

4.13 States are also bound by obligations to co-operate for the protection of the marine environment and the conservation and the management of high seas living resources.[9] The concept of an obligation to co-operate, which is typical of the high seas where no national jurisdiction can be established, is not devoid of legal meaning. It implies a duty to act in good faith in entering into negotiations with a view to arriving at an agreement and in taking into account the positions of the other interested States.

4.14 The Convention requires that conservation measures in the EEZ and on the high seas should be based on the best scientific evidence available to the state(s) concerned. Coastal states must take this into account, whereas on the high seas the requirement is slightly stronger – that conservation measures be designed on such evidence. All states participating in a fishery are obliged to contribute and exchange on a regular basis, through competent international organizations, available scientific information, catch and effort statistics, and other relevant data.[10]

4.15 The Convention standard for conservation measures is that they be designed to maintain or restore populations of harvested species at levels which can produce the maximum sustainable yield (MSY), as qualified by relevant environmental and economic factors. This standard has been criticised in a some quarters for focusing on maintaining production rather than protecting the ecosystem, and for not being sufficiently precautionary in the light of the many uncertainties about population and catch levels.[11]

Dispute settlement under UNCLOS

4.16 The Convention established a unique system for international dispute settlement that has since been adapted for use in other agreements. It offers states a menu of options for settling disputes, but in the end they must submit to compulsory, binding procedures in most situations. In contrast, most other international treaties require parties' consent to dispute procedures, which are not therefore compulsory.

4.17 Dispute procedures can be applied to the protection and preservation of the marine environment. However, there are some exceptions. Coastal states have discretion regarding fisheries conservation and management laws and regulations (for example, setting a limit on total allowable catch).[12] Other fisheries matters are subject to conciliation only. There is also an optional opt-out for disputes arising from law enforcement activities linked to, *inter alia*, this exception.[13] However, in circumstances where serious harm to the marine environment may result, the court or tribunal dealing with the dispute may prescribe provisional measures to prevent such harm. There are safeguards to ensure that provisional measures are not delayed in urgent situations. The parties to the dispute must comply with them.[14]

4.18 The Convention relies on links to other international agreements, particularly in the field of marine environmental protection. It subjects a coastal state to compulsory and binding procedures when it is alleged to have acted in contravention of applicable international rules and standards for marine environmental protection established by the Convention or through a competent international organization or diplomatic conference in accordance with the Convention.

4.19 UNCLOS has not, however, prevented overfishing within the EEZs or on the high seas.[15] The Straddling Stocks Agreement (4.20) strengthens the requirement that coastal states base conservation measures for these stocks on the best scientific evidence available.

UN STRADDLING STOCKS AGREEMENT

4.20 The full title of this agreement is the United Nations Agreement for the Implementation of the Provisions of the United Nations Convention on the Law of the Sea of 10 December 1982 relating to the Conservation and Management of Straddling Fish Stocks and Highly Migratory Fish Stocks.[16] Its aim is to ensure the long-term conservation and sustainable use of straddling fish stocks and highly migratory species. In particular, it requires states to co-operate so that there is compatibility between national and high seas measures.[17]

4.21 The Agreement was adopted in 1995, entered into force in 2001 and the EU became a party (of which there are now 52) in 2003. The Agreement includes commitments to:

- adopt measures to ensure the long-term sustainability of straddling fish stocks and highly migratory fish stocks;

- ensure that measures are based on the best scientific advice available and are designed to maintain or restore stocks at levels capable of producing maximum sustainable yield;

- apply the precautionary approach;

- assess the impacts of fishing and other activities target stocks and associated or dependent species;

- minimise pollution, discards, bycatch, etc; and

- protect biodiversity in the marine environment.

4.22 The Agreement contains additional provisions for fisheries conservation, for example, Article 5(h) to "take measures to prevent or eliminate overfishing and excess fishing capacity". This provision is part of the same package of obligations listed above and needs to be implemented in a consistent way. This type of integrated approach to fisheries conservation and management is recognised in the UN Food and Agriculture Organization (FAO) International Plan of Action on Managing Fishing Capacity adopted in 1999.[18] The Plan of Action calls for the reduction ('management') of fishing capacity consistent with the conservation provisions contained in the UN Straddling Stocks Agreement as well as the UN FAO Code of Conduct for Responsible Fisheries. Paragraph 9.IV of the Plan of Action states:

"The management of fishing capacity should be designed to achieve the conservation and sustainable use of fish stocks and the protection of the marine environment consistent with the precautionary approach, the need to minimize by-catch, waste and discards and ensure selective and environmentally safe fishing practices, the protection of biodiversity in the marine environment, and the protection of habitat, in particular habitats of special concern."

4.23 The Agreement applies to EU fisheries in the north-east Atlantic for albacore tuna, blue whiting, Atlanto-scandian herring and other straddling and highly migratory stocks both inside and outside waters under EU jurisdiction. It also applies to Mediterranean fisheries for tuna, swordfish and other migratory species and EU distant water fisheries for tuna, swordfish, and other highly migratory species in the Central and South Atlantic, the Indian Ocean and the Pacific Ocean, as well as straddling stock fisheries in various parts of the world.

4.24 Some of the provisions of the Agreement are already reflected in EU legislation and commitments under the Common Fisheries Policy, the Habitats Directive and the Agreement on the Conservation of Small Cetaceans in the Baltic and North Seas. However, EU legislation governing fisheries on straddling and highly migratory fisheries will have to be further modified to bring them into line with the obligations established in the UN Straddling Stocks Agreement.[19]

DISPUTE SETTLEMENT UNDER THE STRADDLING STOCKS AGREEMENT

4.25 One of the most important elements of the Agreement is its provision for settling disputes. This should be of great help in enforcement. For example, if the European Community feels that the Agreement is not being implemented by other states, it has recourse to the binding arbitration procedures, including the International Tribunal for the Law of the Sea, provided for under UNCLOS. However only states may appear in contentious cases before the International Court of Justice in The Hague.

4.26 The dispute procedures may also open the way for stronger conservation action. UNCLOS Article 290 allows for the application of 'provisional measures' in the event that there is a threat of 'serious harm' to the marine environment. This could include a legally binding decision to halt the activity in question, pending the final outcome of the arbitration process. It is possible that a strong argument could be made that fishing that involves a high risk of fish stock depletion or species extinction would constitute a threat of 'serious harm' to the marine environment.

4.27 In our view, it would be odd to say the least, if the EU set a lower standards for conservation and management of North Sea fisheries, or for EU boats fishing off West Africa, than the standards by which the EU is bound under the UN Straddling Stocks Agreement.

FAO CODE OF CONDUCT ON RESPONSIBLE FISHERIES

4.28 Although the provisions of the UN Straddling Stocks Agreement are only binding on fisheries for straddling and highly migratory fish stocks, Articles 5 and 6 of the Agreement set international standards for the conservation and management of fisheries in general. These are reinforced by the FAO Code of Conduct for Responsible Fisheries.[20]

4.29 The Code, adopted in 1995, is voluntary, and aimed at everyone involved with fisheries and aquaculture, irrespective of whether they are located in inland areas or in the oceans. It consists of a collection of principles, goals and elements for action. The code advocates that countries should have clear and well-organised policies in order to manage their fisheries and that these should be developed with the cooperation of all groups that have an interest in fisheries. The Code calls for consideration of the environmental and social impacts of fishing. It also calls for use of the best possible scientific information, while taking traditional fishing knowledge into account, and for more cautious limits to be set in the absence of adequate scientific information.

FAO *INTERNATIONAL PLAN OF ACTION TO PREVENT, DETER AND ELIMINATE ILLEGAL, UNREPORTED AND UNREGULATED FISHING*

4.30 The FAO Code of Conduct has a number of International Plans of Actions (IPOAs). The UN FAO, in 2001, adopted the International Plan of Action to Prevent, Deter and Eliminate Illegal, Unreported and Unregulated (IUU) Fishing.[21] This calls on all states to prevent the import of fish caught in contravention of the regulations established by regional fisheries treaty organizations and to discourage companies within their jurisdiction (e.g. insurers and equipment suppliers) from doing business with fishing vessels engaged in IUU fishing.

4.31 Separate IPOAs were developed to reduce incidental catch of seabirds in longline fisheries,[22] for the conservation and management of sharks[23] and for the management of fishing capacity.[24]

WORLD TRADE ORGANISATION *(WTO/GATT)*

4.32 The General Agreement on Tariffs and Trade (GATT) was agreed in 1947 as the main international agreement to encourage trade between States, though the International Trade Organisation of which it was envisaged to be a part did not materialise. It was not until 1993 that a World Trade Organisation was established, with the GATT one of its covered agreements. The rules set down in the Agreement prohibit restrictions on imports or exports with some exceptions set out in Article XX. These require exceptions to be applied in a way that is not a means of arbitrary or unjustifiable discrimination on trade. Measures to protect human, animal or plant life and health and measures related to the conservation of exhaustible natural resources (which must also apply to domestic production) are permitted. There is no explicit exemption for measures to deliver environmental protection, although this is an issue in the current renegotiation of the rules in the Doha round.

Dispute settlement in the WTO

4.33 The WTO has a Dispute Settlement Body that interprets the rules where trade resolutions are disputed. The US has been involved in several disputes involving trade restrictions related to marine conservation and fishing. In the Mexico-US tuna-dolphin dispute the GATT panel (predecessor to the WTO Dispute Settlement Body) noted that states had no right to impose discriminatory restrictions on trade on the basis of process and production methods (such as fishing technology) in other countries, or to protect natural resources outside their own jurisdiction.

4.34   But in a later case, the Dispute Settlement Body ruled that restrictions on shrimp imports imposed by the US were legitimate in principle under Article XX of the GATT. This was despite the fact that the turtles that the measures sought to protect were outside US territorial jurisdiction at the time (although being migratory they could well have been within US waters at other times), and that the measures were triggered by other countries' process and production methods. However, the Settlement Body found that the way in which the measures had been applied by the US (including failing to consult adequately and applying US standards without taking sufficient account of different conditions in other countries) constituted 'unjustifiable and arbitrary discrimination', and therefore was not consistent with Article XX.

4.35   Nevertheless, the case represented the first time that an international judicial body had recognised the link between fisheries freedoms and biodiversity controls. There is an ongoing case involving Spain and Chile in which the Spanish distant water fleet is catching swordfish on the high seas and landing them in Chile for airfreight to the EU. Unlike the earlier cases this relates to landing (transit) rights as a lever to influence compliance with international conservation standards on the high seas. This operation is currently in contravention of EU regulations and a temporary settlement to this matter is being sought.

BIODIVERSITY-RELATED TREATIES

UN Convention on Biological Diversity

4.36   The UN Convention on Biological Diversity was adopted at the 1992 Earth Summit and entered into force in December 1993. It was the first treaty to provide a legal framework for biodiversity conservation, and established three main goals: the conservation of biological diversity, the sustainable use of its components, and the fair and equitable sharing of the benefits arising from the use of genetic resources.

4.37   In 1995, the Conference of Parties to the UN Convention on Biological Diversity adopted the Jakarta Mandate on Marine and Coastal Biological Diversity.[25] One of its five major programme areas is the sustainable use of marine and coastal living resources. This is directly affected by the other four: mariculture, alien species and genotypes, integrated marine and coastal area management and protected areas. The UN Convention on Biological Diversity has also led to the development of Biodiversity Action Plans (4.81).

Convention on the International Trade in Endangered Species (CITES)

4.38   The 1973 CITES Convention was developed through IUCN – the World Conservation Union. Its objective is to restrict international trade in commercially valuable species threatened with the risk of extinction. The Convention has three appendices listing species that receive different levels or types of protection.

4.39   A number of marine species found in the north-east Atlantic are listed in the appendices. These include marine dolphins and basking sharks, the latter having been proposed by the UK government. Commercially important fish species are not listed and the FAO has resisted suggestions to include them.

UN WORLD HERITAGE CONVENTION

4.40   The Convention Concerning the Protection of the World Cultural and Natural Heritage (the World Heritage Convention) was adopted by UNESCO in 1972. To date, more than 170 countries are Parties to the Convention. It seeks to protect both natural and cultural heritage, and has been applied to marine areas such as the Great Barrier Reef.

WORLD SUMMIT ON SUSTAINABLE DEVELOPMENT

4.41   The World Summit on Sustainable Development has emphasised the need to implement a number of important international agreements for the conservation and protection of the marine environment. The 2002 World Summit on Sustainable Development led to targets for developing a coherent network of marine protected areas by 2012[26] and to restore depleted fish stocks to maximum sustainable yields by 2015.

INTERNATIONAL WHALING COMMISSION

4.42   The International Whaling Commission (IWC) was set up under the International Convention for the Regulation of Whaling established in 1946. The Convention's purpose is to provide for the proper conservation of whale stocks and make possible the orderly development of the whaling industry.

4.43   The main duty of the IWC is to review and revise the measures laid down in the Schedule to the Convention that governs the conduct of whaling throughout the world. These measures, among other things, provide for the complete protection of certain species; designate specified areas as whale sanctuaries; set limits on the numbers and size of whales which may be taken; prescribe open and closed seasons and areas for whaling; and prohibit the capture of suckling calves and female whales accompanied by calves. The compilation of catch reports and other statistical and biological records is also required. In addition, the IWC encourages, co-ordinates and funds whale research.[27]

AGREEMENT ON THE CONSERVATION OF SMALL CETACEANS OF THE BALTIC AND NORTH SEAS (ASCOBANS)

4.44   ASCOBANS was concluded in 1991 under the auspices of the UN Convention on Migratory Species (also known as the Bonn Convention) and entered into force in 1994. ASCOBANS covers the marine environment of the Baltic and North Seas and has been signed by Belgium, Denmark, Finland, Germany, The Netherlands, Poland, Sweden and the United Kingdom. It aims to promote close cooperation amongst signatories with a view to achieving and maintaining a favourable conservation status for small cetaceans. A Conservation and Management Plan forming obliges Parties to engage in habitat conservation and management, surveys and research, pollution mitigation and public information.

REGIONAL AGREEMENTS

CONVENTION ON THE CONSERVATION OF ANTARCTIC MARINE LIVING RESOURCES (CCAMLR)

4.45   CCAMLR applies to the Antarctic, which, while it is outside the geographical focus of this study, has been included here because it articulates a number of principles that are widely applicable. For example, it was one of the first conventions to adopt an ecosystem based approach to fisheries management. This is enshrined in three principles of conservation set out in Article 3. These aim to:

- prevent decreases in the size of any harvested population to levels below those which ensure its stable recruitment;

- maintain the ecological relationships between harvested, dependent and related populations of Antarctic marine living resources; and

- prevent changes, or minimise the risk of changes, in the marine ecosystem.

4.46   The CCAMLR Commission also adopted the first detailed description of how to apply a precautionary approach to fisheries management. It restrains harvesting so that a fishery does not develop more quickly than the information necessary to ensure that it can be conducted in accordance with CCAMLR's ecosystem conservation principles. This approach was triggered by a proposal in the 1990s for a new crab fishery. Today, it applies to new fisheries and to existing fisheries for which there is insufficient information to estimate potential sustainable yield and the impacts of fishing on other system components.

4.47   To implement the approach, the CCAMLR Scientific Committee must prepare and annually update a plan identifying data needs and how to collect the data. The plan may specify location, gear, effort, and other restrictions on the fishery. A precautionary limit is set on the harvest at a level slightly above that required to obtain the data and conduct the evaluations. Those engaged in the fishery are responsible for submitting an annual research and fishery operations plan. This must conform to the Scientific Committee's data collection plan and describe fishing methods, including an assessment of the likelihood of impacts on dependent and related species. It is reviewed by the Committee and the decision-making Commission.[28]

4.48   During 2003, the difficulties of enforcement in Antarctic waters were highlighted by the pursuit on the high seas by a number of states of a vessel fishing for Patagonian toothfish, a highly sought after cold water deep-sea species.[29]

CONVENTION ON THE INTERNATIONAL COUNCIL FOR THE EXPLORATION OF THE SEAS (ICES)

4.49   The 1964 Convention sets out the constitution for ICES, although the organisation was established as far back as 1902. ICES's role is to promote marine research in the North Atlantic, including adjacent areas such as the Baltic and North Seas. It acts a forum for national fishery laboratories of member states and other states in the area.

4.50 Following a number of internal conventions, its role was extended to include the provision of advice on fisheries and fish stocks to three international fisheries commissions and the European Community. ICES has a system of Working Groups and Advisory Committees comprised of scientists from member countries. Working Groups report to the Advisory Committees, which are responsible for formulating unbiased and non-political advice.

4.51 ICES promotes the study of all aspects of the marine ecosystem in order to understand and advise on physical processes, water chemistry, pollutants, fish and fisheries, seabirds and marine mammals.

OSLO-PARIS COMMISSION (OSPAR)

4.52 The area covered by the 1992 OSPAR Convention for the Protection of the Marine Environment of the North East Atlantic is the focus of this report. The Convention requires contracting parties to prevent and eliminate pollution, and to take the necessary measures to protect the maritime area of the north-east Atlantic from the adverse effects of human activities, so as to safeguard and conserve marine ecosystems. An additional annex to the Convention was adopted in 1998[30] to protect and conserve the ecosystems and biological diversity of the maritime area. OSPAR's proposals to establish marine protected areas are discussed in paragraph 4.58.

4.53 Although OSPAR has taken a number of important measures on pollution and conservation, it has little influence over fishery management.

INTERNATIONAL COMMITMENTS ON MARINE PROTECTED AREAS AND RESERVES

4.54 This section provides more detail on the legal framework for establishing a particular type of management tool, known as marine protected areas (MPAs). The practical scope for using MPAs is considered in chapter 8.

4.55 MPA is an umbrella term that covers a wide variety of designations but in this report we adopt the definition provided by the IUCN:

> 'An area of land and/or sea especially dedicated to the protection and maintenance of biological diversity, and of natural and associated cultural resources, and managed through legal or other effective means'.[31]

4.56 IUCN has defined several categories of MPA and it is important to be clear about the definition and management objectives of the different categories. MPAs that benefit from a greater degree of protection are known as marine reserves. These are usually protected from all fishing and for this reason are sometimes known as fishery no-take zones (see chapter 8 and appendix L).

4.57 MPAs and marine reserves have been shown to be effective in helping ecosystems and fish populations to recover from the effects of overfishing. The groundwork for an MPA network in the north-east Atlantic has already been laid through initiatives by OSPAR, the EC Natura 2000 process, the EC Habitats and Birds Directives. In addition, at the Fifth North Sea Conference, Environment Ministers called for a network of marine protected areas to be established in the North Sea by 2010. This is known as the Bergen Declaration.

4.58 The commitments under the Bergen Declaration were reaffirmed by the joint OSPAR/HELCOM ministerial meeting in Bremen 2003, which committed the OSPAR states to designate 10% of territorial waters as marine protected areas to form an ecologically coherent network.[32] Under this process, it was agreed to:

- consider, in arrange for the evaluation in 2004 and 2005 of the areas reported by Contracting Parties...as components of the OSPAR Network of Marine Protected Areas ("the OSPAR Network"). This evaluation will be to see how far the purposes of the Recommendation on a Network of Marine Protected Areas have been achieved;

- evaluate in 2006 whether the components of the OSPAR Network that have been selected by that date will be sufficient to make that network an ecologically coherent network of marine protected areas for the maritime area;

- consider whether any action by the Commission, or concerted action by the Contracting Parties, is needed to support efforts by Contracting Parties to achieve the institution of management measures by an international organisation for any component of the OSPAR Network;

- consider reports and assessments from Contracting Parties and observers on possible components of the OSPAR network and on the need for protection of the biodiversity and ecosystems in the maritime area outside the jurisdiction of the Contracting Parties, in order to achieve the purposes of the network...; consultation with international organisations..., how such protection could be achieved for areas identified...and how to include such areas as components of the network;

- identify any gaps which need to be filled in order to achieve the OSPAR Network by 2010 and maintain it thereafter, and take steps towards filling any such gaps;

- create and maintain a publicly available database of the OSPAR Network;

- develop practical guidance on the application of the Guidelines for the Management of Marine Protected Areas in the OSPAR Maritime Area;

- develop guidance on, and make arrangements for, assessing how effectively the management of the components of the OSPAR Network of Marine Protected Areas is achieving the aims for which those areas were selected;

- in 2010 and periodically thereafter, assess whether an ecologically coherent network of well-managed marine protected areas in the maritime area has been achieved.

4.59 Parties will decide how the MPA network will be developed in their own waters, and EU member states may look to Natura 2000 to provide the necessary framework.

4.60 As mentioned in 4.41, in 2002 WSSD called for the use of MPAs and reserves in areas of national jurisdiction and on the high seas. The UK, along with other countries, committed to 'the establishment of marine protected areas consistent with international law and based on scientific information, including representative networks by 2012'.

EUROPEAN MEASURES

EC HABITATS AND BIRDS DIRECTIVES

4.61   The UN Convention on the Conservation of European Wildlife and Natural Habitats (the Bern Convention) was adopted in Bern, Switzerland in 1979, and came into force in 1982. The principal aims of the Convention are to ensure conservation and protection of all wild plant and animal species and their natural habitats (listed in Appendices I and II of the Convention), to increase cooperation between contracting parties, and to afford special protection to the most vulnerable or threatened species listed in Appendix 3 of the Convention.

4.62   To implement the Bern Convention in Europe, the European Community adopted the Birds Directive in 1979,[33] and the Habitats Directive[34] in 1992. The Directives provide for the establishment of a European network of protected areas (Natura 2000 sites) to tackle the continuing losses of European biodiversity on land, at the coast and in the sea to human activities.[35] Following a judgement of the UK courts in 2000, it is clear that under the Habitats and Birds Directives, the UK's conservation obligations extend to its internal waters, territorial seas, areas within British Fishery limits, and certain areas of the continental shelf beyond those limits. The UK is also required to select and protect Special Areas of Conservation and Special Protected Areas.

4.63   Four marine habitat types listed in Annex I of the Habitats Directive are known to occur, or to potentially occur, in UK waters:[36]

- sandbanks which are slightly covered by seawater all the time – these occur in UK offshore waters off north and north east Norfolk, in the outer Thames Estuary, off the south-east coast of Kent and off the north east coast of the Isle of Man;

- reefs – these occur in the English Channel, Celtic Sea, Irish Sea and west and north of Scotland extending out into the North Atlantic. Deep-water coral (*Lophelia*) reefs occur in the North Sea;

- submarine structures made by leaking gases – 'pockmarks' containing carbonate structures deposited by methane oxidising bacteria may fit within this definition;

- submerged or partially submerged sea caves – none yet identified in UK offshore waters.

4.64   Several areas of sandbank and reef extend into the offshore areas of other Member States and inshore into UK territorial waters.

4.65   Special Areas of Conservation are also proposed to offer protection to marine mammal species; grey seal, common seal, bottlenose dolphin and harbour porpoise. There are already Special Areas of Conservation in place to protect breeding colonies and other sites for the two seal species and three for bottlenose dolphins and harbour porpoise within territorial waters.

4.66   Marine Special Protected Areas are classified under the Birds Directive Annex I and for migratory species. These include breeding colonies, inshore areas used in non-breeding seasons (e.g. for divers, grebes and seaduck) and marine feeding areas. Although the

Directive was adopted in 1979, the UK is still working to develop its suite of marine Special Protected Areas which are being considered for 56 bird species.

4.67 Up to the time of the recent closure of the Darwin Mounds to bottom trawling (a candidate Special Areas of Conservation), selection of these sites had only taken place on land and within UK territorial seas. The emergency protection measures on the Darwin Mounds represents the initial stage in the potential identification and designation of further offshore areas. But it took place under the auspices of the EC Common Fisheries Policy (see 4.83) tending to reinforce the presumption that fishing restraints would not be applied directly under environmental legislation in the European context.

EUROPEAN BIODIVERSITY STRATEGY

4.68 The European Community Biodiversity Strategy[37] was adopted in 1998. In 2001, this was followed by the production of Biodiversity Action Plans for fisheries, agriculture, economic cooperation and development and conservation of natural resources. These sectoral Action Plans define concrete actions and measures to meet the objectives defined in the strategy, and specify measurable targets.

4.69 The fisheries Biodiversity Action Plan[38] identified the following priorities to maintain or restore biodiversity threatened by fishing or aquaculture activities:

- promoting the conservation and sustainable use of fish stocks;

- promoting the control of exploitation rates and the establishment of technical conservation measures to support the conservation and sustainable use of fish stocks;

- reducing the impact of fisheries activities on non-target species and on marine and coastal ecosystems;

- avoiding aquaculture practices that may affect habitat conservation.

EUROPEAN MARINE THEMATIC STRATEGY (EUMTS)

4.70 The European Commission has published proposals for a Marine Thematic Strategy to implement the recommendations of the 6th Environmental Action Plan. The draft thematic strategy aims to provide high-level vision and goals to 'protect and...restore the function and structure of marine ecosystems...to achieve and maintain good ecological status of these ecosystems.[39] The Strategy enshrines the ecosystem based approach and is linked to the development of a set of ecosystem indicators. At the time of writing, the legal status of the Strategy and any associated measures has not been decided. See chapters 7 and 10 for further discussion of the Strategy.

STRATEGIC ENVIRONMENTAL ASSESSMENT (SEA)

4.71 SEA is the subject of a new EC Directive that aims "to provide a high level of protection for the environment and to contribute to the integration of environmental considerations into the preparation of adoption of plans and programmes with a view to promoting sustainable development".[40] Implementing legislation came into force in the UK in July 2004 and will apply to plans started after that date, although some sectors, such as the offshore oil and gas industry have already begun to comply with its provisions on a voluntary basis.

4.72 The Directive applies at the strategic level to plans or multiple projects. It involves putting the plan through a screening process, after which the plan-makers must provide a report on the environmental consequences of their plan. This sets out the plan's objectives and any relevant environmental issues. A baseline study is required, plus an assessment of the environmental impacts of implementing the plan and reasonable alternatives to it. The plan-makers are required to send the draft plan to the relevant authorities and to a number of public consultees. Opinions expressed during the consultation must be taken into account, and the availability of the final plan publicised.[41]

4.73 Individual projects within a plan may require an additional Environmental Impact Assessment (EIA) under the 1985 EIA Directive. The EIA Directive has an appendix listing the specific areas to which it applies. These include fish farms, but not capture fisheries. The SEA Directive refers to fishing, but at present the sector falls outside its scope, except in some very limited cases where fishing activities interact with either the EIA or the Habitats Directive. This is an issue we return to in chapter 10.

CURRENT UK LEGISLATION ON MARINE CONSERVATION

4.74 In the UK, conservation policy has focused mostly on terrestrial species and habitats, resulting in restrictions on the use of private or common land. However, the same policies have also created opportunities to protect coastal or marine environments. For example, the Wildlife and Countryside Act 1981 made specific provision for marine reserves. Relevant types of designations are given table 4.1.

Table 4.I

Legal designations that can be used to provide area-based protection[42]

Type of designation	Acronym	Coverage
Sites of Special Scientific Interest/ Areas of Special Scientific Interest	SSSIs/ASSIs	UK
National Nature Reserves	NNR	UK
Marine Nature Reserve	MNR	UK
Local Nature Reserves	LNR	UK
Special Protection Areas	SPAs	EU
Special Areas of Conservation	SACs	EU
Fisheries closures		EU
Marine Protected Areas	MPAs	OSPAR north-east Atlantic
Ramsar Sites		Global
Particularly Sensitive Sea Areas	PSSAs	Global

Figure 4-I
Territorial coverage of UK designations[43]

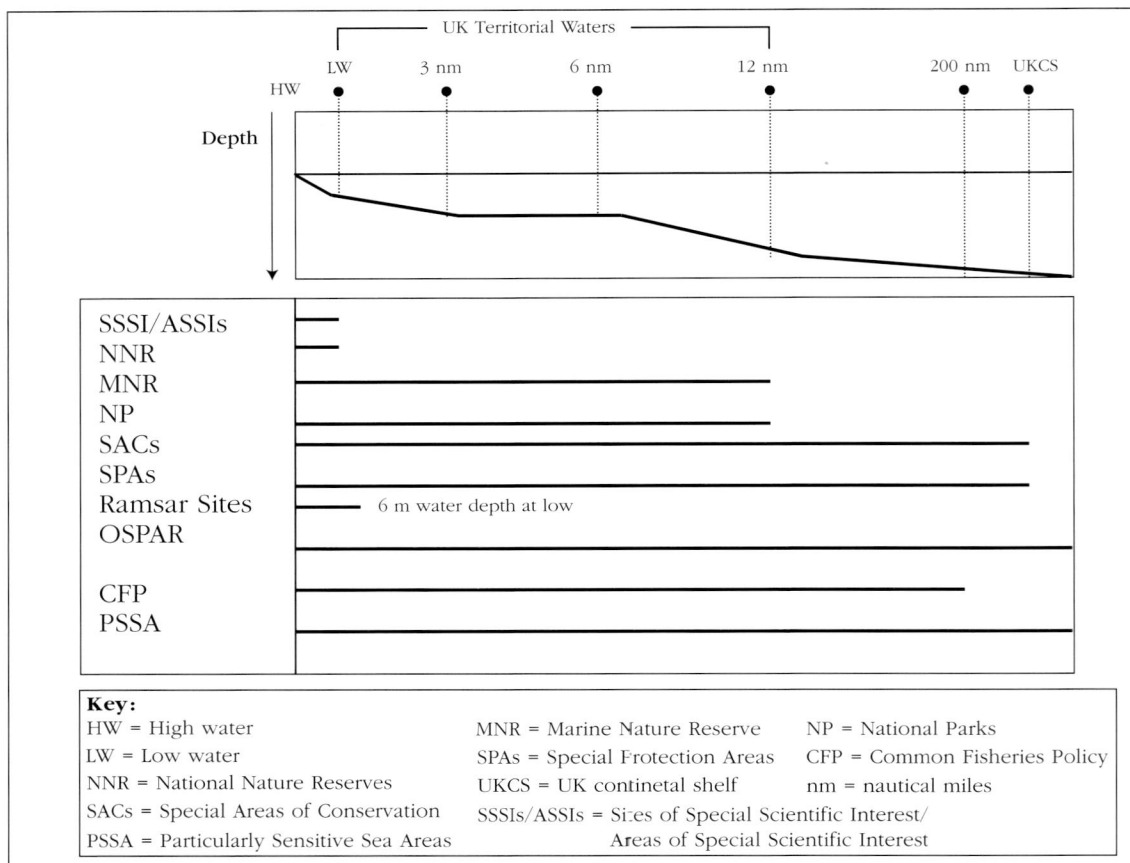

4.75 A number of nature conservation designations can be applied to the marine environment, given sufficient patience and ingenuity. However, Special Areas of Conservation and Special Protection Areas are the only ones that are applicable throughout the UK's territorial and offshore waters, and the UK Continental Shelf (figure 4-I). Even then, Special Areas of Conservation offer only limited possibilities for species that range widely. The territorial coverage of these sites contrasts with Marine Nature Reserves, which are limited in geographical scope, extending to 12 nm at best.

4.76 Most of these designations lack way of protecting sites from fishing, even if they are of national importance. In the UK and Europe, marine areas are only weakly protected, with conservation effort focusing on a small-scale individual approach with a high level of management complexity. In the UK, Marine Nature Reserves, have been rendered largely impotent as a designation tool, due to the legal and political processes involved in their establishment, combined with a lack of legal requirement to designate sites. There are only three Marine Nature Reserves in the territorial waters of UK, and only one of these incorporates a fisheries no-take zone (chapter 8).

4.77 In addition, although UK nature conservation designations offer some species and habitat a degree of protection, the scope for protecting ecological processes (for example, tidal fronts, eddies and hydrothermal vents) is less clear. Coherence and connectivity issues are only addressed by the European designations.

4.78 The only route that seems to be currently available to give full protection from fishing activities is the Common Fisheries Policy, which offers a range of possibilities for closing areas and restricting fisheries, as has been the case for the Darwin Mounds. However, these can only be agreed on a case-by-case basis through the Council of Fisheries Ministers. Securing Council agreement can be an up-hill struggle, since a number of Member States are likely to resist the use of Common Fisheries Policy measures if their fishing industry's short term interests are directly affected. However, there is a requirement on the Council of Ministers to adopt measures in support of an ecosystem-based approach, and specifically to agree recovery plans.

4.79 Formal management schemes are to be introduced for certain designations, including marine Special Areas of Conervation, Special Protected Areas and Ramsar sites. The latter are linked to the UN Convention of the same name, which is implemented in the EU under EC Birds Directive. New legislation has progressively introduced greater byelaw-making powers for use in a nature conservation context, increasing the scope for marine site protection. However, the overall effectiveness of site management frameworks is questionable. Where management schemes identify the need for fisheries measures to reduce environmental impacts, action will normally depend on the decisions taken by the Fisheries Council.

4.80 In addition, any attempt to secure UK offshore marine protected areas will at present need to rely on the limited provisions of the EC Habitats and Birds Directives, which are not best suited for this purpose (chapter 8).

UK Biodiversity Action Plan

4.81 In 1994 the UK Government launched the UK Biodiversity Action Plan. This national strategy identifies broad areas for conservation work over the next 20 years, and establishes the principles for future conservation of biodiversity.[44] Separate strategies for England and Northern Ireland were published in 2002, and for Scotland in 2004. A biodiversity strategy for Wales is under in development.

4.82 The UK Biodiversity Action Plan contains grouped plans for commercial marine fish (including cod, hake, herring, mackerel, plaice, saithe, sole in the North Sea) and deep-water fish (including blue ling, roundnose grenadier and orange roughy). There are also individual plans for species such as basking sharks and skates.

The Common Fisheries Policy (CFP)

4.83 The provisions of the Common Fisheries Policy (CFP) are laid down in the European Community's founding treaty. Although there is no specific fisheries chapter, the CFP has the same general objectives as the Common Agricultural Policy (CAP), laid out in Article 33, which are to:

● increase productivity by promoting technical progress and by ensuring the rational development of production and the optimum utilisation of the factors of production, in particular labour;

- thus ensure a fair standard of living for the [fishing] community, in particular by increasing individual earnings;

- stabilise markets;

- assure the availability of supplies;

- ensure that supplies reach consumers at reasonable prices; and

- ensure the principle of non-discrimination.[45]

4.84 Article 6 of the Treaty stipulates that environmental protection requirements must be integrated into Community policies, in particular with a view to promoting sustainable development. Moreover, Article 174 requires that Community policy on the environment shall be based on the precautionary principle. The Common Fisheries Policy also has to take into account consumer protection requirements, the objectives of economic and social cohesion and development co-operation.[46]

LEGAL COMPETENCE

4.85 The Common Fisheries Policy sets fisheries policy at a Community level and thus limits the extent to which member states can develop their own fishery measures. The way in which powers are shared between member states and the Community (i.e. their competency) is discussed below.

4.86 Community competence over fisheries is determined by the CAP/CFP provisions of the EC Treaty (Articles 32 to 38) as amended and by the UK Treaty of Accession, and, *inter alia*, Regulation 2371/2002 on the conservation and sustainable exploitation of fisheries resources under the CFP.

4.87 Community competence over fisheries is exclusive where conservation measures are concerned, and shared in relation to all other fisheries matters; Community competence in both the exclusive and the shared area extends to setting the criteria for sanctions, but neither the CAP/CFP provisions nor the Accession Treaty entitles the Community to insist on the imposition of criminal penalties.

4.88 External Community competences in relation to issues extending beyond Community waters follow broadly the same model: international agreements are negotiated and entered into by the Community on matters where Community competence is exclusive, by Member States or the Community where it is not, or by both in the case of 'mixed agreements' where some provisions are within exclusive Community competence and some are not.

4.89 Where there is exclusive Community competence (i.e. in relation to conservation measures), member states can take unilateral action only (a) to preserve the prior position where otherwise there would be a genuine legislative hiatus or (b) where specifically so permitted by Community legislation.

4.90 Key areas of importance to the UK where the CFP permits the UK as a Member State to act independently include the:

- ability to take non-discriminatory measures for the conservation of stocks and to minimise the impact of fishing on marine ecosystems up to 12 nm offshore, provided they are in line with the CFP, and to restrict fishing by non-UK vessels in waters up to that limit to vessels that traditionally fish in those waters;

- right to take temporary emergency measures where there is evidence of a serious and unforeseen threat to the conservation of living aquatic resources or to the marine ecosystem resulting from fishing activities;

- ability to take conservation and management measures in all waters provided they are applicable solely to UK flagged vessels.[47]

4.91 Where there is shared competence (i.e. on fisheries matters other than conservation), member states can take unilateral action (a) where the Community has not yet legislated and, once again, (b) where specifically so permitted by Community legislation. Areas where the UK attaches importance to its freedom to act independently include:

- the right – subject to general Community principles on the right of establishment – to fly the UK flag and so be entitled to share in the UK total allowable catch;

- measures for the management of fleet capacity;

- representation of UK Overseas Territories in Regional Fisheries Organisations.

4.92 The new EU Constitutional Treaty preserves these distinctions.

REFORM OF THE COMMON FISHERIES POLICY (CFP)

4.93 In March 2001, the European Commission launched a Green Paper[48] on the future of the CFP. It acknowledged that 20 years after is inception, the CFP was facing significant problems and had failed to deliver sustainable exploitation of resources. The European Commission judged that many stocks were outside safe biological limits, and that if trends continued many stocks would collapse. At the same time, the capacity of the Community fishing fleet was far too high and the enforcement of rules needed to be improved. Enlargement and environmental considerations posed further challenges, along with the need for better scientific advice.

4.94 In May 2002, the European Commission presented its first reform proposals, and the Fisheries Council adopted new measures in December 2002. Rules on conservation and the sustainable exploitation of fisheries resources entered into force on 1 January 2003. Complementary measures also encouraged a reduction in the fishing fleet, including decommissioning of vessels. Measures were also included on the environment, on action to combat illegal fishing and discards and on improving the sustainability of aquaculture.

4.95 Another significant strand of reform is the move to regionalise the management of CFP through the establishment of seven Regional Advisory Councils. These will provide a way of consulting stakeholders such as the fishing industry and other interested parties on the management of their areas. For a fuller discussion of the RACs and their possible implications for the environment, see chapter 10.

4.96 A key change enshrined in the new basic CFP Regulation[49] is the adoption of a stronger commitment to the protection of the marine environment as a fundamental objective of the CFP. Application of the precautionary approach to management is laid down in the objectives, together with sustainable exploitation, minimisation of the impacts of fishing on the marine ecosystem, and progressive implementation of an ecosystem-based approach to management. The new basic Regulation thus provides a clear legal basis for future measures intended to reduce the negative impacts of fishing on the marine environment.

LIMITING ACCESS TO FISHERIES AND CONTROLLING CATCHES

4.97 Since January 1995, all vessels fishing in Community waters and EU vessels operating outside Community areas have required a licence. Fishing effort can be regulated through the allocation of special fishing permits stating the terms of access, time and specific fisheries. The Council of Ministers decides which fisheries require such permits and the conditions attached to fishing.

4.98 The CFP sets maximum quantities of fish that can be caught and landed every year. These maximum quantities, called total allowable catches (TACs), are divided among Member States. Each country's share is called a national quota. TACs are fixed on an annual basis. Following scientific studies on the main stocks, the Council of Ministers decides on fishing allocations for the following year.

4.99 The level of the TACs is based on scientific advice. The major source of information is ICES, which brings together information on the state of the stocks. The European Commission also consults its own Scientific, Technical and Economic Committee for Fisheries made up of national experts. Negotiations also take place with non-Community countries that have an interest in the same fishing grounds or stocks and relevant regional fisheries organisations. The final decision regarding TACs, quotas and any related measures is currently taken by Fisheries Ministers at their end-of-year Council meeting.

4.100 TACs are divided into national catch quotas based mainly on past catch records, under an established allocation mechanism that gives each member state a fixed percentage each year (although a number of international quota swaps are regularly undertaken). This mechanism is still used today, on the basis of what is known as the principle of 'relative stability', which ensures member states a fixed percentage share of fishing opportunities for commercial species. Each member state is then free to determine the means for allocating its quotas and for regulating quota uptake.[50] Member states must monitor quota uptake and close fisheries as and when quotas have been caught.

4.101 The perceived inability to define a TAC, and difficulty in forecasting supply in general, has allowed the catching sector to pressure regulatory bodies into increasing TACs resulting in overfishing. In addition, EU national-level and European Commission fisheries advisory groups are often dominated by interests from the fishing industry through powerful lobbying and this has often led to scientific advice being ignored or compromised. TACs are limits on landings and do not reflect the numbers of fish killed when fish are discarded at sea. Despite this, the quota system remains popular with politicians and fisheries managers as it is readily accessible and lends itself to bartering in the process of political compromise. In the case of the CFP, sometimes the TACs are identical to the levels of fishing recommended by ICES, but in most cases they are not. On average, the chosen TACs have been about 30% above the scientifically advised levels. For a single year, 2000, the average deviation was about 50%, reflecting the inevitable compromise between scientific advice and political pressure.[51] The lack of confidence of the catching sector in the forecast of fish stocks also has a strong negative impact on compliance and the resulting levels of fishing mortality.

4.102 The 'roadmap' for the reform of the Common Fisheries Policy[52] notes the intention to move to a multi-annual framework for conservation of resources and management of fisheries. These plans will set targets for management of stocks in terms of population size and fishing mortality rates. These will be backed up by catch and fishing effort limits designed to meet these long-term goals. The intention is that these long-term management plans will be based on the most up to date information about the state of stocks.

MANAGING FISHING EFFORT

4.103 Previously, the main tool for managing effort was the set of multi-annual guidance programmes (MAGPs), which allowed for planned development of each Member State's fishing fleet, and the reduction of effort. To date, four separate MAGPs have been implemented.

4.104 MAGP IV (1997-2001) specified a reduction in capacity and effort by fleet segment, with the amount of capacity/effort reduction required based on the degree of overexploitation of the fish the fleet segment targets. Each quota stock was classified as either fully exploited, overfished or at risk of collapse. Effort reduction targets for the fish at risk and overfished were established as 30% and 20% respectively. For all other fish, the target was for no effort increase rather than an explicit effort decrease.

4.105 With around half of the Member States adopting the option of effort control (i.e. days at sea restrictions) rather than vessel reduction in at least some fleet segments, the capacity of the fleet has not substantially decreased under MAGP IV. Instead, capacity utilisation has decreased (as the existing vessels are prevented from being fully utilised), resulting in a less efficient use of the capital resources tied up in the fishing fleet. The overall result was that the MAGP IV had no significant impact on the degree of overcapacity in the Community fleet.[53] In spite of these difficulties, MAGP IV was extended for an additional year with revised effort reduction targets for the end of the extension period.[54]

4.106 Since the end of MAGP IV, effort reduction measures have been more closely tied to specific stocks. One of the first tests of support for effort reduction measures under the reform of the CFP will come with the cod recovery plan, which advocates effort reduction as the principal tool for achieving recovery (box 4C).

BOX 4C	EFFORT REDUCTION AND THE COD RECOVERY PLAN[55]

In 2003, the European Commission proposed the establishment of a long-term recovery plan for a number of commercial fish species stocks threatened with imminent collapse, including North Sea Cod.[56]

Measures were agreed under Council Regulation 2287/2003[57] to impose a spatial management plan for threatened species based on a set of closed or semi-closed areas. The Regulation limits fishing effort in these areas, with specific control and monitoring rules to ensure implementation. The stocks covered by the plan include cod in the Kattegat, the North Sea including the Skagerrak and the Eastern Channel, the west of Scotland and the Irish Sea.

The 2003 agreement is the first multi-annual plan and is expected to be part of a series of long-term recovery and management plans addressing a number of severely depleted stocks. The European Commission has already made proposals to revise the 2003 plan. The long-term strategy aims to achieve its overall goals within five to ten years.

The plan makes recommendations on long-term TAC setting, but its central pillar is the limitation of fishing effort. Fishing effort is calculated in kilowatt-days (i.e. the engine power of a vessel multiplied by the days spent fishing). Effort limitations for particular categories of vessels are set directly by Council legislation.

Kilowatt-days can be redistributed among vessels within, but not across, the geographical areas containing vulnerable cod stocks. They will be fully transferable and usable at any time throughout the year. Fishing for other species, such as haddock, continues within the Cod Protected Areas.

A €32 million 'scrapping' fund has been established to help meet the required reductions in fishing effort under the recovery plans, but this has not been activated. Once operational, it would add to the funds already available for decommissioning vessels under the Financial Instrument for Fisheries Guidance for the 2000 to 2006 period.

TECHNICAL MEASURES

4.107 Technical measures are generally defined by geographical area and include minimum net mesh sizes, the use of selective fishing gear, closed areas and seasons, minimum landing sizes for fish and shellfish and limits on by- or incidental catches. Their aim is to avoid or limit the capture of immature fish to allow them to contribute to stock renewal as adults, unwanted fish because of their lack of commercial value or fish for which fishermen have no more quotas, and marine mammals, birds and other species such as turtles.

4.108 While the aim of technical measures is clear, their drafting and implementation are extremely complex. European fisheries, particularly for demersal species tend to be mixed in nature, meaning that restrictions such as those on net mesh sizes are often a compromise. Fishers also maintain that several different net sizes may be needed for any fishing area to allow catch of different species. This complicates enforcement.

4.109 Temporary closures have also been used in the North Sea for particular stocks. Examples of such areas include the Plaice Box (38,000 km² in area), the Norway Pout Box (95,000 km²) and the Mackerel Box (67,000 km²) (see figure 4-II). Herring spawning grounds are also protected in the North Sea and in the UK fishing for bass is restricted in their estuarine nursery areas. None of these is a fully protected no-take zone for fishing.

CFP SUBSIDIES

4.110 The EU fisheries sector continues to attract substantial amounts of direct and indirect subsidy from both EU and national sources to support and develop the sector. The main sources of aid are the Financial Instrument for Fisheries Guidance (FIFG), providing structural assistance to the sector, and the EU payments for fishing access to third country waters. These subsidies are under increasing scrutiny for a number of reasons, including the inclusion of fishing subsidies on the WTO agenda and the issue of coherence with other EU policies, such as sustainability, development and environmental protection.[59]

Figure 4-II[58]

Seasonal area closures around the UK. Closed areas have been popularly advocated as a vital tool that fisheries managers have been slow to utilise. In fact, there are a large number of areas where fishing is banned or regulated, such as the Plaice Box protecting juveniles plaice and sole in the North Sea, the Mackerel Box in the western Channel, and closed herring spawning grounds.

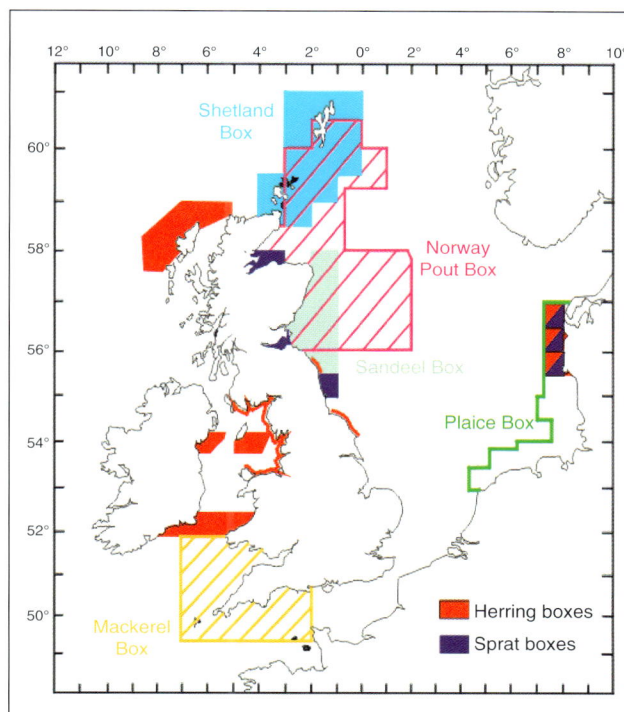

4.111 Despite this, funding to the sector remains high. Current budgets include around €1.2 billion per year for the fishing sector through FIFG, national matching funding, state aid, fishing access agreements and the European Fisheries Guarantee Fund. While some contributions are decreasing, figures for FIFG for the period 2000-2006 suggest a total commitment of almost €5.6 billion, of which €3.7 billion is supplied by the EU, almost twice as much as the €1.8 billion committed in the period 1994-1999. This increase underlines the growing dependence of the fisheries sector on EU subsidies, particularly in light of depleting stocks.

4.112 In recent years, there has been an added impetus towards more environmental considerations in the sector. In 1999, the FIFG was subject to major reform as part of the EU's Agenda 2000 process that placed a greater prominence on "environmentally sustainable and economically viable exploitation of fisheries resources". Consequently, changes were made to the funding criteria. For example, public aid for the entry of new capacity is now conditional upon Member States meeting their annual objectives under the Multi-Annual Guidance Programme and that at least the equivalent capacity is withdrawn without public aid. In addition, aid is not made available for permanent transfer of vessels to third countries identified as permitting fishing that jeopardises the effectiveness of international conservation measures. So, in principle, the FIFG 2000-2006 programme must take environmental impacts into consideration, and ensure compliance with the fleet reduction programme.

4.113 Over the period 1 January 2003 to 31 December 2004, to introduce capacity with public aid, a certain capacity must be permanently withdrawn without aid according to the following ratios:

- 1 Gross Tonne (GT) withdrawn for 1 GT introduced, for vessels up to 100 GT;

- 1.35 GT withdrawn for 1 GT introduced, for vessels between 100 and 400 GT.

From 1 January 2005 aid is restricted to vessel modernisation.

4.114 However, environmentally 'perverse' subsidies, such as capital investment in vessel building when the total EU fleet is estimated to be between 40 and 60% over capacity, still exist. Funding also continues to promote permanent transfer of vessels to non-member countries, with the exception mentioned above. Some aid, however, is used to support fisheries management objectives, notably aid for vessel decommissioning, and, to a far lesser extent, support to environmentally sensitive practices.

4.115 In spite of moves to increase transparency in public spending, it is still difficult to trace the subsidies to the fishing industry and calculate total amounts, making it even harder to analyse the impacts on the environment. This is partly due to the large variety of direct and indirect fisheries subsidies that exist. Transfers are administered by different parts of the European Commission, some as part of broader programmes, for instance funding allocated under the European Social Fund. In addition, funding programmes will often include different types of transfers, supporting, for example, a combination of capital investment and research.

4.116 The proportion of FIFG funding to each Member State seems to be more or less the same in the current round as in the previous period from 1994-1999. Spain still receives more than 45% of the EU transfers and Italy around 10%;[60] both are budgeted figures only.

4.117 The European Community PESCA initiative was set up to address the problems of areas particularly dependent on fishing and ran until 2000. PESCA enabled eligible areas to access structural funds for specific schemes aimed at lessening their dependence on fishing. PESCA was not renewed in 2000, but similar assistance was made available to areas dependent on fishing. Most of these areas were regarded as regions facing economic and social reconversion problems, giving them access not only to FIFG, but also to the European Regional Development Fund and to the European Social Fund.

CFP ENFORCEMENT

4.118 CFP measures are binding on the Member States and individual operators. Normally they take the form of regulations, which are directly applicable without requiring implementing legislation. But Member States are still responsible for proper enforcement of Common Fisheries Policy measures in the waters and territories under their jurisdiction. Boats over 10 m have to keep a logbook which includes details of the quantities of quota species caught and retained on board, and the time and location of capture. Enforcement is carried out through inspection at ports and at sea as well as by protection vessels and aircraft.

Satellite monitoring will apply to vessels over 18 metres from 1 January 2004 and to vessels over 15 m from 1 January 2005. In some cases observers are put aboard vessels.

4.119 In both 2000 and 2001, the Department for Environment, Food and Rural Affairs (Defra) recorded 250 infringements of regulations (10% and 13% respectively of the number of inspections carried out). Half related to inaccuracies in recording catches, the remainder mainly to breaches of technical measures such as net sizes or of licensing and registration requirements. However, it was noted in the report that the number of infringements recorded did not give the full picture of compliance.

4.120 In 2003, the National Audit Office reported that there was a very low probability (less than 1% chance) that on any day of fishing a vessel would be subject to a physical inspection at sea and around six per cent chance of being inspected on land.[61] There was however a much higher probability (60 – 70%) that submitted documents would be cross-checked against each other, fish available on the market and other information such as sightings at sea or satellite information.

4.121 Fishers were recorded in the report as saying that they knew when they were going to be inspected and that there were numerous places to hide illegal catches of which inspectors were unaware. Fines were also found to be inadequate, being on average 1.7 times the value of the infringement, despite the low probability of detection and prosecution. The possibility of introducing administrative penalties (such as license restrictions) is being investigated.

4.122 During 2003, the European Commission issued a letter of formal notice to the UK and Spain advising them of serious shortcomings in inspections and monitoring of fishing and landing activities, validating and cross-checking of data, follow up of infringements and applying deterrent sanctions against wrong-doers. This issue is ongoing.

4.123 In 2004, the European Commission announced[62] that the new EU Fisheries Inspection Agency would be located in Spain, one of the leading fishing regions of the Union. The aim of the new agency is to ensure a level playing field in fisheries enforcement. Its main task will be to ensure operational co-ordination of the deployment of the pooled national means of inspection and surveillance. Inspections will be assured by multinational teams according to inspection strategies, including the setting of benchmarks and common priorities, to be adopted by the European Commission.

4.124 In September 2003, Defra announced a review of marine fisheries and environmental enforcement arrangements in England and Wales, and in 2004 wrote to vessel owners and others with proposals for improvements (see chapter 9).

ACCESS AGREEMENTS

4.125 The EC has bilateral fisheries agreements with fifteen developing countries, which involve paying for access to surplus fishing in the exclusive economic zone of the country in question. The access agreements are intended to protect the interests of the European distant water fleet and to ensure that conditions for sustainable fisheries are strengthened in partner countries.

4.126 Between 1993 and 1997 almost 2.7 million tonnes per year of fish were caught through access agreements, i.e. about 40% of the total European Community catch.[63] Over the same period, the agreements enabled an average of 2,800 vessels to operate solely or partially in third country waters or on the high seas. Almost 2,100 boats operate to the north of the EU, while about 800 operate in the waters to the south, mainly in the waters of developing countries (figure 4-III and table 4.2). The agreements represent nearly 41,000 jobs, over 80% of which depend on 'southern' agreements.

Figure 4-III

Number of vessels operating solely or partly outside Community waters, 2000[64]

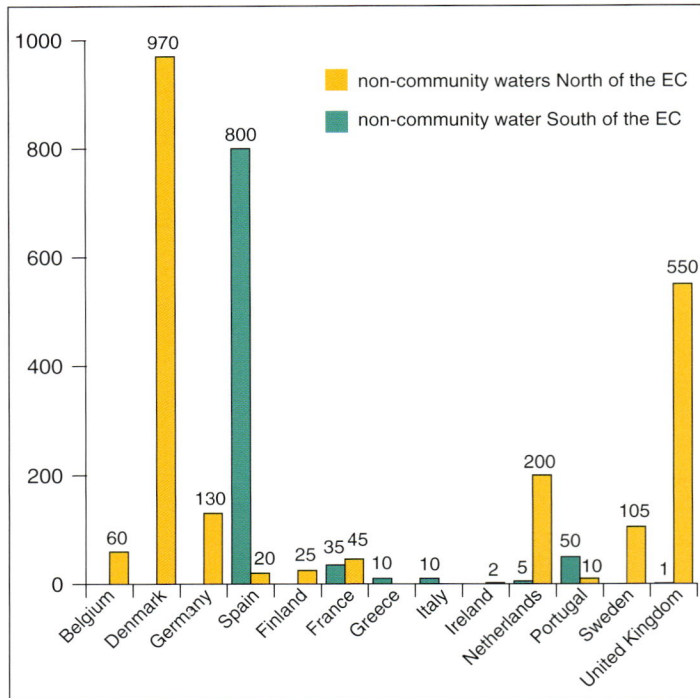

Table 4.2

Third party commitments under fisheries access agreements in 2000[65]

Figures are for countries where the value of the agreement is around €1 million or more. Total value of agreements was €137 million in 2000

Third party	€ million, 2000
Mauritania	54
Senegal	12
Guinea Bissau	7.4
Guinea	3.0
Seychelles	3.5
Angola	14
Cote d'Ivoire	1.0
Greenland	38

Table 4.3

Average benefit from access agreements per EU country[66]

EU Beneficiary	% of fisheries resources under agreements	Allocation in 2000 (million €)
Spain	59%	81
France	24%	31
Portugal	8.9%	12
Italy	5.0%	6.9
Greece	4.0%	5.5
UK	0.1%	0.2
Ireland	0.2%	0.2
Total	**100%**	**137**

4.127 Spain and France benefit significantly more than other EU countries (table 4.3). Community fishers also benefit more from the agreements than the third party countries since the value-added is estimated to be greater than the cost of the agreements, at a total of €294 m per year on average.

4.128 Access agreements have been strongly criticised in some quarters, for transferring fishing pressure from European waters elsewhere, for allowing EU boats to out-compete local fishers and for their potential environmental impacts[67,68] (chapter 10). Particular difficulties have also been identified with the enforcement and management of the fleet fishing in these waters.

4.129 It has been recognised that non-Community countries do not always have sufficient means to enforce the inspection arrangements in agreed protocols. Member Sates are therefore required to monitor the operations of their vessels in non-Community waters.[69] The measures include requirements for vessels to keep a logbook in which they note their catches, and to make a statement to their country of registration concerning products landed or transhipped in non-Community ports or onto non-Community ships. This information should be given to the European Commission on quarterly basis. However, problems have been identified in the collation and formatting of this data as well as concerns about reliability.[70]

UK ORGANISATIONS INVOLVED IN MANAGING FISHERIES

FISHERIES DEPARTMENTS

4.130 The UK Fisheries Departments comprise Defra, the Scottish Executive Environment and Rural Affairs Department, the National Assembly for Wales Agriculture Department, and the Department of Agriculture and Rural Development for Northern Ireland. Departments in the Isle of Man, Jersey and Guernsey are responsible for administering fishing activity in their respective areas.

4.131 Within the UK, fisheries is a devolved matter and the devolved administrations regulate fishing activity in their waters, but the power to legislate can be exercised concurrently by the Westminster Government. In EU negotiations, the Defra Minister leads for the UK, following a line agreed among the Ministers responsible for fisheries in the UK, who may attend in support. Licensing and quota management are devolved. Apart from enforcement carried out by Sea Fisheries Committees in their areas (up to 6nm from the coastal baselines), enforcement off England and Wales is carried out by Defra's Sea Fisheries Inspectorate (with support from the Royal Navy and, on aerial surveillance, from a private contractor). Enforcement is carried out off Scotland by the Scottish Fisheries Protection Agency and off Northern Ireland by the Northern Ireland Fisheries Inspectorate.

PRODUCER ORGANISERS

4.132 The UK quota management system is an entirely informal arrangement between government and industry. Bodies known as Producer Organisations have been established under to Common Fisheries Policy to share out the quota. While catch track records are nominally attached to licences, no quantitative catch rights are specified in the licences.

4.133 While the government retains overall responsibility for the uptake of quota at the national level, the twenty Producer Organisations may decide on the means that they use to manage their quota allocation. They may also swap quota both with other Producer Organisations and with companies that receive individual allocations of pelagic quota. This means that there are now several market-based elements present in the distribution of UK quota.

4.134 Producer Organisations may broadly be divided into two categories: those that have retained quotas for the collective benefit of local or regional industry groups, and others that have been formed by smaller numbers of vessel operators wanting to work wholly under an Individual Quota regime.[71]

4.135 The current system of quota allocation on the basis of track record has led to problems, as this provides an incentive to fish to the quota limit and in some cases to over-report catch.

LICENSING

4.136 Under EU law it is the responsibility of the flag Member State to license fishing vessels. The UK operates a restrictive licensing system for its commercial fishing vessels. That is to say, no new licences are issued by fisheries departments, and the owners of new vessels must obtain a licence from one or more existing licensed vessel(s). Such licence transfers are permitted provided that the capacity, tonnage and engine power of the recipient vessel are no greater than those of the donor vessel(s). In the majority of cases a capacity penalty is also payable to offset the increased catching efficiency of the newer vessel.

4.137 Licences contain restrictions on the types of fish and, in the case of vessels not in Producer Organisations, the quantities of fish that can be landed. These may be varied throughout the year as quotas are exhausted or increased. Licences also contain various conditions relating to fishing which must be observed, for instance the requirement to land into designated ports or to give prior notification of landing, and the requirement for vessels to maintain a genuine economic link with the UK. Owners and masters of fishing vessels found fishing without a licence or breaching the conditions of their licence are liable on summary conviction to a fine of up to £50,000 plus a fine of up to the value of the fish. Courts also have the power to disqualify a person from holding a licence.

SEA FISHERIES COMMITTEES

4.138 There are twelve Sea Fisheries Committees that regulate[72] local sea fisheries around virtually the entire coast of England and Wales out to 6 miles. Sea Fisheries Committees (SFCs) were established in the last century and are empowered to make byelaws for the management and conservation of their districts' fisheries. In 1995 their powers were widened to include the control of fisheries in their districts for environmental reasons (table 4.4).

Table 4.4

Legislation relevant to Sea Fisheries Committees[73]

Act/Regulation	Implications for Sea Fisheries Committees
Sea Fisheries (Wildlife Conservation) Act 1992	Requires SFCs to recognise conservation needs and balance these with other considerations when making decisions
Conservation (Natural Habitats & c.) Regulations 1994	SFCs must protect wildlife within Special Areas of Conservation, and/or Special Protection Areas by ensuring that any fishing activities they authorise do not cause significant damage to the sites
Environment Act 1995	States that SFC members should include one (or more) "environment" member(s), to provide an environmental input to meetings. It also enables fisheries byelaws to be made or withdrawn for environmental purposes
Wildlife & Countryside Act 1981, amended by Countryside & Rights of Way Act 2000	Gives extra protection to sites of Special Scientific Interest (SSSIs). SFCs must obtain permission from English Nature before they authorise fishing activities or change fisheries management within SSSIs.

MANAGEMENT OF FINFISH AQUACULTURE

4.139 Chapter 6 examines the aquaculture industry in more depth – including the finfish and shellfish sectors. Some of the key regulatory aspects are set out below.

EUROPEAN POLICY AND LEGISLATION

4.140 EU aquaculture is governed by the CFP, and its reform has resulted in a new strategy for aquaculture.[74] The Strategy aims to increase employment in aquaculture by 8,000 to 10,000 full-time job equivalents over the period 2003-2008. This will be mainly be achieved by developing cage and mollusc farming in fisheries-dependent areas, to compensate for jobs lost in the capture sector. It also aims to increase the aquaculture production growth rate in the EU to 4% per year, with a particular focus on mollusc farming, diversification into farming new species, organic production and environmentally certified production.

4.141 The strategy recognises that conflicts over space are currently hindering the development of aquaculture in some areas. It endorses moves to enhance inland closed systems and to move more cage farming further offshore. The Strategy also identifies shortcomings in current aquaculture governance.

4.142 The EC Directive on Environmental Impact Assessment (EIA)[75] is an integral part of the process of determining applications for marine fish farms (although it does not apply to shellfish farming). Marine fish farming falls within the types of projects listed in Annex II to the Directive. Such developments must therefore, be subject to EIA whenever they are likely to have significant effects on the environment. This includes changes or extensions to existing developments that may have significant adverse effects on the environment even where the original development was not subject to EIA. The Regulations also apply to renewal of existing leases.[76]

UK REGULATORY FRAMEWORK FOR AQUACULTURE

4.143 The Crown Estate is responsible for the management of the territorial seabed and most of the foreshore between high and low water mark except for special arrangements in Shetland and the Orkneys under the Zetland and Orkney County Council Acts 1974. Anyone wishing to establish a marine fish farm must apply to the Crown Estate for a lease of the seabed (and foreshore where appropriate) within which the marine fish farm will operate. The Crown Estate monitors marine fish farm operations to ensure compliance with lease conditions. It also maintains a register of marine fish farm leases.

4.144 The industry is essentially Scottish. The Scottish Executive Environment and Rural Affairs Department (SEERAD) as a statutory consultee under the EIA Directive, advises the Crown Estate on the implications for disease control, existing fishing interests and the inshore marine environment of applications for marine fish farm leases, and is consulted by the Scottish Environment Protection Agency (SEPA) on discharge consent applications. SEERAD's Fisheries Research Services carries out a wide range of marine fish farm research and offers advice on aspects of production and disease control. SEERAD is also the formal point of contact for statutory notifications of escapes of farmed fish.

4.145 Development proposals that extend below the mean high water mark on spring tides require a licence under part II of the Food and Environmental Protection Act 1985, issued by SEERAD.

4.146 SEERAD is responsible for statutory measures under the Diseases of Fish Acts and related EC Fish Health legislation, to prevent the introduction and spread of serious pests and diseases of fish and shellfish which may affect farmed and wild stocks. All fish farms must be registered with the Department for disease control purposes. Certain diseases must be notified to the Department and there are procedures laid down for the treatment and disposal of infected stock.

4.147 Under the Control of Pollution Act 1974, consent is required for the discharge of effluent from marine fish farms to coastal waters from SEPA. SEPA consults other regulatory authorities and is a relevant and competent authority under the Conservation (natural habitats and conservation) Regulations 1994. Conditions designed to prevent, minimise, remedy or mitigate adverse environmental effects may be attached to discharge consents. SEPA is responsible for ensuring that appropriate monitoring of the aquatic environment is undertaken and this is achieved by applying specific consent conditions and by its own audit monitoring.

4.148 Planning authorities have the lead role in advising the Crown Estate on marine fish farm proposals under the interim arrangements. They prepare statutory development plans that provide the basis for making decisions about planning applications on land. These may include the landward developments needed to support offshore operations, or freshwater farms, which are currently subject to planning control. National Planning Policy Guideline 13: Coastal Planning notes that in areas where the potential for new or expanded fin and shellfish farms is recognised, planning authorities should consider the preparation of non-statutory Framework Plans, which would guide the location of new off- and on- shore facilities. It also notes that the involvement of the industry as well as local and environmental interests in the preparation of these Framework Plans is essential.

4.149 There are also requirements overseen by the Scottish Executive Development Department (SEDD) to ensure that works in tidal waters do not constitute a hazard to navigation. Consent for the installation of marine fish farming equipment in sea areas must be obtained from SEDD.

4.150 Salmon fisheries management in Scotland has been devolved to district salmon fishery boards.[77] These boards may do such acts, execute such works and incur such expenses as may appear to them to be expedient for the protection or improvement of salmon fisheries, the increase of salmon and the stocking of the waters of the district with salmon. It is an offence for a person intentionally to introduce salmon or salmon eggs into inland waters in a salmon fishery district for which there is a board, unless he/she has the written permission of the board, or the waters constitute a fish farm within the meaning of the Act.[78]

4.151 Scottish Natural Heritage is consulted on aquaculture applications, particularly in relation to areas designated for natural heritage purposes, such as SACs, SPAs, SSSIs and NNRs; areas such as Marine Consultation Areas which, although not designated, deserve particular distinction in respect to the quality and sensitivity of their marine environment and where the scientific information available substantiates their nature conservation importance; direct or indirect impacts upon biodiversity, protected under the UK Biodiversity Action Plan; possible conflicts with potential predator species arising from proximity to seal haul-out areas, and otter and fish-eating bird populations; the risk of introducing alien species and the likely consequences for wild animal and plant communities; the risk of genetic contamination of native stocks, particularly of Atlantic salmon; impacts upon the character and special qualities of Scotland's landscapes, and their enjoyment, including potential impact on wild land.

4.152 In Shetland[79] the Council has powers to licence works in coastal waters, which it exercises in conjunction with its powers as planning authority. Under these powers, the Council has developed policies for the development and regulation of salmon and shellfish farming. Anyone wishing to undertake marine fish farm development within the Shetland coastal waters must obtain a works licence from the Council. Applicants and objectors enjoy the right of appeal to Scottish Ministers against the Council's decision. Under the Orkney County Council Act 1974, the Council exercises works licensing powers within certain designated harbour areas. In the event a Works licence is granted the applicant must also apply to the Crown Estate for a lease in the usual manner.

CONSTRAINED AREAS

4.153 Scottish Ministers introduced a presumption against further aquaculture development on the north and east coasts in 1999, taking into account the Salmon Strategy Task Force's recommendations, the 'Resolution By The Parties To The Convention For The Conservation Of Salmon In The North Atlantic To Minimise Impacts From Salmon Aquaculture On The Wild Salmon Stocks' (NASCO – 'the Oslo Resolution') and ICES reports of the threats salmonid stocks face throughout their North Atlantic range.

4.154 In addition to the presumption against further development on the east and north coasts, Scottish Ministers proposed three categories, based on the level of nutrient loading and benthic impact within an area (see chapter 6).

4.155 Particular arrangements must be applied when considering any proposals that might affect Special Protection Areas and Special Areas of Conservation. Any proposal that is likely to have a significant effect on the interests for which the site was designated, must be subject to an appropriate assessment. If this assessment cannot demonstrate that the proposal will not adversely affect the integrity of the site it can only proceed in very exceptional circumstances.

MANAGEMENT OF SHELLFISH FISHERIES

4.156 Under European legislation, all shellfish production waters must be tested and classified. There are three grades of classification – Grade A, Grade B and Grade C. Shellfish gathered from or farmed in Grade A waters are suitable for direct sale and consumption. Shellfish from Grade B waters must be either re-layed in Grade A waters or cleaned by depuration (placing the shellfish in purified water). Re-laying and depuration allows the shellfish to filter clean water through their system, flushing out any bacteria that may be harmful to human health if consumed. Shellfish from Grade C waters are not suitable for human consumption.

4.157 UK retail multiples have elected to be more cautious than stipulated by European legislation and will only sell UK product that has been grown or collected from Grade A waters and depurated. They will not stock any shellfish produced in UK Grade B waters even after depuration. They do, however, stock product from outside the UK that has grown in Grade B waters and depurated. This practice by UK multiples effectively excludes much of the UK shellfish production from being sold directly in the UK. Continental buyers will accept product from Grade B waters that has been depurated.[80]

ALTERNATIVE MANAGEMENT APPROACHES

4.158 This chapter has concentrated on the aspects of the legal framework that are most relevant to the OSPAR area, and within that, to the EU and the UK. However, other nations have adopted different management approaches to similar problems. Below, we set out some examples relating to particular themes. Many of these are also discussed in more detail later in the report.

FISHERIES MANAGEMENT

United States

4.159 The US Magnuson-Stevens Fishery Conservation and Management Act[81] sets out a number of principles for both fisheries and conservation management. The operation of the Act is overseen by eight Regional Fisheries Management Councils, which are required to develop regional fishery management plans and to allow for public hearings on these plans so that any interested party can express their views.[82]

4.160 During our visit to the US we learnt that the system appeared on paper to build in requirements for environmental assessments. But in practice these were weak and could be circumvented where there was a lack of will to give sufficient emphasis to them in determining policy. In particular the predominant role given to those with direct connections to the fishing industry in the regional fisheries management councils, which set quotas, meant that many quotas had been set too high. This has led to continuing problems of sustainability and damage to the marine environment. Such issues led to the setting up of a Presidential Commission on Ocean Policy, which reported in draft in April 2004. It recommended radical change to the governance of marine policy and an ecosystem based approach to the management of marine resources including significant use of marine spatial planning and the separation of fishery assessment and allocation decisions (chapter 10).

New Zealand

4.161 In New Zealand, fishing rights have become permanent and transferable through the formal introduction of an Individual Transferable Quota system. New Zealand fisheries have also produced conservation initiatives such as industry-generated catch sampling and industry-generated codes of practice. ITQs would generally be expected to make these initiatives easier to introduce but there is no evidence that they are inevitably linked.

4.162 The Quota Management System (QMS) was introduced in 1986 and controls catches from all main commercial fish stocks in New Zealand waters. Total Allowable Commercial Catches (TACC) are set annually. Since 1991, under the ITQ system fishers own the right to fish a set proportion of the TACC. Pre-1991 ITQs were tonnage based, thus the government had to buy or sell quota tonnage to fishermen depending on the absolute level of TACC set each year. The transferability of ITQs allow fishers to adjust their holdings of quota, through purchase or leasing, so allowing the system to smooth out fluctuations in the catch of different species over time.

4.163 This system is particularly important in a multi-species fishery where there are fluctuations between catches of different species. Quota requirements therefore change often. Flexibility minimises illegal activity and discarding relating to catches of over-quota or non-quota species. New Zealand fisheries also attempt to minimise discarding by allowing fishers to pay a set 'deemed value' to government to allow landings of over-quota or non-quota species, thus effectively buying them the right to land such fish legally. The deemed value is set at a level which aims to encourage fishers to land such fish rather than discarding at sea which is illegal, but at the same time not allowing fishers to generate any significant economic benefit from landing and selling such catch. If the fisher goes on to purchase or lease quota for those species in the same year, the deemed value payments are returned to the fisher.[83]

Iceland

4.164 Iceland has also moved to managing its stocks using an Individual Transferable Quota (ITQ) system. Quota holders pay an annual fishing inspection fee and vessel owners pay for transfer of quota between vessels. Also, fishery-monitoring charges are collected through a charge on quota issue amounting to US$2m annually. The Icelandic fishing industry also contributed to a decommissioning scheme between 1990 and 1997 in which vessels were decommissioned from commercial fishing activity by having their licenses revoked but did not have to be physically scrapped. The decommissioning program was financed by a loan from the government to the Fisheries Development Fund, to be repaid with interest. The Fund issued grants to vessel owners who wished to decommission with preference being given to smaller vessels.

Faeroes

4.165 The Faeroese introduced vessel licensing in 1987, then TACs in 1994, followed in 1996 by effort management (individual transferable days at sea). TAC management was rejected because it caused discarding and high grading. The fisheries management system in operation there includes some technical measures, including seasonal spawning area closures, and real-time closures for small fish.

4.166 Since introducing effort management at Faeroe in 1996, fisheries managers have not followed ICES' advice, which has recommended cuts in effort even when stocks are above precautionary biomass limits. It has been claimed that the Faeroes effort system is successful because it allows fishermen to follow the natural fluctuations in stocks, and because landings of demersal fish have since increased.

4.167 Effort management in the Faeroes may appear to be successful to fishermen because of recent improved catches, but since 1996, when it was introduced, the average fishing mortality of cod and haddock has actually increased.

Norway

4.168 Norway's management of fisheries in relation to concentrations of juvenile fish also appears to have much to commend it. Fisheries patrol vessels can request vessels to move elsewhere if there is a high proportion of juvenile fish in the catch. Norway has also introduced special protection measures were announced to protect cold water coral reefs off its coast. In 2003, a ban on bottom trawls was introduced in an area approximately 43 km long and 6.8 km wide.[84] Other similar reef structures have also been granted protection.

ECOSYSTEM APPROACH AND MARINE PROTECTED AREAS

Canada

4.169 Canada's Oceans Act integrates many marine management objectives into one overarching framework.[85] The Act is split into three parts; Part I – Recognizing Canada's Oceans Jurisdiction, Part II – Oceans Management Strategy, Part III – Consolidation of Federal Responsibilities for Canada's Oceans. The Management Strategy (introduced in 2002) is designed to implement the Act and is based on the premise that oceans must be managed as a collaborative effort among stakeholders. The Act promotes integrated management and provides for the use of some basic management tools, including Marine Protected Areas, marine environmental quality guidelines and the development of management plans, including integrated coastal zone management. The Act is founded on the implementation of an ecosystem based approach to management and recognises clear roles for improved ocean science and for international leadership by Canadian authorities (chapter 10).

Australia

4.170 In 2003, the Great Barrier Reef Marine Park Authority carried out a rezoning exercise that established large-scale marine reserves, closed to fishing and covering 33% of the park. The primary goal was to protect representative examples of the entire range of habitats and biodiversity within highly protected areas[86] (chapter 8).

CONCLUSIONS

4.171 At the European and UK level, the current legal framework for protecting the marine environment is weak in a number of areas. First, areas of conservation importance are offered only very limited protection from the effects of fishing. Second, the area given over to conservation is vanishingly small and largely designed to protect specific areas and habitats, rather than interconnected parts of the ecosystem. Third, it does not provide a formal basis for newer management tools, such as marine protected areas, that may be required to help the recovery of commercial fish species and the wider ecosystem.

4.172 Overall, it appears the current system is unlikely to deliver reforms on the scale that are necessary to secure the future of the marine environment. New legislation will therefore be necessary to deliver the changes outlined later in our report, and innovative examples of how these problems can be tackled are already to be seen in operation in other countries.

4.173 The European legislative framework limits considerably the scope of what can be done by the UK. Major change will require action at EU level as well as at national level. We therefore look to the European Commission, the European Parliament and the Council in framing our advice as well as to the UK government and devolved administrations.

Chapter 5

THE LEGACY OF OVERFISHING

Commercial fisheries have expanded massively, shifting marine ecosystems into new states that are less desirable and may be difficult to reverse. Do we face a future of collapsed fisheries and the extinction of many marine species?

INTRODUCTION

5.1 Innovations in fishing technologies have had a continuous impact on marine ecosystems from ancient times to the present day. Archaeological evidence has shown that technological advances, such as the introduction of nets, have been associated with dramatic reductions in biodiversity and progressive depletion of fish populations. There are documented examples such as the collapse of the now-extinct Scania herring population in the Western Baltic which came about as a result of improvements in netting in the middle ages.[1,2]

5.2 Historical changes, however, are of a limited and local nature by comparison with the global changes being wrought by modern commercial fishing. Improvements in technological capability, such as the use of sonar and bigger trawling gear, have accelerated the rate of change over time and the magnitude of the over-exploitation. Modern fishing vessels can compensate for reductions in fish biomass by continually improving catching technology (Chapter 3) and fishing for 24 hours a day in almost all weather.

5.3 Reviewing 40 years of catch data compiled by the UN Food and Agriculture Organization (FAO), researchers calculated that the mean trophic levels of landings (2.17) is falling as a result of overfishing, while top predators such as sharks, cod, tuna, and swordfish are in decline. Many target fish populations are 10% or less of what they were 50 years ago (5.42). As a result only organisms lower down the trophic chain are prevalent, and we are now starting to fish these more intensively – a trend sometimes referred to as 'fishing down the marine food web'.[3]

5.4 The effects of overfishing and damage to the environment from fishing practices over the past 50 years will continue to destabilise and reverberate throughout marine ecosystems for decades or centuries to come. The problem with the current level of fishing is simply the removal of animals and the destruction of their habitats. This reduction in biodiversity has led to a loss of resilience in the system, which is likely to compound the effects of climate change (2.22-2.27).

THE IMPACTS OF FISHING

5.5   The sea is subject to a range of impacts from human activities, but no other activity has such a large direct impact on the whole marine ecosystem as fishing. Technological advances have enabled overfishing to the degree that the overall global catch has levelled off and started to decline despite increased capacity and effort. The adverse impacts of excessive fishing on target fish populations are apparent in almost all fisheries. According to FAO figures, 47% of the world's populations of commercial marine species are fully exploited, while 18% are overexploited and 10% are severely depleted or recovering from depletion. Only 25% of such populations are under- or moderately exploited.[4] The proportion of commercial fish populations in the area covered by the International Council for the Exploration of the Sea (ICES) that is inside 'safe biological limits' has declined (from 26% to 16% between 1996 and 2001). The status of EU fish populations has also deteriorated over the last thirty years. The proportion of those that are overexploited or severely depleted has increased (Figure 5-I and appendix E).[5]

Figure 5-I
The status of EU commercial fish populations from 1970 to 2000 as calculated by the Prime Minister's Strategy Unit on the basis of ICES stock assessments[6]

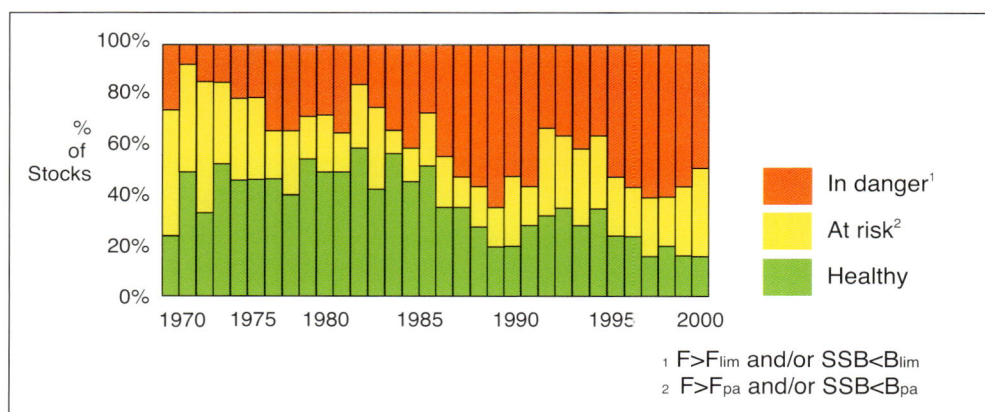

5.6   In the OSPAR region, total landings of major gadoid species (i.e. soft-finned fish of the family *Gadidae*, such as cod) have declined from over 3.5 million tonnes in 1970 to around 1.5 million tonnes in 2001 (excluding industrial landings for aquaculture and agricultural feedstocks), reflecting a decline in population size over this period (see figure 5-II). It is likely that the original prefishing or 'pristine' fish populations were much larger than 1970 levels, as the most rapid depletion of commercial fish populations occurred early in the history of commercial fisheries. In 2002, of the main assessed commercial fish populations of importance to the UK, only 29% were deemed to be within safe biological levels on the basis of population size and fishing rates, and only 47% were above precautionary biomass levels (the biomass level below which management action should be taken).[7] The tonnage of fish landed in England and Wales has decreased by 23% from 1996 to 2003,[8] reflecting declines in population levels.

Figure 5-II

Landings of major gadoid fish species in the North East Atlantic and the Baltic Sea excluding industrial fishing landings of Norway Pout and Blue Whiting, 1970-2001[9]

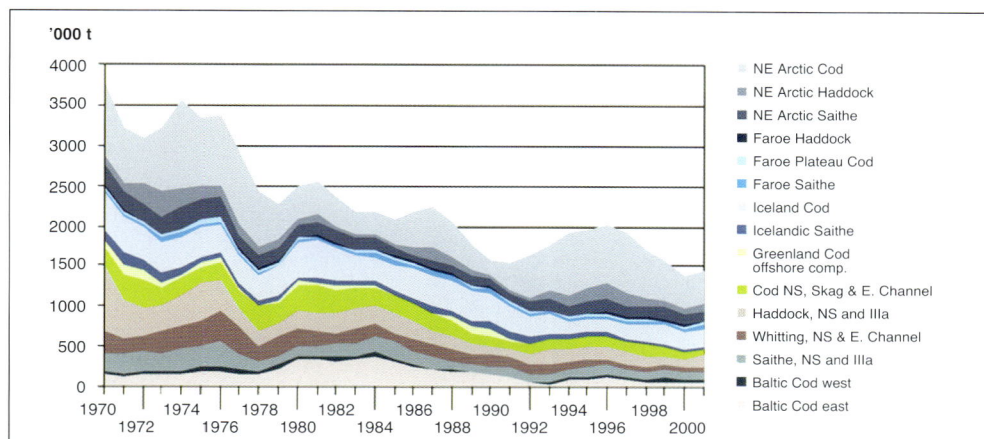

5.7 Previous concerns about fishing have tended to focus on its impact on commercial fish species and on cetaceans, birds and turtles caught as by-catch. While these may represent some of the severest, direct effects, it is clear that fishing can damage entire ecosystems by disrupting food webs and changing multi-species population dynamics. There is a growing appreciation of these wider and subtler effects.[10] In this chapter we examine the profound changes to ocean and coastal ecosystems that widespread overfishing appears to be provoking.

THE ECOSYSTEM EFFECTS OF FISHING

5.8 Although much is still unknown, scientists are starting to understand the full scale and scope of the impacts of overfishing on the marine environment.[11] These include the direct removal of target species,[12] changes in size structure of target populations,[13] alteration in non-target populations of fish and benthos,[14] alteration of the physical environment through seabed disturbance,[15] alterations in the chemical environment, including nutrient availability,[16] effects on trophic interactions[17] and altered predator-prey relationships.[18] The combined effect is to reduce the overall stability of marine ecosystems.[19]

5.9 The ecosystem effects of fishing are widespread and global, though populations of fish on continental shelves have perhaps been more affected than those in the very deep ocean thus far. Some species are particularly vulnerable; generally those that are long-lived and only start breeding after a relatively long period of immaturity. Sharks, rays and skates, and many species of fish in deep water fall into this category. In general, seabed communities in heavily trawled areas have changed from long-lived to more opportunistic species. The sheer variety of gears used and areas fished means that the footprint of the industry on the marine ecosystem is large.[20]

MORTALITY OF BY-CATCH AND EFFECTS ON NON-TARGET SPECIES

5.10 As mentioned in Chapter 3, fishing kills many more fish than are landed as a result of discarding and bycatch (figure 5-III). Globally, the proportion of fish caught and discarded amounts to about 27% of the overall catch.[21] If industrial fisheries, where all the catch is landed, are not included in this calculation, the proportion rises to over 50% in many individual fisheries (e.g. for shrimp and prawn fisheries discard rates represent over 80% of the catch). ICES have estimated that in 1990, around 260,000 tonnes of roundfish, 300,000 tonnes of flatfish, 15,000 tonnes of rays, skate and dogfish, and 150,000 tonnes of bottom-dwelling invertebrates were discarded in the North Sea alone. This amounted to about 22% of the declared landings.[22]

Figure 5-III
The bycatch from a scallop dredge, including flatfish[23]

5.11 For every kilogramme of North Sea sole caught by beam trawl on the seabed, up to 14 kg of other animals are killed.[24] Often this includes species of commercial importance that are either below the minimum size for landing or for which the boat has no quota allowance, and therefore cannot be landed legally. For example, every tonne of Dublin Bay Prawn landed from the Irish Sea trawl fishery just under half a tonne of whiting are discarded as bycatch.[25] But much discarding is also done for purely economic reasons. In the UK whitefish trawl fishery over two-thirds of the discarded catch consists of commercial species, of which undersized fish make up the majority.

5.12 In the North Sea demersal fishery the average proportions of discarded cod and haddock are estimated at 22% and 36% respectively by weight, representing 51% and 49% by number. Only half of the plaice caught by beam trawl are usually retained, decreasing to 20% in shallower inshore grounds.[26] The estimated cost (in terms of loss of future catch) of this discarding is over 40% of the total annual landed value of the whitefish trawl fishery.[27] A feature of heavily exploited fish populations is that the mean size and age of the fish in the population decreases (5.25). This results in increased discard rates because a higher proportion of the fish are too small to be legally landed.

5.13 The impact of fishing on some non-target marine species is severe, as some species are particularly sensitive to the mortality caused by fishing (5.40). Consequently, some species have been made locally extinct in parts of their ranges and diversity among fish populations is also declining as a result of fishing pressure. For example, studies have demonstrated clear downward trends in diversity of both commercial fish species and non-target fish off Plymouth since 1913.[28]

BY-CATCH OF CETACEANS AND SEABIRDS

5.14 Fishing with trawls, seine nets, set nets, trammel nets, long lines and gillnets can result in by-catch of cetaceans, seals and seabirds. While all gear presents all marine animals with hazard, some gears are particularly risky for some animals. Bottom-set gillnetting particularly affects harbour porpoises, some pelagic trawling practices affect dolphins; longline hooks catch seabirds (such as albatross) and turtles; and drift nets catch both birds and dolphins.

5.15 **Cetaceans:** On a global scale, by-catch in fishing gears is believed to be the biggest single threat to cetacean populations. Extrapolation of US figures suggests a global annual cetacean by-catch rate of 65,000 to 85,000 animals with at least 26 different species dying from entrapment in fishing gear.[29] By-catch of North Atlantic right whale (*Eubalaena glacialis*) in gillnets that has a population of around 380 individuals, is believed to be a major factor in inhibiting the recovery of this protected species.[30] The EC Habitats Directive (4.61) requires Member States 'to establish a system to monitor the incidental capture and killing of all cetaceans, and in the light of the information gathered to undertake further research or conservation measures to ensure that incidental capture and killing does not have a significant negative impact on the species concerned'. The UK and some Member States have entered into commitments beyond those in the EC Habitats Directive under the Agreement on the Conservation of Small Cetaceans of the Baltic and North Seas (ASCOBANS) (4.44). This obliges signatory states to pursue a precautionary objective of reducing annual by-catch to 1% of the total population.

5.16 Data on by-catch of cetaceans is inadequate for management and research purposes and generally does not meet the standards required by European law. Some good information has, however, come from independent discard observer schemes and there have been some studies of marine mammal by-catch. Observers on European pelagic vessels have recorded the accidental capture and drowning of many mammal species in pelagic trawls, including harbour porpoises (*Phocoena phocoena*), short-beaked common dolphins (*Delphinus delphis*), white-sided dolphins, white-beaked dolphins, striped dolphins, bottlenose dolphins, Risso's dolphins, killer whales, long-finned pilot whales and grey seals.[31] Pelagic trawl by-catches of dolphins are widespread in the Bay of Biscay, the Western Approaches and the Celtic Sea.[32]

5.17 Estimates of harbour porpoise by-catch in the North Sea, of around 8,000 a year (from a population of about 280,000[33]) indicate that their numbers are certainly constrained by bycatch and probably depleted.[34] Off the south-west of the UK, Irish and English gillnetters fishing for hake were estimated to catch around 2,000 porpoises each year in the mid 1990s (from a population of about 38,000).[35,36] Some Member States have established systems to monitor cetacean by-catch, including the UK, but most have done little or nothing and commitments to reduce cetacean bycatch are not being fulfilled.[37] Since 2000, the UK has conducted observer monitoring to estimate the level of by-catch in UK pelagic fisheries. The only area where by-catch of common dolphins and other cetaceans has been observed regularly in UK waters is in the Western English Channel. The by-catch was taken by boats, primarily from the UK, pair trawling for sea bass.[38] The UK government has recently undertaken steps to halt this practice (9.56). High dolphin by-catch rates have also been recorded in the Dutch mackerel and horse mackerel single-trawl fishery in the north-east Atlantic and Irish tuna albacore pair-trawl fishery in the Celtic Sea.[39]

5.18 The UK government (led by the Department for Environment, Food and Rural Affairs (Defra)) has funded research to investigate potential by-catch mitigation measures, including acoustic deterrent devices (pingers) and exclusion devices. Defra has also consulted the public and the fishing industry on a *small cetacean by-catch reduction strategy*, which includes the compulsory use of pingers, an effective observer scheme, continued research into mitigation methods in pelagic fisheries and better surveys of cetacean abundance and distribution.[40] The European Commission is also proposing a new regulation to address cetacean by-catch, but there are concerns that these efforts maybe insufficient to reduce by-catch levels substantially.[41] It has been suggested that the proposals place too much emphasis on pingers, at the expense of other mitigation measures and more selective fishing methods, the suggested level of observer coverage is inadequate, cuts in overall fishing effort are needed to effectively reduce by-catch and the proposed regulation fails to set any objectives or targets for reducing cetacean by-catch.[42]

5.19 The greatest by-catch issue in the Southern Hemisphere comes from vessels fishing illegally; many of which appear to be controlled from companies based in OSPAR member states, but until recently there has been little effort to bring them under control (4.31). Overall there are insufficient data on by-catch to assess impact and to be able to propose appropriate management measures. In addition, there is also evidence of substantial cetacean by-catch occurring in the EU's considerable distant water fisheries such as the Dutch pelagic trawl fisheries off Mauritania, indicating a lack of regulation of these distant water fleets (4.125-4.129).[43]

5.20 **Seabirds:** The ICES working group on seabird ecology recently stated that gillnets and longlines are responsible for drowning thousands of seabirds each year.[44] North sea populations of northern fulmars (*Fulmaris glacialis*), common guillemots (*Uria aalge*) and razorbills (*Alca torda*) are all believed to be affected by fishing-gear induced mortality (figure 5-IV). There have been very few studies of bird by-catch, however, in European waters. Reviews in the 1980s indicated that some nearshore salmon gillnets were catching substantial numbers of auk species (guillemots, razorbills, puffins, etc.) from nearby colonies around Scotland and Ireland; but most of these fisheries in Scotland have now closed for salmon conservation reasons.

5.21 Coastal gillnet fisheries have been implicated in the decline of several European populations of auk species.[45] There are also reliable reports of these species being caught in both pelagic trawl gear and trawls for sandeels, but catches have not been quantified in either.[46] Gillnet entanglements of shearwaters (*Puffinus* species), red throated divers (*Gavia stellata*), Leach's petrel (*Oceanodroma leucorhoa*), gannet (*Sula bassana*), shag, razorbill, great northern diver (*Gavia immer*), Slavonian grebe (*Podiceps auritus*), scaup, common scooter, long-tailed duck (*Clangula hyemalis*) and guillemot have all been recorded in European waters.[47] The impacts of discards on seabird populations are discussed below (5.34-5.36).

Figure 5-IV
Dead gannet caught in a fishing line[50]

5.22 One study of catches on longline fisheries in the OSPAR area, showed a relatively high by-catch of northern fulmars (approximately 20,000 per annum) by the Norwegian longline fisheries to the north of Scotland.[48] There are indications that the by-catch of Cory's shearwater (*Calonectris diomedea*) in southern OSPAR waters is also relatively high.[49] The effect of Southern Ocean longline fisheries on albatross populations has attracted much attention. In general, there is sufficient information on the legal fisheries in the Southern Ocean, and good mitigation measures have been put in place to ensure that this catch is minimised. The same mitigation measures that apply to Southern Ocean longline fisheries (including setting the line at night, use of streamer bird scaring lines and optimal weighting of line to ensure it sinks quickly) could be introduced for the longline fleets in the Northern Hemisphere, but as yet there is no legal requirement to do so.

STRUCTURAL CHANGES IN EXPLOITED MARINE COMMUNITIES

5.23 Communities of marine animals have been substantially altered, largely as a result of consistent fishing pressure over the last century. The main change in structure is that the numbers of small fish are greater and numbers of large fish are severely reduced both within and across species (5.25). There is a clear link between fishing and reductions in biodiversity (5.75-5.77).

5.24 There has been a consistent decline in the mean trophic level of the global landings from top predators to organisms lower down the food web. Decreases in the trophic level of fish communities are expected to be even greater than decreases in the trophic level of landings because fishers tend to target species selectively at higher trophic levels. Many species from higher trophic levels and large or slow growing species with late maturity have become scarce following intensive exploitation (5.60-5.62). The extensive exploitation of marine fish has led to substantial reductions in the abundance of some target species and changes in the structure and species composition of fish communities.[51,52,53,54]

5.25 Studies in the North Sea have also shown that the differential effects of fishing on organisms with different life history parameters determine changes in the structure of the fish community, such that there has been a shift towards small organisms with higher turnover rates, with long-lived, large organisms with low reproductive rates becoming locally extinct as fishing effort has increased.[55] This size shift has occurred both within and across species, such that individuals of commercially exploited species are smaller and mature earlier as a result of genetic selection for these traits (5.58), and species that already have such characteristics have increased in comparative abundance.

5.26 This may be because shorter life histories enable these species to sustain higher instantaneous mortality rates or they may suffer lower mortality rates because they are less desirable and less accessible targets in a selective size fishery.[56] However, if fishing mortality is high enough even populations of relatively small and early maturing species may be depleted. For example, long-term trends in catch rates suggest that the biomass of some sole populations has been reduced by 90% in the twentieth century.[57] Populations of some shellfish (such as shrimp and Norway lobster or Dublin Bay prawn) have proved more resilient to depletion, partly because a proportion of the population (particularly females with eggs) are in burrows at any one time and therefore not available to be fished. However, it is possible that growth in populations of shellfish may often be a consequence of the loss of large fish such as cod, which consume large numbers of benthic crustaceans in some areas (5.32), and this in turn will have affected food webs of bottom-dwelling organisms.[58]

5.27 Although studies of fishing impacts have shown changes to the structures of marine ecosystems, they may under-estimate the extent of these changes because analyses are based on datasets that began after the onset of exploitation and knowledge of the structure of unexploited fish communities is usually lacking.[59] Even where such data are available changes in climate mean that cessation of fishing will not necessarily lead to reversion of historic states.[60] In order to fully comprehend the impacts of fishing on the structure of marine ecosystems, it is necessary to predict what would be the structure of such ecosystems under present climatic influences if they were not impacted by fishing.[61]

5.28 Recent studies using primary production measurements and predator-prey body mass ratios have modelled what the total fish abundance and size structure would be if the North Sea were unexploited. These studies suggest that the biomass of large fish in body mass classes 4-16 kg and 16-66 kg in the North Sea is 97.4% and 99.2% lower than it would be in the absence of fisheries exploitation.[62] The assumption that the predator-prey mass ratios are primarily influenced by environmental variation rather than fishing underpins the model used to derive these figures. However, it should be noted that such size based approaches cannot inform us of the unexploited levels of some of the largest fish, which may feed at the lowest trophic levels where biomass is most abundant, such as basking sharks that feed on plankton or tuna feeding on small planktivorous fish. We further consider how knowledge of the unexploited marine communities could be used in the management of the marine environment in Chapter 7.

5.29 This method relies on the relationship between abundance-body mass relationships, predator-prey mass ratios and transfer efficiency of energy between trophic levels to make the appropriate predictions. Predator-prey mass ratios in the North Sea have previously been estimated using information from nitrogen stable isotope analysis in order to analyse the size composition in relation to trophic position. This analysis indicated that the trophic level of the North Sea demersal fish community has decreased as a result of fishing exploitation, a factor that is related to overall size reduction.[63,64]

5.30 Fishing has a great many direct and indirect effects on the food web. It may affect populations of predators, prey and competitors. The impact will depend on how strong inter-species linkages are within the community or ecosystem. For example, changes in food availability as a result of fishing of a prey species will severely affect those species unable to switch prey easily, and result in a predator population decline, even if the prey species is fished at 'sustainable' levels.

5.31 This is worrying because knowledge of the ecology and relationships between species is limited and effects are unpredictable, particularly against a background of natural variability. An example of this is the effect

Figure 5-V
Green sea urchins (*Psammechinus miliaris*) grazing kelp. This species is commonly found round the coasts of the UK, at depths of 16-70 m. Its distribution frequently coincides with that of sugar kelp (*Lamnaria Saccharina*)[66]

of northern cod fisheries off the coast of Newfoundland. Reduction in cod populations has led to an increase in lobster survival. When fishing effort was redirected to the lobster populations, lobster numbers declined, sea urchin survival improved, leading to an explosion in their numbers and overgrazing of kelp forests occurred (figure 5-V).[65]

5.32 There have been several cod fishery collapses in recent decades that have led to changes in the food webs of associated ecosystems. Where cod is the dominant predator, the collapse of cod populations often leads to large increases in benthic crustaceans, and pelagic fish,[67] both a major food source for cod. A serious effect of this is to inhibit recovery of cod populations due to intense predation on cod eggs and larvae. For example in the Baltic Sea, the collapse in cod populations has meant a growth in planktivorous species that were once heavily predated upon by cod. Now, these species, herring and sprat (*Sprattus sprattus*), consume large quantities of cod eggs,[68] thus suppressing the ability of the cod population to recover. This situation could continue even in the absence of fishing pressure and the presence of a large abundance of prey.

5.33 Changes have occurred in the North Sea, where echinoderm (starfish and urchin) larvae have dominated the plankton since the early 1980s. This period of domination coincided with the increase in beam trawling, which started in 1970 and continues to the present day. Such a shift may well be the result of fishery pressure because common starfish suffer less mortality during trawling than many other species, although other echinoderm species such as burrowing urchins and brittlestars (figure 5-VI) may suffer higher mortality rates. Fishing may also have reduced predation on some species of echinoderm by reducing the abundance of fish predators.[69]

Figure 5-VI
Two red brown brittlestars of the species *Ophiura albida*. This species is found on a variety of muddy sediments. Brittlestars are a common component of the diet of flatfish such as plaice[70]

5.34 Fishing activities may directly support scavengers by providing dead, discarded food. Populations of many species of scavenging seabird have increased in size over the past century (appendix E). It is difficult to prove cause and effect in this area as there has been inadequate long-term assessment of diets and feeding relationships and changes are not necessarily immediate. The reversibility of these changes is also unknown.

5.35 As an example, within the last hundred years, the population of great skuas in the North Sea area has increased by 200 times their historic abundance due, it is believed, to increased availability in their mixed diet of sandeels and discards from fishing vessels. Both sandeel abundance and discards have now decreased in the North Sea area as a result of overfishing and changes in fisheries management, and consequently great skuas have switched diet. Unfortunately this switch has been into seabird predation and black-legged kittiwakes in particular have taken the brunt of this.[71]

5.36 On the basis of trophic modelling it has recently been claimed that the number of levels in the North Sea food web halved between 1880 and 1981.[72] Biomass estimates from these modelling techniques (Box 7B and appendix F) suggest that total fish biomass in the North Sea has decreased from about 26 million tonnes in the 1880s to 10 million tones by 1991.[73] It is clear that the complexity of marine ecosystems is being reduced, and with the change from complex food webs to simple food webs, resilience in the marine ecosystem is also being lost. Fast turnover of exploited organisms at low abundance (5.24-5.25) will lead to greater instability in biomass production and increased sensitivity of populations and communities to environmental change.[74]

THE MYTH OF FECUNDITY

5.37 There is a common misconception that marine species are more resilient to catastrophic mortality events than terrestrial ones. This idea was articulated by Thomas Huxley in 1883, who, impressed by the great fecundity of broadcast spawning fishes (box 5A) such as the cod, stated 'that the cod fishery, the herring fishery, the pilchard fishery, the mackerel fishery, and probably all the great sea fisheries, are inexhaustible; that is to say nothing we do seriously affects the number of fish'. At the same meeting, however, another biologist, Sir Ray Lankester argued that the millions of young produced by marine fish were not superfluous and that the fish in specific area of the ocean were in equilibrium with their predators, such that "those that survive to maturity in the struggle for existence merely replace those that have gone before".[75]

5.38 This is now one of the tenets of population ecology: when an animal population is at or near equilibrium natural selection favours individuals whose reproductive strategy allows them to produce enough offspring to replace themselves.[76] Unfortunately, in the case of fisheries science, Huxley's incorrect hypothesis prevailed and most fisheries scientists have until recently continued to assume that very fecund fish species such as the gadoids can be fished to low population levels and recover. This argument still features in fisheries management debate as one of the drivers for over-fishing, that the reproductive capacity of fecund fish species is an adaptation to increased mortalities as a result of environmental fluctuations in the oceans thus allowing fish populations to recover from low levels caused by high fishing mortality.[77] Undoubtedly, some clupeid species (fish species from the taxonomic order *Osteichthyes* that are mostly planktonic feeders) do have directly naturally fluctuating population sizes in response to environmental factors. These include the pacific sardine (*Sardinops sagax*) and northern anchovy (*Engraulis mordax*), which have populations that have fluctuated by factors of six and nine respectively, over a period of two millennia.[78] There is, however, little empirical or theoretical basis for the hypothesis that highly fecund fish species are less at risk of decline as a result of fishing pressure than those of low fecundity and the use of fecundity in estimating reproductive potential is flawed (box 5A).[79]

5.39 Comparative analyses indicate that large size, long life, late maturity and low rates of natural increase render many species of larger fish particularly vulnerable to fishing. Many long-lived marine teleosts (fish species with a bony skeleton) share a suite of life history characteristics, including delayed sexual maturity, long reproductive life-span, sporadic recruitment and the repeated production of offspring at intervals throughout the life cycle. These are adaptive responses and predictable 'trade-offs' to low probabilities of successful reproduction (box 5A), due to high egg mortality.[85]

Box 5A	BROADCAST SPAWNING

Broadcast spawner species are defined as those that release both sperm and eggs into the water column, with fertilisation taking place externally and producing planktonic offspring that can potentially drift long distances on ocean currents. Many marine invertebrates are broadcast spawners but lead a sessile or semi-sessile life as an adult. The larvae may remain for only hours or up to months in the pelagic zone, depending on the species, where they may drift with water currents; some are capable of travelling great distances. These dispersal systems produce high variability in breeding success, as demonstrated by the difficulties in predicting recruitment of species (the number of individuals successfully progressing from the larval to adult stages) in any given year. It may be that both local and remote recruitment are important in the same population at the same time. The duration of the pelagic phase (which is often shorter for invertebrates) is an indication of the possibility of long-distance dispersal.[80]

Although broadcast spawners produce large numbers of eggs, they have high mortality rates. Compared with species that produce small numbers of eggs, the chances of survival to a reproducing adult can differ by orders of magnitude. Empirical support for this presumption comes from studies of fecundity and variance in reproductive rate in different species of marine fish. These suggest that species with high fecundity are no more likely to produce high levels of recruitment than those with relatively low fecundity.[81] Using meta-analysis of maximum reproductive rates it has been shown that reproductive capacity in marine fish populations is surprisingly low and uniform, generally ranging between 1 and 7 replacements per year.[82] It should also be noted that long-lived species also typically undergo years of low recruitment interspersed with occasional high levels of recruitment when oceanographic conditions are right. Thus, reproductive success is quite low averaged over an individual's lifespan and longevity is an important component of this 'bet hedging strategy'. Fishing leads to the truncation of the age structure in such long-lived species and may contribute to severe population declines in such species (5.40).[83] The suggestion that the fecundity of marine fish can be used as a means for assigning level of threat is therefore inconsistent with a precautionary approach to fishery management and conservation of marine biodiversity.[84]

5.40 Many long-lived fishes also show complex social structures and mating systems (e.g. cod, box 5B), and removal of dominant males can reduce reproductive success. At low population density it may not be possible to locate a mate and fertilisation success may be reduced (the 'Allee effect' or dispensation).[86] In addition, the success of larval offspring of some fish species appears to be related to the age and size of the female parent. Older larger female fish appear to produce more eggs and larval offspring with higher rates of survival. Thus, the loss of large individuals is likely to have considerable impact on the overall reproductive ability of the population.[87] A high rate of mortality means that the populations are more heavily dependent on recruiting year classes (the number of juvenile fish joining the population each year). This creates more variability in the system, as variability in recruitment from year to year has a bigger effect when there is a lack of a 'buffer' provided by sufficient numbers of large adult fish. Similarly, some species of marine organisms change sex during their life cycle, such that the older larger individuals are all males, or in the case some shrimps and crab species female.[88] Obviously in such cases, the loss of larger older individuals will dramatically affect the reproductive viability of the population.

5.41 Nearly all species targeted in demersal fisheries have become depleted. The assumption has often been that relieving fishing pressure on depleted populations through moratoriums or other measures will inevitably allow these populations to recover. Some fish populations, however, have not even recovered when fishing pressure is reduced or removed, as changes in predator-prey interactions and in food web structures may shift ecosystems into alternative stable states (5.32). Although the relaxation of fishing pressure almost always leads to rapid, and potentially long-lasting increases in community biomass, the trajectories of individual species are more difficult to predict.[89]

5.42 Studies have provided evidence that most marine fish species are not resilient to large population reductions, with the possible exception of herring and related short-lived pelagic species (clupeid species) that mature early in life and are fished with highly selective equipment (5.38). Analysis of ninety other populations shows that for some gadoids (cod, haddock) and other non-clupeid species (for example, flatfish) the time taken to recovery may be considerable after 45-99% reductions in biomass (North Sea Cod has been reduced by 90%). Of the populations either collapsed or severely depleted over the past several decades, most have not recovered after 5, 10 or 15 years of management measures designed to promote their restoration, such as for cod on the Grand Banks (box 5B), possibly because management measures have been ineffectual in reducing fishing mortality.[90]

5.43 Nonetheless, there remain the grounds for cautious optimism, as some species have shown recovery where fishing effort has been drastically reduced and appropriate management regimes enforced,[91] as on Georges Bank where some areas have been closed to all fishing and overall fishing effort greatly restricted (8.25). The slow recovery of some populations is perhaps not surprising, because fish species are no more resilient to rapid declines in numbers than any other animal taxa. In addition, many management strategies for recovery only provide partial protection, such as species-specific fisheries closures where fishing is still permitted in the area but not for that species, so the remaining fish can still be caught as by-catch by other fisheries using the same area. In addition to ineffective protection measures, the ecosystem consequences of the exploitation such as changes in community structure and food webs (5.23-5.36) are likely to be equally important factors in the slow recovery of these populations.

Box 5B   **COD COLLAPSE ON GRAND BANKS**

Cod have been caught off Newfoundland since the early 1500s, but in 1992, after almost 5 centuries of fishing, these were fished to commercial extinction. Cod were largely caught with hook and line until the early 1900s, and although the fishery supported many fishers and provided good economic returns, the replenishment of the cod populations was fast enough to sustain the fishery and the fishery was developing. In later years, more effective fishing techniques such as bottom trawling, traps and gill nets were used. Landings of the so-called northern cod increased from 100,000 – 150,000 t per year from 1805-1850 to 200,000 t at the end of the nineteenth century. Factory freezer trawlers from Europe started fishing northern cod in the 1950s and catches rapidly increased from 360,000 t in 1959 to a peak of 810,000 t in 1968. In 1977 Canada extended jurisdiction to 200nm and took over management of the cod populations.

It was already overexploited and landings were now 20% of 1968 level. Under Canadian management the stock continued to be fished at low but increasing levels and in 1992 the six Canadian populations of Atlantic Cod (*Gadus morhua*) collapsed so severely that the Canadian government closed the fishery. The size of the spawning stock had fallen from an estimated 1.6 million t in 1962 to 22,000 t in 1992 (since this it has risen to more than 50,000 t).

This collapse was a direct result of the setting of excessive quota levels that led to overfishing.[92] The management system was over-reliant on population modelling and prediction, and the catch per unit effort data used for these models (derived from information given by fishers) provided an inaccurate picture of the state of the population, resulting in a severe and consistent underestimation of fishing pressure. The leader of an inquiry into the validity of the population assessments during the crisis likened fisheries science to the 'Ptolemaic model of the solar system, where, when observations did not fit a theory an additional layer of complexity was added, rather than questioning the basic theory'.[93]

A number of scientists (and the inshore fishers) did query the calculated fishing mortalities and highlighted the uncertainties involved, but were usually dismissed as mavericks supporting political agendas. When attempts were finally made to limit the catch, it was seen as an admission that the scientists had got it wrong, which in turn meant that the science could no longer act as an arbiter between the conflicting resource claims. Offshore fishers disputed there was any evidence that populations had fallen and insufficient efforts to reduce catch limits led to a complete collapse of cod populations within a couple of years. An apparent 'systems failure' prevented scientists, fishers and politicians from responding to existing information or extracting themselves from the situation.[94]

The core problem was overfishing, but it was compounded by a lack of knowledge about the basic biology of cod, in particular their mating behaviour. No attempts were made to limit the fishing of spawning aggregations, severely affecting the reproductive capacity of the population.[95] The mating behaviour of cod is far more complex than has previously been assumed.[96] Mate competition, mate choice and other components of mating systems affected population growth rate deleteriously during and after periods of intense exploitation, increasing the rate of population decline and diminishing the rate of recovery. For example, there was a change in migratory behaviour to spawning grounds that was previously led by older dominant males, which were all removed from the population by high fishing levels.

Rates of cod larvae survival are also affected by environmental changes and corresponding fluctuations in plankton levels (2.26). Once cod populations are fished to low levels these environmental changes have much greater effect, exacerbating the effects of the collapse and impeding recovery.

Decline in large predators

5.44   Globally, fisheries have previously over-exploited large predatory fishes, such as tuna or cod, on which to a large extent they rely, with little or no regard for the consequences as a result of assumptions about fecundity. Management schemes are usually implemented well after commercial fishing has begun, and at best only serve to stabilise fish biomass at low levels. The removal of the majority of these large predators, which comprise a large proportion of the vertebrate biomass of the oceans is likely to have a widespread ecosystem effect, possibly difficult to reverse because of the global nature of the decline. Analysis of data (box 5C) from fisheries in four continental shelf and nine oceanic systems shows that large predatory fish populations are depleted by about 80% during the first 15 years of industrial exploitation, which is usually before comprehensive scientific monitoring has taken place.[97] Overall, the analysis suggests that 90% of large predatory fish have been lost from the global oceans and that there is pronounced decline in entire communities across widely varying ecosystems.

Figure 5-VII
The extent of declines in fish populations. The charts each represent one fish population, portraying the near universal decline, in the last decades, of abundance of commercial fishes in the North Atlantic in terms of biomass (blue lines), and the increase of the fishing mortality to which they are subjected (red lines). The time scale for each chart is 1950 to 2000. The status of most populations in the 1990s can be seen to have worsened rapidly from what were already depressed levels[98]

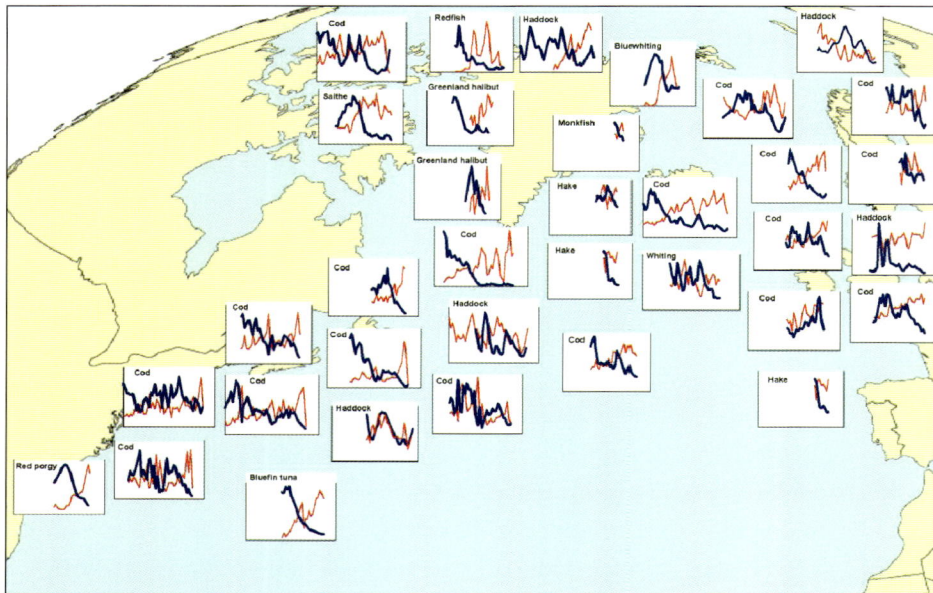

5.45 The targeted capture of marine mammals in past centuries resulted in significant declines in their populations (box 5D). The earlier removal of the large whales from much of the north-east Atlantic would also have caused substantial damage to the marine ecosystem and its dynamics. Many marine mammals and seabirds would be more abundant in the absence of human impacts and, as these species feed predominately on smaller fish classes, they could be expected to compete with resources that would otherwise be utilised by large predatory fish.[99] At present population levels and distributions, however, whales and marine mammals consume less than 1% of primary productivity in areas that overlap with fisheries.[100]

5.46 With some exceptions in the OSPAR area (whaling off Norway, Iceland and the Faeroe Islands, and illegal killing of dolphins in Iberian waters), the targeted capture of marine mammals has now ceased, but it will be a considerable time before their numbers recover sufficiently to fulfil their previous role in marine ecosystems, particularly given continuing mortalities as by-catch and from disease outbreaks.

Box 5C	DISPUTE OVER TUNA NUMBERS

A 2003 study[101] claims that populations of large fish often 'stabilise' at about 10% of their pristine size, although this has been strongly disputed by fisheries population assessment modellers. One particularly contested example is that of the Japanese pelagic longlining fishery, which removed 200,000 large bluefin tuna off Brazil in its first 15 years, but then caught none in the following 15 years with a similar amount of fishing effort. Some fisheries scientists, in defence of existing fisheries management regimes, have questioned the reliability of the data used for these studies, particularly how catch per unit effort data should be statistically analysed. The study is faulted both for summing catching and effort over spatial cells to produce a ratio estimate of catch per effort and making hidden assumptions about abundance trends in spatial cells that were not fished at all early or late in fishery development.

The first fault places much more weight on data from those cells that were heavily fished than on cells for which catches and efforts were low, whether or not the heavily-fished cells were representative in any way of relative abundance in lightly fished or unfished cells. The second fault is to apply the correct formulae for stratified sampling, but only for those cells that were actually fished in a given time period. These mistakes are liable to make it appear that the population size has declined more than it actually has.[102] Tuna scientists have also pointed out that the area in which tuna fishing was defined as taking place was too small, that longline fishing catches only older, larger fish living in the deeper cooler water layers and that the majority of tuna fishing is now done on younger fish in shallower water layers using purse-seining, the data from which were not included in the study.[103]

The fact remains that numbers of large tuna off South America appear to be very low and the purse-seining of the younger fish would predict that they are unlikely to be replaced in the near future. The overall picture of massive declines on a global basis in the populations of predatory fish is impossible to dispute. Further analysis of independent studies has shown that a reduction to 10% is likely a general, and in many cases a conservative, estimate of the depletion of large predatory fishes and higher biomass is only seen in areas with particularly tightly managed fisheries such as the Faeroe Islands, Iceland or the Gulf of Alaska.[104]

Extinction risks in marine organisms

5.47   Human influence has overwhelmed the natural ecological processes of extinction, resulting in a rate of extinction at least four times greater than that seen in the fossil record.[105] The process of exploiting the oceans began with the hunting of large vertebrates such as seals, manatees, turtles and whales (box 5D).[106] People first exploited large animals, moving from one to another and from place to place as populations were depleted and eventually made locally extinct.[107] Exploration gave way to exploitation and the pattern went from small-scale to commercial fisheries, from high- to low- value species, and from abundance to scarcity.[108] An early example of marine extinction due to hunting was that of the Steller's Sea Cow, which was widely distributed throughout the northern Pacific Rim through the Late Pleistocene. Humans apparently hunted these animals to extinction at the end of the Pleistocene and beginning of the Holocene in most of their range. The last population of sea cows persisted in some unpopulated areas of the Aleutian Islands until the mid-eighteenth century, when they were finally wiped out by Russian fur traders.

5.48 The sea otter on the Pacific coast of North America almost suffered a similar fate. Hunted to low levels by the aboriginal Aleuts, they were brought to the point of extinction in the 1800s by fur traders before receiving legal protection in the twentieth Century. The otters performed a vital role in kelp forest ecosystems by predating sea urchins that grazed on the kelp, and the decrease in their numbers resulted in overgrazing and a collapse of these ecosystems.[109] It has also been estimated that at the time of Columbus's voyages to the Caribbean in the 1490s there may have been as many as 33 million green turtles there, the majority being eliminated before the nineteenth century. Green turtles closely crop turtlegrass, but in the near absence of green turtles today, turtlegrass beds grow longer blades that baffle currents, shade the bottom, start to decompose *in situ*, and provide suitable substrate for colonisation by the slime molds that cause turtlegrass wasting disease. Large vertebrate species such as these often play vital roles, so many marine ecosystems had already been negatively affected by their removal prior to the advent of large scale fisheries.

Box 5D	WHALES BEFORE WHALING IN THE NORTH ATLANTIC[114]

Although it has been known for some time that commercial whaling reduced the population size of all baleen whale species, the true extent of this reduction is only now being revealed by the genetic profiling of present populations. A small population will have little genetic variation, as genetic differences are eliminated by inbreeding, but a large population will have a correspondingly large amount of variation. Unexpected variation in small populations indicates they were once part of a much larger population. In the case of whales, the amount of variation indicates that populations were considerably bigger than historical estimates based on whalers' log-books, about 10 times larger than expected in the case of humpback whales. Based on models for mitochondrial DNA sequence variation, the genetic diversity of North Atlantic whales suggests original population sizes of approximately 240,000 humpback, 360,000 fin and 265,000 minke whales. These estimates for fin and humpback whales are far greater than previous ones, and 6-20 times higher than present-day populations; for example the 2003 population of humpback whales is about 10,000. The data starkly underline how far away the International Whaling Commission is from realising its goal of allowing whale species to recover fully from relentless exploitation. The data also raises a number of questions about how the removal of whales has affected marine ecosystems, for example baleen whales are major consumers of krill and small fish, as well as themselves being prey of killer whales and sharks. The genetic data also support conclusions from archaeological and ecological research that the past abundance of large consumer species such as whales, turtles, sharks, and pelagic fish was much greater than more recent observations.[115] The suggestion that vast cetacean populations may have existed before the advent of hunting by humans raises fundamental questions for our perceptions of the world's oceans.

5.49 As yet, there have been only a few documented cases of commercial fishing leading to biological extinction or near extinction of the target species, such as the white abalone (*Haliotis sorenseni*).[110] The extinction of sensitive populations and species is, however, a threat at present levels of fishing pressure.[111] Because the majority of fish species are not managed, local extinctions of marine fishes and invertebrates as a result of by-catch or changes in the ecosystem tend to be overlooked until long after they have occurred. Studies have documented 133 local, regional and global extinctions of marine populations, with the Wadden Sea in the OSPAR area being a particular extinction hot spot. Fishing was found to be the primary cause of extinction, followed by habitat loss; some extinctions have been linked to other factors such as invasive species, climate change, pollution and disease.[112] Habitat loss and degradation can be caused by fishing itself in the form of trawling (5.70-5.80).[113] However, although this may have severe local effects on benthic communities, the patchiness of the fishing effort (5.81-5.83) makes it difficult to ascertain whether larger scale extinctions have been caused.

5.50 There is an increasing awareness of how overfishing can drive species to ecological extinction, that is, depleting populations of fish to such an extent that they no longer play a functional role in the ecosystem. One of the effects of over fishing is the increase in non-target or less vulnerable species due to release from predation or competition. Consistent patterns of compensatory increase and decline have been seen in most pelagic communities, and also in some demersal communities.

5.51 Although attributes such as high fecundity or large-scale dispersal characteristics do not make marine species less vulnerable to extinction than terrestrial organisms, attributes such as long life-spans and late maturity may greatly increase it, just as is the case for large terrestrial species with late maturity.[116] Sharks, skates, sawfishes and rays are believed to be particularly vulnerable to overexploitation due to a combination of large size, associated large offspring size, slow growth and late maturation (figure 5-VIII).[117]

5.52 Large and vulnerable species have been severely depleted in British waters.[119] For example, the common skate (*Raja batis*) was made locally extinct in the Irish Sea in the 1980s, and four North Sea skate species have undergone severe declines and now exist only in localised pockets (figure 5-IX). By contrast some of the smaller species of skate such as the cuckoo (*Raja naevis*), spotted (*Raja mantagui*) and starry (*Raja radiata*) rays, less than 85 cm in length, have increased in abundance suggesting a competitive release mechanism.[121,122] The species compensation is, however, often reversed in a decade or less, most probably as a result of changes in targeting by fishers.

Figure 5-VIII

Dead Angel Shark (*Squatina squatina*), this species is now locally extinct in the Irish Sea, English Channel and Bay of Biscay[118]

Figure 5-IX

A thornback ray (*Raja clavata*), which has diasppeared from the south-east coast of the North Sea and has undergone an approximately 45% decline in abundance in the Irish Sea between 1988 and 1997. It remains the commonest skate species in the Irish sea and is present on the Thames coast of the North Sea. Even if fishing pressure is reduced, as skates exhibit only limited seasonal movement within geographical ranges and lay benthic eggs, recolonisation of areas in which they are extinct is unlikely[120]

5.53 Spurdog (*Squalus acanthias*), for example, are slow-growing, long-lived shark species that have a 22-month gestation period and give birth to 20 live young (figure 5-X). This species is therefore prone to over-exploitation and long-lasting depletion. Total biomass has fallen by 95% in the north-east Atlantic since fisheries for this species began over one hundred years ago, and it has been proposed for listing on Appendix II of the Convention on International Trade in Endangered Species.[123]

Figure 5-X
A spurdog[124]

5.54 Shark species have been shown to be in severe decline as a result of direct targeting and from by-catch by longlining pelagic fleets.[125] Total populations of sharks such as scalloped hammerheads (*Sphyrna tiburo*), whites (*Carcharodon carcharias*) and threshers (*Alopias vulpinus*) in the north-west Atlantic have each fallen by over 75% in the past 15 years, largely due to the expansion of longline fishing fleets out into the open ocean in the last 50 years.[126] Oceanic whitetip sharks (*Carcharinus longimanus*), once the most common large shark in the world (and possibly the most common large animal in the world), have become almost completely extinct in some areas within the last 50 years, an event that has gone almost unnoticed.[127] The barndoor skate (*Raja laevis*) has been driven to extinction in parts of its large geographic range, despite attempts to protect it.[128]

5.55 The conservation status of less than 5% of approximately 27,600 fish species has been assessed according to the World Conservation Union's (IUCN) Red List of Threatened Species.[129] Short-term economic interests in relatively few commercial species tend to prevail over long-term conservation interests in non-targeted species and ecosystem health. These conflicting economic and conservation goals were highlighted by the controversy surrounding the listing of the Atlantic cod and halibut on the IUCN Red List.[130,131] It is often argued that 'economic extinction' of exploited species will occur before 'biological extinction', but this is not the case for non-target species caught in multi-species fisheries or species with high commercial value, especially if this value increases as the species become rare. It should also be noted that prices for an exploited species, the bluefin tuna (*Thunnus thynnus*) have soared as high as $178,000 per fish as populations have declined, making it economically viable to use aircraft to direct boats to single individuals.[132]

5.56 This raises the possibility that it may well be economically viable to hunt large-sized species to extinction if the right technology is available such as side-scan sonar and high technology autobaited longlines. Case studies of terrestrial species modelling the economics of extinction also found that, in a multi-species context, neither privatisation, nor price/cost shifts with scarcity were sufficient to avert extinction.[133]

5.57 There needs to be a much more proactive approach to the prediction of vulnerability, estimation of extinction risk and the prioritisation of species of particular conservation or management attention, although this will require more information about the biology of species, as well as geographical range and endemism (areas to which particular species or subspecies are restricted (box 5E).[134,135] The risk of extinction for certain long-lived deep-sea species may be particularly acute (box 5H).

Loss of genetic diversity

5.58 Since fishing is selective with respect to a number of life history traits, which are at least partially heritable, traits may evolve in response to size-selective harvesting. Fisheries can be viewed as large-scale experiments on life history evolution of target and by-catch species.[136,137,138] Evolution occurs through changes in gene frequencies resulting from interactions between genotypes and their environment. Hence environmental changes (including fishing) can be expected to influence both the ecology and evolution of the genotypes of affected marine organisms.[139]

5.59 In fisheries management, the objective is to apply relatively moderate levels of mortality to species continuously in a manner that ensures a sustainable harvest in perpetuity. But because it is beset with the problems of predicting the immediate response to fishing, such management has not dealt with the evolutionary consequences of exploitation.[140] Two general classes of threat to genetic biodiversity may result from declines in marine organism populations sizes:

- extinction of populations or species, which leads to complete and irreversible gene loss.

- reduction in genetic variability in populations due to the selective effects of fishing or due to a decrease in population size that results in inbreeding. It seems likely that marine species may be vulnerable to erosion of genetic diversity when undergoing fishery-induced declines.

5.60 Evidence from phenotypic traits (observable traits of an organism resulting from the interaction of genes with the environment) has identified potential cases where fishing may have driven selection. Large phenotypic changes are taking place in fish populations such as the gadoid populations in the North Atlantic, North Sea, Baltic Sea and Barents Sea. To disentangle the genetic component of variation, direct environmental effects need to be taken into account. However, it is difficult to separate out the direct selective effects of fishing from indirect and other environmental factors, such as changes in the benthic environment caused by dragged fishing gear or changes in water temperature resulting from climate change. Nonetheless, evidence from such studies shows that a substantial proportion of changes in phenotypic variability in fish, such as growth rate, length, age at sexual maturation and fecundity are evolving as a direct result of fishing selection pressures (figure 5-XI).[141,142] When fishing mortality is relaxed, the surviving genotypes in the fish population will be those with reduced fitness under normal environmental conditions (small, slow-growing, early-maturing fish) but high fitness under fishing selection, leading to slower population recovery times.

Figure 5-XI
Cod showing variability in length and age at sexual maturity[143]

5.61   Where DNA-based studies have been carried out, they have shown decreases in genetic diversity as a result of fishing pressure. Studies have estimated the genetic diversity of populations of New Zealand snapper (*Pagrus auratus*) using data based on DNA extracted from fish samples spanning a 36-year period (1950-1986). Most commercially exploited species are fished wherever they occur, so comparisons between exploited and unexploited populations are usually not possible. One of the populations studied was only commercially exploited from the beginning of this period, thus a baseline for the diversity of a natural spawning population biomass level could be set. The data provide evidence for a loss of genetic diversity in overexploited fish populations, despite an estimated minimum census size of three million fish, the effective (or genetic) population size being an order of magnitude lower than the actual number of fish. The results suggest that commercial fishing may not only result in selective genetic changes in exploited populations, but also in reduced diversity caused by genetic drift (appendix E).

5.62   Genetic diversity has usually been considered to be unaffected by commercial fishing, mainly because diversity is significantly reduced only in very small populations and even 'collapsed populations' may consist of far too many individuals to show decline in diversity measurable with feasible sample sizes. For example, the collapsed Grand Banks cod population (box 5B) still consists of several million fish, so there would appear to be little cause for concern.[144,145] It was assumed that a harvested population retains the capacity to grow back to its equilibrium state over a very short time and that repeated bouts of harvesting can go on indefinitely without changing the genetic structure of a population.

5.63   However, the number of fish in a population is often much larger than the genetically effective population size, which determines the genetic properties of a population (5.61). In marine organisms, high fecundity, a strong bias in reproductive success, large variations in year class strength (the number of fish recruited to the population annually), and size dependent fecundity may reduce the genetically effective population size by several orders of magnitude. This evidence shows that millions of individuals may be equivalent to an effective genetic population size of only hundreds or thousands.[146,147] The result of this loss of diversity is to reduce the future capacity of fish populations to adapt to a changing environment.

5.64   DNA-based studies have also examined the levels of genetic diversity in one of the four cod subpopulations in the North Sea, off Flamborough Head in Yorkshire. Although this study did not provide information about the levels of genetic diversity in natural populations of this species as this was a population that had been commercially fished for a long time, it was able reveal the effects of increasing levels of exploitation over a 50 year time period. Genetic diversity was compared between 1954, 1960, 1970, 1981 and 2000 using DNA from archived biological samples.

5.65   The results showed a marked decrease in genetic diversity indicating a substantially reduced effective population size up to 1970; however, after 1970 there was an influx of novel alleles from other populations of cod in the North Sea as a result of increased immigration. By 1998, the Flamborough Head population was largely composed of immigrants from neighbouring populations, indicating the effective loss of the original population, which comprised one-tenth of total North Sea cod prior to the 1970's decline.[148] This may be significant in terms of loss of adaptations to local conditions and indicates how fishing pressure may be affecting species that consist of large numbers of individuals but are subdivided into smaller, genetically isolated sub-populations (box 5E).

BOX 5E **SUB-POPULATIONS AND DISPERSAL**

More information is required concerning rates of recovery in severely depleted populations, particularly negative population growth as a result of the spatial dynamics and connectivity of sub-populations. Fisheries management units generally comprise multiple biological (i.e. genetic) populations units and are in essence mixed-stock fisheries. This can result in over-exploitation of vulnerable populations. Census population sizes also generally refer to multiple biological populations. There is a need for fisheries management to be based on biological production units and not on political and other sometimes arbitrary boundaries. In addition to characteristics outlined in paragraph 5.51, high extinction vulnerability will be found in species with small geographic range size and ecological specialisation, including feeding modes, habitats and migrations. Marine species are often assumed to have large geographical ranges, but populations of some species may be far more geographically restricted than has been previously assumed, often forming discrete sub-populations, especially in the case of broadcast spawners (box 5A).[149]

Such dispersal mechanisms have led to the perception that marine species are more widespread than those on land and hence at less risk of extinction.[150] It should be noted, however, that, as many fish populations consist of subpopulations that may be adapted to particular marine habitats, they are at a greater risk of extinction. There is little chance of repopulation in such cases, as neighbouring populations are also likely to be adapted for specific ecological niches.

For example, western Atlantic cod and herring populations comprise small geographically discrete sub-populations that have been serially depleted; this may have disproportionately reduced the resilience of the entire aggregate population.[151, 152] Studies mapping the geographic range of more than 3,000 species of fish, corals, snails and lobsters that inhabit coral reefs have shown that the majority of species have relatively small ranges.[153] Some broadcast spawning species are also presently threatened with extinction (5.49-5.57).[154,155]

5.66 ICES has recently formed a working group, the working group on the application of genetics to fisheries and mariculture (WGAGFM), to consider these issues and provide advice to managers. This will include providing recommendations on the applications for the estimation of effective population size in wild populations of marine fish and shellfish, evaluating the evolutionary and genetic effects of selective fishing, producing list of species for which there is reason to be concerned for loss of genetic variation and a list of species for which there is good genetic information from which to advance management advice.[156]

5.67 Scientific justification for conserving genetic biodiversity is clear. It includes: maintaining adaptability of populations, the future utility of genetic resources for medical and other purposes, and changes in life history traits and behaviour that influence the dynamics of fish populations, energy flows in the ecosystem, and ultimately sustainable yield.[157] Equally, it is clear that fishing is affecting the evolution of various traits in populations of marine organisms. There remain, however, a considerable number of questions that require answers if fish populations are to be managed on a long-term basis that avoids the deleterious effects of fishing selection pressures. For instance, altering the patterns of fishing will not readily reverse genetic changes caused by fishing such as changes in maturation, as it is easier to select for early maturation than late maturation. These questions can only be answered by the incorporation of an evolutionary perspective into management of the marine environment.[158,159,160,161]

DAMAGE TO SEABED HABITATS BY FISHING GEARS

5.68   Demersal trawling scrapes and ploughs the seabed, reducing the diversity of benthic communities (box 2D), either directly by inflicting unsustainable rates of mortality on species with slow rates of population increase or indirectly by modifying and removing the habitats that support diverse communities, as well as disrupting and resuspending sediment and effecting biogeochemical cycling. Most demersal fishing activity takes place in shallow seas on the continental shelf at depths of less than 200 m, although deeper water trawling off the continental shelf is rapidly increasing (5.92). Bottom trawling is a source of chronic and widespread disturbance in shallow seas,[162] estimates of the area trawled range from 50-75% of the global continental shelf.[163,164] The estimates of the percentage of seabed impacted depend on the scale of the analysis, since analyses at large spatial scales such as 0.5 degree grid squares imply a far greater area of impact than analysis at smaller scales (5.86). If 50% of the continental shelf is trawled every year, as one estimate indicates, this is the equivalent of 150 times the rate of forest clearcutting on land.[165,166]

5.69   Figures from 1989 suggest that beam trawling has affected approximately 323,000 km² of the central and southern North Sea seabed in this year, and that otter and pair trawls have scarred 99,000 km² and 108,000 km², respectively, in the entire North Sea. Industrial trawling has been implicated in approximately 140,000 km² of damage over the entire North Sea. In the southern North Sea, beam trawling is the most common form of fishing, mostly involving Dutch (80%) and Belgian (but Dutch-owned) fleets. Figures indicate that some 171,000 km² of the North Sea, approximately 40% of the area, is fished by Dutch beam trawlers, about 80% of the total beam trawl effort in the area.[167] The UK has only a small beam trawl fleet and most of UK waters are subjected to otter trawling rather than beam trawling.

5.70   Trawling modifies the diversity, community structure, trophic structure and productivity of macrobenthic invertebrate communities.[169,170] Field studies have shown that the total biomass of macrobenthic infaunal and epifaunal species, such as corals, bryozoans (sessile colonial animals with plant like appearance), hydroids (soft colonial animals) and crinoids (feather stars, filter feeding sessile members of the starfish family), is lower in trawled areas. Fragile species with larger body sizes and slow life histories are generally more vulnerable than smaller species to the effects of trawling.[171] As a result, trawled communities are increasingly dominated by small infaunal species with short life histories (usually various small hard shelled bivalve, starfish and crustacean species).[172] Trawling and particularly scallop dredging are massively damaging to long-lived sessile benthic invertebrates (e.g. soft corals) (figure 5-XII). Beam-trawling disturbance has inflicted unsustainable rates of mortality on large bivalve species in the southern and central North Sea, leading to regional reductions in diversity.[173,174]

5.71 Not surprisingly, the weight and width of the gear, and the nature of the benthic substrate to a large extent determines the degree of impact. The most destructive techniques being used are beam trawling, rock hopper otter trawls and shellfish dredging. Beam trawling involves the use of tickler chains which plough into the seabed to a depth of 8 cm, while otter trawl boards may penetrate as deep as 15 cm in soft sediments and the width of the otter trawl board tracks left may vary between 0.5 and 6m.[176] The net and roller gears on the footrope of rock hopper trawls, impact surface living organisms over a far greater area.

Figure 5-XII

A furrow left by scallop dredge, which has removed or destroyed all the surface benthic fauna[168]

Vulnerability of different benthic habitats to trawling

5.72 The scale of damage done by trawling depends on the vulnerability of the habitat and habitat distribution (2.28 and figure 5-XIII). Benthic communities experience continual disturbance at various scales in time and space, with the shallow continental shelf sea environments experiencing more frequent disturbances than deeper sea environments that are not exposed to wave action and strong currents. Theoretically, habitats can be ranked in terms of vulnerability to a specified fishing impact or gear (9.36). The heaviest gears in use are beam trawls. These are used primarily in waters shallower than 50 m over sandy or muddy-sand seabeds in order to catch flatfish. Sandy seabeds are relatively resilient to impact as they are naturally relatively mobile, but some muddy-sand seabeds (as opposed to solely mud sediments) can show trawl traces for many months after a single impact. Trawling over naturally less disturbed mud or gravel seabeds can have even longer-term impacts of years, with trawling in rocky reef habitats being the most destructive of all.[177]

5.73 As habitat stability increases, so do the potential effects of fishing damage.[179,180] Studies have suggested that the effects of typical fishing intensities on the less vulnerable habitats are not dramatically different from the effects of natural disturbance, and recovery times may be on a scale of months or years (5.81-5.83). The effects on the most vulnerable habitats however, are irreversible on time scales of decades to centuries. Moreover, fauna living in more mobile seabed habitats are likely to be well adapted to continual disturbance and theoretically more resilient to the effects of trawling (figure 5-XIV).

Figure 5-XIII

An untrawled area with intact benthos, including pink sea fans (*Eunicella verrucosa*). This hydroid species is a charismatic component of the benthos and one that illustrates slow growth and poor rates of recovery if lost[175]

111

Nonetheless, it should be noted even in more mobile seabed habitats the effects of disturbance are additive and can thus exceed the background level and frequency of natural disturbance and become ecologically significant. In the Dutch Wadden Sea, for example, dredging led to the loss of reefs of the calcareous tube-building worm (*Sabellaria spinulosa*) and their replacement by communities of smaller non-reef building bristleworms (polychaetes).[181] Tube building worms play an important ecosystem role in such habitats by stabilising the sediment and providing sites for the establishment of mussel beds (box 2D).

Figure 5-XIV
Plaice on a sandy sea bed[178]

5.74 As 'recovery' is often only measured at the site of impact, apparent recovery may simply involve the relocation of fauna from adjacent unfished areas. In this case, large-scale and chronic fishing impacts may reduce the probability of local recovery. Another problem in making these sorts of estimates is that trawling modifies the communities to dominance by resilient species. Even in high energy (continually disturbed) environments, it is possible that real recovery may take many years. While the current fauna of sandy seabeds is not particularly vulnerable to trawl effects there is some evidence (from palaeontological records) that unfished sandy habitats did contain sessile epibenthos. For example, the unfished north-west Australian shelf is reported to have large expanses of hydroid and sponge assemblages even in shallow water exposed to ocean swell. Although it can be accepted that some areas of mobile habitat already damaged by fishing could continue to be fished it should not be assumed these areas never contained vulnerable fauna.[182] It is also clear that in deeper areas with less natural disturbance, fishing is likely to account for a significant proportion of total disturbance, and most habitats can be vulnerable if fishing is intensive, as is the case in the southern North Sea.[183]

Effects of trawling on marine organisms

5.75 The effects of fishing on habitat are most significant in areas of relatively low natural disturbance and on structurally complex and delicate biogenic habitats such as cold-water coral reefs or maerl beds (Box 2D). Both of these increase three-dimensional habitat diversity and consequently have a relatively high diversity of associated species.[184,185] Although a significant positive relationship between fish biomass and topographical complexity has only been shown for coral reefs, they are likely to exist in other biogenic habitats. It is by effecting changes to biogenic structure that demersal fishing is most likely to influence the benthic communities of marine systems. Sponges and hydroid/bryozoan turf are likely to be most important fish habitats in the OSPAR area.

5.76 Trawling disturbance in areas where complex biogenic habitats are found will dramatically change the habitat structure and the abundance and diversity of associated species.[186, 187] The role of such habitats is only just beginning to be understood. Cold-water coral reefs (figure 5-XV) are slow growing (about 2.5cm per year) and provide habitats for diverse communities of animals that are not found on the surrounding seabed. The recovery of reefs impacted by

Figure 5-XV
Cold water corals and glass sponges at 850m, off Ireland[188]

trawling is expected to take many decades or centuries. In most cases, the first pass of trawl gear is sufficient to damage or destroy some areas permanently.

5.77 The removal of the organisms themselves can also affect the nature of the habitat, and species that act as 'ecosystem engineers' are believed to play an important role in marine ecosystems.[189,190] For example, the chlorophyll a concentration (a measure of phytoplankton biomass/nutrient enrichment) in Canadian estuaries is best predicted by the biomass of mussels, rather than by any measurement related to nutrient biofluxes or concentrations, as sufficient biomass of these shellfish will filter out any increase in phytoplankton occurring as a result of increased nutrient input. The destruction of mussel beds by trawling could have a significant impact on the nutrient dynamics in such systems.[191,192]

5.78 Short- and long-term effects of the environmental impacts of trawling have been investigated in a number of studies.[193] Most investigations have examined the immediate or short-term effects of the passage of the gear on seabed individuals and communities, effects that can be reliably quantified. Many of these short-term studies have examined effects on what are already considerably altered environments from which vulnerable species have been made locally extinct and outcomes for existing benthic fauna are predicted with no knowledge of the original composition of the fauna (5.73). Data are often lacking at a scale or over a period sufficient to ascertain the disturbance history of an area and it is hard to demonstrate convincingly that towed bottom-fishing activity has been responsible for all the changes in bottom fauna and habitats. For example, it has been estimated that 25% of the bivalve fauna (in terms of number of species) recorded in the southern North Sea in the first half of the twentieth century are no longer present.[194] Fewer longer-term studies have been conducted because of the general lack of historic data on communities (except fish) and the difficulty in differentiating fishing effects from other long-term influences (e.g. climate change).[195] It has been argued that this lack of data can only be resolved by closing large areas of the North and Irish seas to fishing for many years to determine the long term effects of trawling on the benthic marine ecosystems present.[196]

5.79 In most cases, the first pass of a trawl over an unfished benthic habitat will cause the greatest damage,[197] although given the lack of truly pristine habitats, it is difficult to know what the effect would be on these untouched habitats. In most areas of the continental shelf, damage is being inflicted on communities already substantially changed by fishing. The magnitude of the immediate response of organisms to fishing disturbance varies significantly according to the type of fishing gear used, the habitat in question and amongst taxa. The short-term outcome of disturbance, however, is of less ecological importance than the potential for recovery or restoration. Heavily-fished habitats suffer reductions in habitat complexity as well as increases in the relative abundance of smaller free-living species and individuals. Within and among species, mortality is generally size dependent, thus larger bivalves and attached epifauna suffer very high mortality while smaller bivalves and polychaetes may suffer lower mortality, possibly because lighter animals are pushed aside by the pressure wave in front of the gear.[198] Not only are larger species more likely to suffer high mortality, but the mortality rates they can withstand will be lower because they tend to have lower potential rates of population increase and slower growth.[199,200]

Box 5F   **ESSENTIAL FISH HABITATS**

While it is fairly simple to identify those habitats that might be considered essential to the life history of some fish, for example spawning and nursery areas, and thus protect them, other equally important secondary habitats may be patchily distributed. Secondary habitats are important for the acquisition of food and predator avoidance at various stages in the life histories of fish species; their importance may be related to habitat complexity and structure. Habitat complexity is a product of the surface features of the seafloor and the animal life that inhabits the seabed. Subtle features such as sand ridges and pits created by the feeding or burrowing action of benthic fauna may provide shelter for bottom-dwelling fish species. In general, juvenile fish often survive better in structurally complex habitats than in simple ones, because such habitats offer refuge from predators and better feeding conditions.[201,202]

It is estimated that benthic invertebrates can supply as much as two thirds of the annual food requirement of demersal fish. Bottom fishing is capable of greatly reducing habitat complexity by direct modification of the substratum or removal of the fauna that contribute to surface topography and can lead to changes in associated fish assemblages.[203] For example, certain species of sole preferentially live in uniform sandy areas, a habitat which is maintained by towed bottom fishing gear, as opposed to dab and plaice, which require more varied ecosystems. The effects of trawling disturbance on ecosystem processes are not well known, but in general terms, beam trawling appears to have created a system where small fish feed on small food items.[204,205] This may have minimal effects on the growth and production of some species of flatfish such as sole, but the size structure, biomass and total production of the infaunal and fish communities are fundamentally different from those in the unfished state.[206]

5.80 Research has shown that in the North Sea, the size spectrum of the infaunal community is determined by the over-riding effect of trawling disturbance rather than of sediment size and depth.[207] Therefore, under conditions of repeated and intense bottom-fishing disturbance a shift from communities dominated by relatively high-biomass species towards dominance by high abundance of small size organisms is likely to occur.

Patchiness of fishing effort

5.81 The presence of some unimpacted areas has an important influence on the overall level of impact because the first effects of fishing on an unfished area are consistently greater than the subsequent effects.[208,209] The development of new gears and techniques for fishing on previously unfished grounds causes great damage to the marine environment. Examples in recent years include the development of rockhopper gear, the development of more powerful trawlers able to fish in deeper water and the development of accurate navigation and trawl handling gear enabling small patches of suitable habitat to be targeted. Fishing boats have gear that now penetrates into areas previously considered too rough for fishing, and trawls are able to move rocks measuring up to 3 m in diameter and weighing 16 t.[210]

5.82 Although, damage to benthic communities by trawling is widespread, the degree of damage varies because fishing effort on local scales can be very patchy, with some areas of seabed fished many times and others not fished at all (figure 5-XVI). This distribution also probably changes over time, so present fishing pressure on little fished areas may have been higher in the past before they were fished out. The net result of this patchiness is that the overall impacts of fishing on habitat are less than would be assumed if the effort were allocated uniformly in space and time. At present, there is no direct evidence for this assertion that patches have been fished in the past, but this is a key aspect of any modelling of recovery dynamics and any attempt to predict the value of areas closed to fishing. What data there are have been used by some to justify the *status quo* – let fishers go where they will and they will avoid some patches.

5.83 As is discussed further in Chapter 8, the use of closed areas as a means of fisheries management can also cause spatial and temporal redistribution of fishing effort. Although the effects of trawling are mitigated in the closed area, the impacts of trawling are exacerbated outside the closed area, which may have been previously little trawled, unless accompanying management measures to reduce trawling effort overall are taken,[212,213] which we discuss in Chapter 9.

Figure 5-XVI

Distribution of trawling effort in the North Sea. International beam and other trawling effort for 1995 shows that some areas of the southern North Sea are very heavily fished while other areas of the central North Sea are rarely visisted[211]

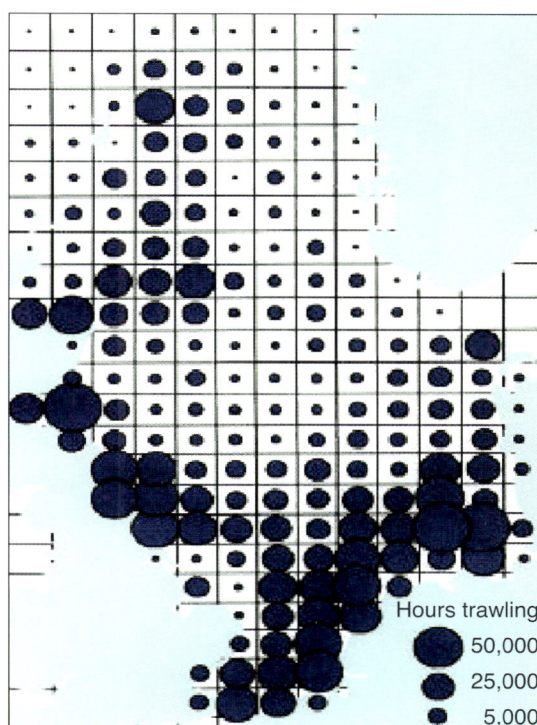

Hours trawling
50,000
25,000
5,000

Effect on nutrient cycling and fluxes

5.84 Direct damage is not the only physical effect of trawling. The seabed habitat can be altered by the stirring of seabed sediments, or the moving or removal of seabed features such as stones and boulders. Nutrient flux will be altered by the stirring of sediments, and the breakdown/alteration of structure may remove vital habitat features. Despite the ubiquity of bottom fishing disturbance in shelf seas, relatively little is known of the effects of this disturbance on sediment community function, carbon remineralisation and biogeochemical fluxes (transfer between the biotic and abiotic parts of the biosphere).

5.85   The physical impact of trawling or dredging, such as scraping and ploughing of the substrate, sediment resuspension and mortality of benthos, disrupts the redox status of the sediment (whether it acts as an oxidising or reducing environment) and hence the microbial activity which controls biogeochemical cycling. The impact may shift the balance of anaerobic to aerobic mineralisation (in the presence of oxygen), which may in turn result in a change in the end-products of mineralisation for both carbon and nitrogen compounds.

5.86   Processes such as sulphate reduction can be stimulated at depth by the input of organic carbon and the physical mixing of organic material deeper into the sediment. As the rate and balance between such sediment processes are mediated by microbial metabolic pathways, carbon input to sediments and oxygen state, trawling activities would be expected to affect bulk sediment biogeochemistry. Further work on the biogeochemical impacts of fishing activities should focus on the large-scale experimental work that is still needed to assess the impacts of entire fisheries.[214][215]

5.87   Demersal trawl fishing also resuspends the surface of sediments and may result in immediate nutrient releases into the water column. Hence, nutrients may be introduced into the water column as a pulse rather than by the usual slower efflux mechanisms. Relatively few studies of the effects of fishing disturbance on nutrient fluxes in soft-sediment communities have been carried out, and yet all the evidence suggests that such effects may have a profound influence on marine ecosystems. They may cause oxygen depletion in the water column as a result of sediment resuspension or increased chemical oxygen demand and may increase or alter sediment nutrient flux rates during periods when natural disturbance is relatively low (e.g. after the spring phytoplankton bloom and when wind stress is low in the Northern Hemisphere).

5.88   These fluxes may provide nutrients to fuel pelagic production. Resuspension may accelerate nutrient recycling and result in an overall increase in primary productivity and organic carbon export rates. The actions of burrowing animals in unfished sediments may allow sediment chemical storage and fluxes to reach equilibrium. This is because these animals consume organic carbon sources and reduce the magnitude of carbon fluxes. Trawling disturbance rapidly depletes populations of the most active bioturbators, and in fished sediments, the biomass of these animals is markedly reduced and larger fluctuations in benthic carbon fluxes and storage are expected.[216,217]

DEEP-WATER DEMERSAL FISHING

5.89   The impacts of fishing are now more widespread because more of the marine environment is accessible to fishing.[218] As traditional populations of fish dwindle, fishers have moved their attention to previously unexploited species in deeper waters. Fishing technology has improved to such a point that bottom trawling can be used around seamounts and on the edge of the continental shelf at depths greater than 1000 m where it has observable impacts (figure 5-XVII). Some vessels routinely trawl to depths of 1,500 m and experimental fishing programmes delve beyond 2,000 m.

5.90 A common feature of many deep-water fisheries is that populations are often exploited before information is available on the biology and productivity of the species present, before time-series data are available for assessments and before an evaluation of the potential impacts of fishing on the environment can be made. Consequently, reliable information on the state of populations, the potential for fisheries and the wider environmental effects of fishing, frequently lags behind exploitation. It is, however, accepted that many deep-sea species display characteristics including extreme longevity, late age of maturity, slow growth and low fecundity (box 5H). Many also form dense aggregations for spawning and/or feeding.

Figure 5-XVII

Extensive trawler scarring of the seabed at depths greater than 500m that has been observed in the Faroe-Shetland channel in DTI surveys of the area. Trawler scarring has also been observed at depths greater than 1000 m in other UK deep-waters such as the Rockall Trough. This is likely to be highly destructive of the sessile benthic communities of these areas[219]

5.91 As a result of these attributes deep sea fish populations are generally unproductive, highly vulnerable to over fishing and potentially slow to recover from the effects of over-exploitation (box 5H).[220] Experience from a range of deep-water fisheries around the world has shown that despite initial optimism about long-term sustainability, based largely on no real knowledge, populations can be quickly depleted and recovery slow. Most of the deep-water fisheries in the north-east Atlantic have been completely unregulated; the latest stock assessments indicate that nearly all exploited deep-water species are being harvested outside safe biological limits and continued exploitation at present rates will result in stock depletion.[221] Many deepwater populations depleted decades ago, such as the orange roughy population off Scotland, have shown no sign of recovery to date.

5.92 The main environmental effects of deep-water fisheries are directly on commercially targeted species and indirectly on the wider fish assemblage and vulnerable habitats. Lost and discarded gill, tangle and trammel nets (ghost fishing) also result in a substantial additional indirect impact in deep sea environments. Deep-water communities, such as those found on coral reefs and mounds, seamounts, hydrothermal vents and in soft muddy bottoms are complex, fragile and exhibit a high local species diversity. There is the potential for destruction of deep-water habitats (seamounts and coral reefs are particularly susceptible) by fishing gears. Observable impacts on deep-water coral ecosystems have been documented from the Darwin Mounds (west of Scotland) and the Sula Ridge (off Norway) (box 5G).

5.93 Deep-sea fisheries are commonly areas of great biological significance, featuring seamounts, steep slopes, such as those of canyon walls, and hard bottoms.[222] Seamounts, a common and abundant feature of the ocean floor, are undersea volcanoes, typically cone-shaped and rising relatively steeply from the seabed. They can be very large features, not only in terms of their elevation but also in area, as some are more than 100 km across at their base (2.35).[223]

5.94 While the biological diversity and vulnerability to exploitation of seamount species and habitats is not well understood, research that has been undertaken shows that some seamounts support a wide variety, and high abundance, of some species, and can be characterised by high rates of endemism (i.e. many species are not found outside that seamount or seamount complex). The tendency for the generally higher abundance of fish around seamounts than in surrounding waters, and for some species of fish to aggregate around seamounts, may be at least partly explained by the apparent increased concentration of plankton associated with some large and relatively shallow seamounts.

5.95 This may be a result of localised upwellings, from the interaction of currents and seamount topography interrupting the flow of water, which bring nutrients to the surface encouraging primary production, or by the trapping of diurnally migrating plankton in eddies that become trapped over seamounts known as circulation cells or Taylor columns.[224] Seamounts can also support dense assemblages of benthic fauna, including suspension feeders such as corals, sponges, hydroids and ascidians. Corals can be particularly abundant with soft (gorgonian), hard (scleractinarian) and black (antipatharian) corals all recorded on seamounts. Again, this appears to result from the interaction of seamounts with currents, creating elevated levels of flow which remove sediments, increase the food supply for filter-feeders and provide hard bottom habitats suited to these fauna.[225,226,227] Feather stars, starfish, brittlestars, sea cucumbers, molluscs and crabs and lobsters have also been reported.[228]

Box 5G | **COLD-WATER CORAL**

The most common cold-water coral is *Lophelia pertusa* which has a global distribution but is most common in the north-east Atlantic (2.37) (figure 5-XVIII). There are other cold-water coral species and patches of coral often include more than one species. One of the most exciting discoveries in UK waters so far has been Darwin Mounds that were found using remote sensing techniques in May 1998 during surveys funded by the oil industry and steered by the Atlantic Frontier Environment Network an industry-government group. The Darwin Mounds are located at a depth of about 1,000 m in the north-east corner of the Rockall Trough, covering an area of approximately 100 km². Some hundreds of mounds lie in two main fields (referred to as Darwin Mounds East and Darwin Mounds West), while other mounds are scattered at much lower densities in nearby areas.[229]

Each mound is approximately 100 m in diameter and 5 m high. Most are also distinguished by the presence of an additional feature visible on side-scan sonar referred to as a tail. Both mounds and tails are characterised by

Figure 5-XVIII
Distribution of cold water coral reefs around the British Isles[230]

LEGEND
AD = Anton Dorhn Seamount
BB = Bill Bailey's Bank
FS = Faroe Shetland Channel
GB = George Bligh Bank
LB = Lorien Bank
PS = Porcupine Seabight
R = Rockall
WT = Wyville=Thompson Ridge

a roughly 15-fold increase in the density of xenophyophores – giant deep ocean protozoa (single-celled animals) – subsequently identified as *Syringammina fragilissima*, which grows up to 20 cm across. The Darwin Mounds rest on seabed composed of deep foraminiferal sand sediments (the microscopic shells from trillions of dead zooplankton) in the northern Rockall Trough. The tails are of a variable extent and may coalesce, but are generally a teardrop shape and are orientated south-west of each mound. The mounds are unusual in that *Lophelia* grow on sand rather than a hard substratum. Prior to research on the mounds in 2000, it was thought that *Lophelia* required a hard substratum for attachment.[231]

Cold-water corals are vulnerable to damage from towed fishing gear; 30-50% of Norway's deep-water coral reefs have been seriously damaged by trawling. Evidence of new damage was visible over about half of the Darwin Mounds East field during summer 2000. This damage was visible as smashed coral strewn on the seabed. A trawler was operating nearby during the surveys. Given that *Lophelia* appears to need (or favour) the elevation provided by the sand volcanoes for growth in this area, it seems likely that this damage will be permanent. Bottom trawl fishing has now been permanently banned in the area under EC Common Fisheries Policy measures.[232]

5.96 The general abundance and common aggregation of fish around seamounts makes them attractive sites for fisheries. Trawl fisheries for a variety of deep-sea species are focused on seamounts (figure 5-XIX). Seamount topography, along with advanced navigation and gear technology, results in a large number of tows of heavy trawl gear over a relatively small area. This creates intense local disturbance. Trawling has been shown to have led to near complete removal of bottom-living communities from exploited seamounts.[233,234,235,236]

Figure 5-XIX

Distribution of seamounts in the north-east Atlantic.[237]

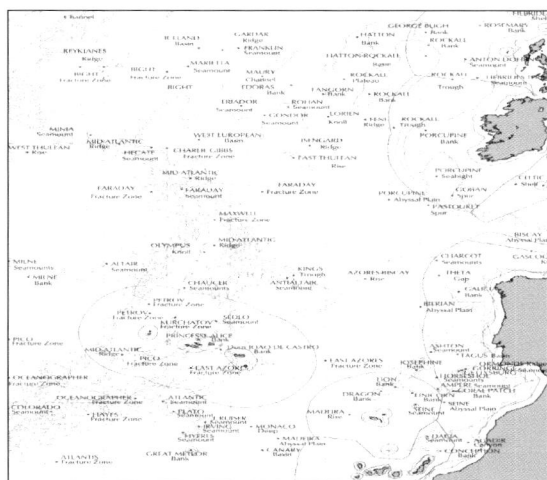

Box 5H	DEEP-WATER FISH SPECIES

The orange roughy is a deep-water fish found in cold waters over steep continental slopes, ocean ridges and seamounts at a depth of 900 – 1,000m. Roughy have slow growth rates and are one of the longest-lived fish, living up to 150 years; they do not mature sexually until their mid-twenties to thirties. Natural mortality and reproductive output are low, as mature fish do not spawn every year. When spawning occurs in the North Atlantic from January to February they gather in large numbers in a small specific area, which is targeted by fishers using bottom trawling.

The orange roughy has only been exploited in the north-east Atlantic since the early 1990s and the species has largely been fished out in waters to the west of Scotland. Although quotas have been set in some areas, ICES considers Atlantic populations of orange roughy already to be outside of safe biological limits, as given the characteristics of this species, recovery times for overfished populations are likely to be high.[238] The North East Atlantic Fisheries Committee (NEAFC) is responsible for management of fisheries in the north-east Atlantic, but this organisation has not yet agreed any regulations on deep-water fisheries management.

Figure 5-XX
Black scabbardfish[241]

The roundnose grenadier is another deep-water species fished in the north Atlantic. It has a depth distribution from about 500-1,800 metres. Although, as with other deep-water species, validation of age is a problem, it is thought that it lives for at least 70 years, maturing at 14-16 years old. The species is the mainstay of the French deep-water fishery with total international landings from waters to the west of the British Isles reaching 11,850 tonnes in 2001. It is likely this species is also being fished unsustainably as populations are acknowledged never to have recovered from Russian exploitation and depletion in the 1960s and 1970s. Deep-water squalid sharks (*Centrophorus squamosus*) with similar life spans of 60 – 70 years are also now being unsustainably exploited.

There are fisheries for other deep-sea species, such as argentines (*Argentina silus*), alfonsino (*Beryx decadactylus and splendens*), ling (*Molva molva*), blue ling (*Molva dypterygia*), tusk (*Brosme brosme*), greater forkbeard (*Phycis blennoides*) and black scabbardfish (*Aphanopus carbo*). Although most of these species are fairly long-lived (20-30 years), some species, black scabbard fish (figure 5-XX) and alfonsino, have comparatively short life spans (12 years) and higher growth rates,[239] which has led to suggestions they could form the basis of a sustainable fisheries, but these species are the exception rather than the rule. Only a handful of deep-sea species are marketable and the majority of species caught are discarded. A common by-catch of these fisheries is Baird's smoothhead, which is a slope-dwelling, deep-water species with a depth range between 700 and 1,800 metres. It can constitute up to 50% of deep-water catches in waters to the west of Scotland at depths of up to 1,000 metres and up to 61% in depths of 1,200 metres or more. Fishermen consider this species as a nuisance fish as they cannot be marketed at all due to the high water content of their flesh. Fishing at these depths means that there will be no survival of catches returned to the sea as discards (this is true for any discarded deep-water species) and any technical by-catch and discard mitigation methods (such as mesh size and selectivity grids) are likely to fail due to the high mortality of escapees.[240]

5.97   The risk of severe depletion, and even extinction, of elements of the benthic seamount fauna is increased by their highly specific habitat requirements, localised distributions and high levels of local endemism.[242] In the Southern Ocean, seamounts that once supported large numbers of invertebrate species have been stripped to the bare rock by a few decades of trawling for orange roughy.[243] Deep-sea fish populations must be considered as non-renewable resources, but at present we are removing species faster than they can replace themselves. In March 2004 more than 1,100 marine scientists signed a statement calling on the UN and world governments to stop bottom trawling of the deep-sea, to stop the unprecedented damage being caused to the deep-sea coral and sponge communities on continental plateaus and slopes, and on seamounts and open ridges.[244] The species found in the deep sea are not amenable to exploitation and the current fisheries are not sustainable in any form.

CONCLUSIONS

5.98 Changes in marine ecosystem structure and function as a result of fishing occurred as far back as late aboriginal and early colonial periods, although these pale into insignificance in comparison with subsequent events. These changes increased the sensitivity of coastal marine ecosystems to subsequent disturbance and thus preconditioned them to collapse. Fishing alters the structure of marine ecosystems in ways that are only just beginning to be understood, such as the removal of long-lived organisms with low natural mortality from the ecosystem (both within a species and within genera and taxa) and their replacement with short-lived organisms with more volatile population dynamics, the degradation of habitats by fishing gear, the reduction of predatory pressure on smaller pelagic organisms, the harvesting of smaller pelagic organisms reducing food for larger piscivores, and reduction in the predictability of exploited marine ecosystems.

5.99 The policy goal of 'sustainable' yields exploited at close to a maximum level is damaging to both fish populations and marine ecosystems. Fishing is the most widespread human exploitative activity in the marine environment; it is estimated that over 20% of primary production is required to sustain fisheries in many intensively fished coastal ecosystems.[245] Previous estimates of the primary production required were much lower,[246] and it was widely assumed that fishing had few fundamental effects on the structure or function of marine ecosystems apart from those on fished species.[247] These views were widely accepted at the time since they were in accordance with the overriding philosophy of many fisheries scientists who based their assessment and management actions upon the short-term dynamics of target fish populations.[248,249]

5.100 Reassessment of the amount of primary production required to produce the number of organisms that are extracted from the marine environment, coupled with empirical evidence for shifts in marine ecosystems, imply that the actions of fishers have important effects on ecosystem function.[250] The ecological effects of the present levels of fishing also preclude the long-term survival of most present-day fisheries. A number of processes working in tandem continue to contribute to the erosion of biodiversity and ecosystem integrity. Fishing acts as a selective force on ecosystems by removing long-lived, slow-growing fish in favour of those with high turnover rates both within and among species. When species or genotypes become extinct, it becomes difficult to restore the ecosystem to its previous state. The fact that the ecosystem has become impoverished will be lost on each new generation of fisheries scientists and historical accounts of former abundance will be ignored or discounted as methodologically naïve. Attempts will only be made to maintain the ecosystem at this impoverished level.

5.101 Additional fishing power is generated in response to population depletion to maintain catches, by investment through loans or subsidies, leading to fleet overcapacity and further population depletion. This ratchets up the continuing spiral of depletion of the marine environment, which is now ubiquitous and unrelenting with few if any refugia remaining

121

for unexploited populations of marine organisms.[251] Acting in conjunction with these ratchet-like processes, competition for scarce resources stimulates the adoption of new technology, and leads to serial depletion of fish populations by geographical area. In addition, management of the marine environment has concentrated on sectoral interests or single issues; only recently has the focus shifted to look at ecosystems as a whole.[252]

5.102 So why has it taken so long to realise that our coasts and seas are in such a degraded condition? There is clear evidence from worldwide studies of changes in community structure in marine ecosystems resulting from fisheries pressure. Such fundamental changes have happened almost unnoticed, and yet fisheries are probably the most studied of the activities affecting the marine environment. In part, it is because problems with the sea are not visible to the public, and changes have occurred slowly but progressively over a long time. Moreover, what we regard as a baseline for judging change is an already degraded system, and this makes it difficult to understand what an environment in good condition would look like.

5.103 The shifting baseline syndrome refers to the incremental lowering of standards with respect to the marine environment. Daniel Pauly introduced the term to describe the way in which each generation perceives ocean life as abundant, even though marine biodiversity has declined slowly and steadily for several hundred years.[253] Each new generation redefines what is 'natural' in terms of personal experience while unaware of earlier declines. The continual decreases in size of fish populations are believed to be normal and these diminished populations are used by managers as the baseline to set limits to the total allowable catch. The next generation makes the same mistake, thus leading to a step-by-step decrease in the standards used to define the optimum status of marine ecosystems.

5.104 The deleterious effects of fishing, the shift in the fish community from large long-lived predatory fish to short lived animals with volatile population dynamics, the reduction in natural predatory pressure by selective removal of predatory species, the simultaneous removal of some the small pelagic fish species that are important food sources for predatory species, and the degradation of marine benthic habitats by various types of fishing gear will all ultimately impact on the fisheries themselves. This depletion in the world's fisheries and the degradation of the marine environment has resulted in the emergence of a new policy emphasis on ecosystem management – formulated and reiterated in several international conventions. As a result the emphasis of marine fisheries research and management has started to shift from fish population studies to ecosystem-based concerns (chapter 7).[254]

5.105 Once depleted, however, affected populations of marine organisms may not rebound rapidly, if at all. The rate of exploitation and loss of marine habitats means that many more species than previously thought are in danger of extinction and the fundamental issue remains that those with responsibilities for maritime issues, outside of the conservation sector, have yet to fully embrace values for the conservation of maritime ecosystems and all that they contain.[255,256] Management of the marine environment will have to change if the constituent parts of marine ecosystems are to be conserved. In the following chapters we suggest a more robust approach to integrating environmental and fisheries policies in order to halt further degradation of the marine environment.

Chapter 6

IS AQUACULTURE THE ANSWER?

What are the environmental impacts of aquaculture and how can they be minimised? Could an expansion in fish farming compensate for a reduction in fish caught at sea? Is this desirable from an environmental perspective?

INTRODUCTION

6.1 At the global level, aquaculture is growing faster than any other means of animal food production.[1] World-wide, aquaculture production is expected to nearly double in the next two decades, climbing from 29 million tonnes in 1997 to 54 million tonnes in 2020.[2] Within Europe, the output of marine fish farming has grown a thousand-fold since 1970.[3] Today, the UK is the largest aquaculture producer in the European Union, producing 30% by volume of the EU's total production; 90% of this effort is concentrated in Scotland, where the industry has an annual turnover of around £500 million and provides up to 7,000 direct and indirect jobs.[4]

6.2 The prospects for further growth within the industry will depend on many factors, but environmental concerns could prove a key constraint. For example, it takes millions of tonnes of wild fish fed to carnivorous fish to support the aquaculture industry. Major concerns include:

- important issues arising from the use of wild fish populations (known as forage fish) to produce fishmeal and fish oil for aquafeed, and other impacts of the supply chain;

- interactions between farmed species and wild populations of fish and shellfish. For example, farmed fish can escape and disturb or breed with wild populations. Aquaculture can also lead to the introduction of exotic species (including disease and parasites) and increased abundance of pathogens;

- nutrient enrichment from faeces, uneaten food and dissolved metabolites that end up in water and sediments;

- chemical pollution from chemicals used to treat disease and parasitic infections; and

- habitat change or destruction. This can include loss of coastal and wetland habitats, the visual impact of fish farms on the landscape and the creation of dead-zones on the bottom of the sea, riverbed or freshwater pond as a result of the build-up of organic matter.

6.3 This chapter analyses each of these concerns and makes recommendations designed to reduce the environmental impacts of the aquaculture industry.

ABILITY OF AQUACULTURE TO SUBSTITUTE FOR WILD CAUGHT FISH

6.4   Some have argued that the aquaculture industry should be encouraged to grow, not purely for economic reasons, but to meet a potential food gap that could develop as the supply of wild-caught fish stabilises or declines, and as world population grows.

6.5   Demand for fish is set to rise. In 2003, a report from the International Food Policy Research Institute (IFPRI) estimated that annual global production of fish for food would need to double by 2020 to keep up with global demand.

6.6   Global consumption of fish has doubled since 1973.[5] IFPRI noted that fish now accounts for 20% of animal-derived protein in low-income, food-deficit countries, compared with 13% in developed countries.[6] Despite recent rapid growth, per capita consumption is still much lower in developing countries, indicating room for further growth. Rising populations and income levels in such countries are likely to mean that their demand for fish will continue to grow for some time, and could be responsible for around 77% of global fish consumption in 20 years' time.

6.7   In developed countries, increasing awareness of the health benefits of eating fish may also contribute to rising demand. For example, in the UK, average consumption of fish would have to increase by 40% to meet the advice of the Food Standards Agency to eat two portions of fish a week (3.42–3.45).

6.8   However as we have seen, the supply from capture fisheries is static, with many fisheries already over-exploited (3.8). This means that growth in aquaculture is expected to play a large part in determining future supplies of fish food and related products. IFPRI's analysis indicates that per capita consumption of fish can be maintained or increased for most parts of the world under plausible scenarios of aquaculture growth to 2020. According to IFPRI, this sector could eventually contribute over 40% of fish for human consumption.

6.9   The largest expansion is likely to occur in production of freshwater fish, crustaceans and molluscs in developing countries. While this trend would place additional demands on forage fish, these would not be as great as those associated with marine fish farming. This is because freshwater fish are largely herbivorous and there is greater scope to substitute vegetable protein for fish protein in their diet. Freshwater aquaculture is of course associated with other environmental problems including nutrient enrichment, water use and habitat loss. There is thus a need for innovation in this sector to combat environmental impacts and increase efficiency.

6.10   Marine fish farming presently accounts for less than 3% of total global supplies of marine fish. It therefore seems unlikely that marine aquaculture (including familiar sectors such as salmon farming in the northern hemisphere) could directly compensate for large-scale declines in wild-caught fish.

6.11   However, there is some evidence that aquaculture can offset regional trends. For example, total aquaculture production in the European Economic Area quadrupled during the period 1970 to 1999, reaching a total of 1.8 million tonnes. Over the same period, capture fisheries fell from 7.6 to 6 million tonnes (although the annual figures have varied considerably). Therefore in terms of production, aquaculture has more than made up for the decline in landings of fish, although there will have been a change in species composition. Europe also continues to rely heavily on imports of wild fish caught elsewhere.[7]

6.12 It is sometimes argued that the forage fish used in aquaculture should be redirected to the human food supply chain, thus meeting some of the global demand for high quality protein. Forage fish fall into three categories: those that are primarily used for human consumption but where the surplus may be used for fishmeal; those that could potentially be used for human consumption but are mainly used for fishmeal; and those that are not used for human food.

6.13 The scope for changing the situation is unclear. The immediate constraints include the cost of getting the fish to human markets in good condition, people's food preferences and income levels in developing countries. We recognise that these factors could change in the future, but in the meantime, in the absence of specific purchasing and/or processing interventions by aid agencies, it appears that many fish will continue to go primarily for fishmeal and oil.

USING WILD FISH FOR AQUACULTURE FEED

6.14 In Europe, marine aquaculture involving carnivorous fish relies on carefully formulated diets to boost production and economic performance, improve flesh quality and reduce local environmental impacts. This means that most farmed fish receive compound feeds based in part on fishmeal and fish oil, which are derived from wild caught fish. As populations of some forage fish are under heavy pressure, and their removal can have significant implications for other parts of the environment, there are concerns that aquaculture encourages unsustainable exploitation of marine resources.[8] Forage fish are, for example, also a source of food for many sea birds, sea mammals and other fish.

6.15 Fishmeal is created by cooking, pressing, drying and milling wild caught pelagic species; fish oil is a by-product of this process. The compound feeds derived from such products are widely used in aquaculture, with salmonids consuming the greatest proportion (table 6.1). Omnivorous species such as tilapia and shellfish also receive a small proportion of aquafeed (around 7% of total aquafeed production)[9] with a lower fish content. Estimates for the amount of forage fish needed to produce enough fishmeal and fish oil to produce 1 kg of farmed salmon vary from about 5 kg to about 2.8 kg (the latter figure applies to newer diets that include a higher proportion of vegetable oil).[10]

Table 6.1

Global use of compound feeds in aquaculture, 2001[11]

	Fishmeal	Fish oil
Total production of aquafeeds, million tonnes	2.6	0.59
Salmonids	29%	65%
Marine fish	23%	20%
Marine shrimp	19%	7%
Carp	15%	–
Eels	6.9%	2.5%

FUTURE DEMAND FOR FISHMEAL AND FISH OIL

6.16 Aquaculture has grown strongly over the past decades and its use of forage fish has increased. The total demand for forage fish has not however changed significantly because demand from the terrestrial livestock feed sector has fallen, partly in response to price changes. There is room for this demand to fall still further since the feed is destined for poultry, pigs and cattle and these animals can use vegetable-based diets more easily than can carnivorous fish such as salmon.

6.17 As a result, the International Fishmeal and Fish Oil Organisation (IFFO)[12] predicts that aquaculture's share of fishmeal demand will rise to nearly half of the total by 2010 (figure 6-I).

6.18 The position is more difficult for fish oil supply, which is likely to reach critical levels before fishmeal (figure 6-II). Aquaculture is predicted to account for 80% or more of demand by 2010 (figure 6-III). Indeed, on some calculations, aquaculture may already use around 75% and this could rise to nearly 100% of the total by 2010.[14] This would represent a tightening of the market for fish oil, and could lead to increased price volatility, especially as the catch of forage fish also varies as a result of El Niño events and other factors. As prices rise, some fear that there may be more pressure on forage fisheries, increasing the risk that they will become over-exploited.

Figure 6-I
Fishmeal demand by sector in 2002 and 2010[13]

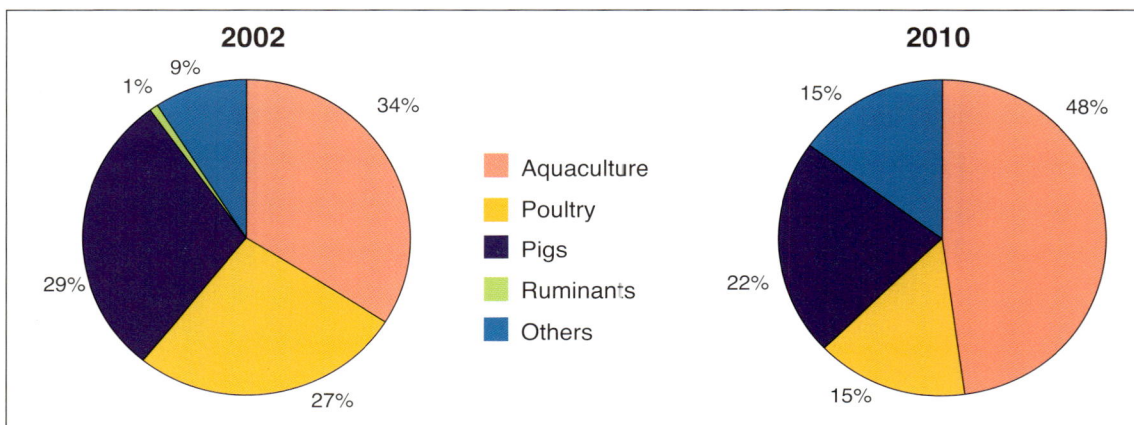

Figure 6-II
Fish oil trends, 1995-2015[15]

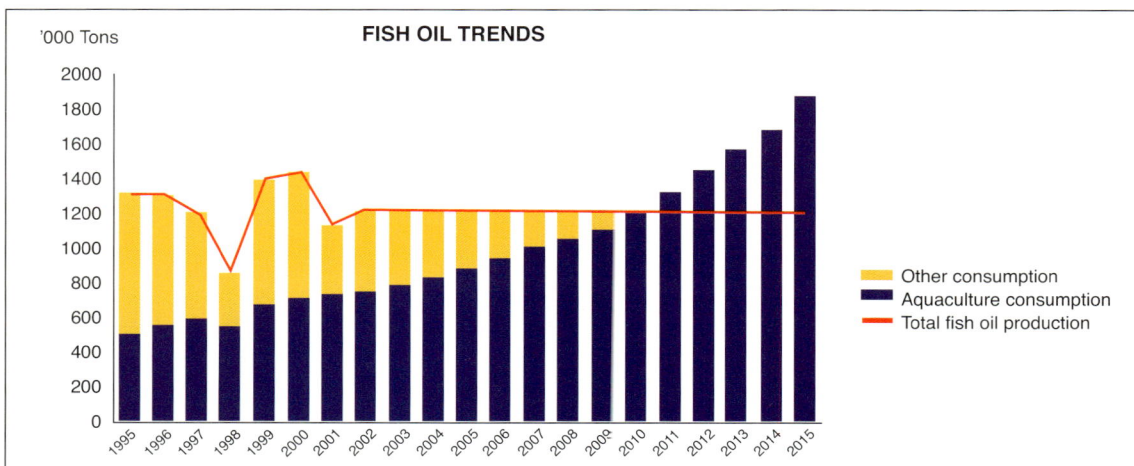

Figure 6-III
Demand for fish oil by sector in 2002 and 2010[16]

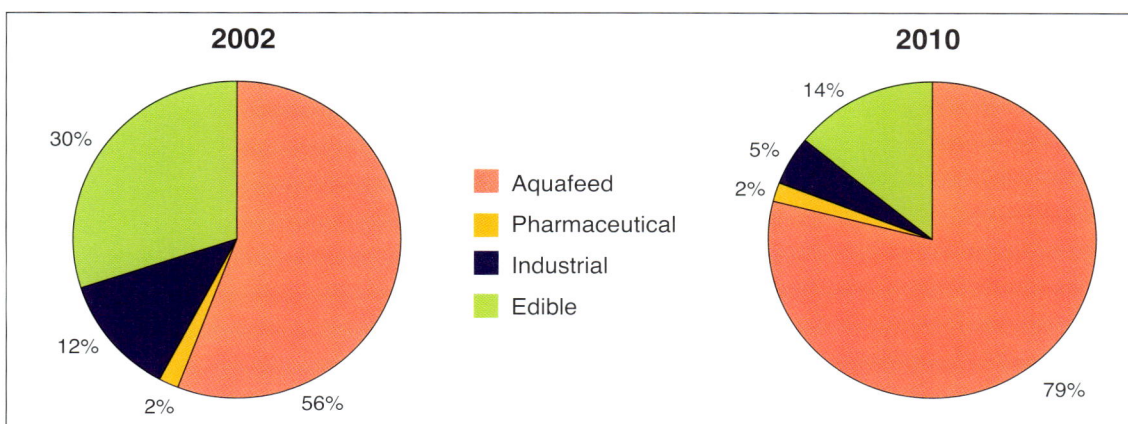

2002

30%

12%

2%   56%

2010

14%

5%

2%

79%

- ■ Aquafeed
- ■ Pharmaceutical
- ■ Industrial
- ■ Edible

STATUS OF FORAGE FISH POPULATIONS

6.19   Around 30% of the total global fish catch is destined for aquaculture, and this rises to 80% for wild pelagic species.[17] As discussed, increases in the price of fishmeal and fish oil may lead to increased pressure on forage fish, with the potential to deplete poorly managed populations.

6.20   Table 6.2 provides information on the status of the main forage fish populations, taken from the EU-funded SEAfeeds project. The report considered that several of these populations were within the safe biological limits established by the International Council for the Exploration of the Sea (ICES). But some are threatened, such as the blue whiting fishery in west Scotland, and the status of others is unknown. In addition, the interactions between species and within ecosystems are often poorly understood.

Table 6.2

Status of forage fish populations[18]

Fishery	Status and management measures
Anchovy Peru	Biomass increasing following last El Niño. Management measures include licensing; satellite tracking; closed seasons; minimum landing size
Anchovy, sardine Chile	Closed seasons; company catch limits
Sandeel North Sea	Within safe biological limits; 2003 class strong; no management objectives
Sandeel Shetland	TAC; safe biological limits not set
Sandeel west Scotland	Multi-annual TAC; closed season; limited access state of stock currently not known
Blue whiting west Scotland	Outside safe biological limits; fishing mortality high; catches above ICES recommendations; no TAC (under discussion); no management objectives; North East Atlantic Fisheries Commission to develop a plan

Fishery	Status and management measures
Norway pout west Scotland	Unknown status; no management; small mesh trawl
Norway Pout North Sea	Within safe biological limits
Herring North Sea	Autumn spawning stock within safe biological limits; spring spawning stock outside safe biological limits
Herring Icelandic	Within safe biological limits
Herring (spring spawning) Norwegian	Within safe biological limits
Capelin Barents Sea	Within safe biological limits
Capelin Icelandic	Healthy, according to Icelandic Ministry of Fisheries
Horse mackerel	State of stock not known
Sprat	Stock in good condition

TAC: Total allowable catch

6.21 The populations of some of these species are also sensitive to climate conditions and respond sharply to El Niño events. They could therefore be vulnerable to the effects of climate change. It can also be difficult to enforce management rules for some high seas populations. Together with rising demand, these factors mean that forage fish are at potentially great risk of over-exploitation. Robust management measures are therefore needed to ensure that they are fished in a sustainable way. Opportunities for substituting fishmeal and fish oil with other sources of protein and improving the efficiency of aquaculture, are also high priorities which we examine in the following sections.

SUBSTITUTION

6.22 Feed typically accounts for between 35-60% of the cost of salmon farming, and is the largest operational cost. Protein sources account for 51% of the cost of high-energy extruded grower diets (which contain 40% protein and 30-35% lipid) for Atlantic salmon. Much of the high cost of the protein fraction is due to the extensive use of South American premium quality fishmeal, which may supply more than 90% of dietary protein in the case of salmon. Other major variables include the cost of fish oil (25%) and astaxanthin colourant (20% of cost at a dosage of 50 parts per million). The costs of the binders and vitamin and mineral supplements are low by comparison.[19]

6.23 A 2003 report suggested that the fishmeal and fish oil content of aquafeeds could be reduced substantially. Research found that for salmonid diets, at least 50% of the fishmeal and 50-80% of the fish oil could be replaced with vegetable substitutes. For marine fish, 30-80% of fishmeal and 60% of the fish oil could come from alternative sources. Salmon are piscivores (carnivores that mainly consume other fish in the wild). It is therefore unlikely that feeds for salmon will ever contain less than 38% crude protein since they rely on it for energy and to provide amino acids, the building blocks of proteins. Proteins contain about 20 amino acids of which ten are essential dietary components. Non-essential amino acids must also be provided in the diet for optimum protein retention. Dietary requirements for most essential amino acids have not been determined for Atlantic salmon.[20]

6.24 Substituting vegetable oils for fish oil will affect product quality. Substantial reductions in fish oil will lower the content of long-chain n-3 polyunsaturated fatty acids (PUFAs) in fish. Since these are beneficial for human health, this may affect consumer attitudes towards the product. The effect can be countered in part by feeding diets that are high in long-chain n-3 PUFA fish oil towards the end of the production cycle. Substitution of fishmeal and oils could also have the effect of reducing the level of some undesirable substances, such as dioxins, that could occur in the final product.

6.25 Several EU-funded projects are looking into the development of alternative feeds. The FORM (Fish Oil and Meal Replacement) network has been set up to discuss new projects on nutrients in fish feed and seafood.[21]

ALTERNATIVE MARINE SOURCES OF PROTEIN AND OIL

6.26 The Australian government has funded research into replacing fishmeal and oil in the diet of salmon. In experimental trials, plant protein meals (soybean meal, pea protein concentrate and lupin protein concentrate) replaced 40% of the protein from fishmeal and resulted in weight gains that were within 10% of the growth achieved by salmon fed fishmeal-only diets. This showed the potential of each of these plant meals to be used in salmon feeds, and was further tested by replacing 25% and 33% of the fishmeal protein in extruded feeds. The parameters used to extrude the feeds were controlled in order to ensure the experimental feeds matched the commercial salmon feeds produced in Australia. The growth on each of the plant proteins and at both inclusion levels was the same as the control, (fishmeal-only) diet. In a final trial, in which salmon were held under commercial conditions of high stocking density and high feed rates, the performance of pea meal and soybean meal in extruded feeds was equal to that of a fishmeal-only control diet as well as a commercial feed formulation. Soybean, lupin and pea meals were shown to have considerable potential as replacements for fishmeal protein in Atlantic salmon feeds.[22]

Discards, by-catch and trimmings

6.27 Discarding is a wasteful use of fish and an additional pressure on fish populations. It has been suggested that a more efficient use of resources would be to redirect discards and by-catch from food-grade fisheries to generate fishmeal. This is already practised in some countries such as Norway, but is not legal in the EU under the Common Fisheries Policy. Any future use of discards or by-catch should not however prejudice measures to reduce by-catch – a key measure in maintaining the sustainability of capture fisheries.

Krill and copepods

6.28   Krill represents a huge resource of marine protein and oil that is highly enriched with carotenoids as well as long-chain n-3 PUFAs. The global population of krill probably exceeds 100 million tonnes. There are, however, technical difficulties both in catching krill (the fine mesh nets require tremendous power to move them through the water) and in preserving them (the small animals begin to degrade very rapidly on capture). Furthermore, krill is rich in fluorine, and current EC legislation relating to animal feeds would preclude its use. Krill is a key species in the food web in Antarctic waters, and any major fishery would need a management regime that took full account of ecosystem effects. The Committee for the Conservation of Antarctic Marine Living Resources has set a precautionary limit of 1.5 million tonnes for krill harvest in the south-west Atlantic sector of the Southern Ocean.

6.29   The copepod *Calanus* (box 2B) is even more abundant, but even smaller and more difficult to catch, preserve and process. Serious research has only just begun on commercial-scale culturing and harvesting of copepods.

Biotechnology

6.30   Biotechnology could generate high quality proteins rich in long-chain n-3 PUFAs from intensive production of bacteria, algae or zooplankton, but costs are currently high and research and development activity limited. Research suggests that prices of bio-fermentation-derived long-chain n-3 PUFAs would have to be reduced by 95% to be competitive with fishmeal and oil. It is known however that long-chain n-3 PUFAs can be synthesised by a variety of micro-organisms in marine sediments. While the culture of, for example, marine worms is unlikely to fill the gap (although it may become an important niche product) biotechnological advances in micro-organisms may have more potential for large-scale production.

6.31   Genetic modification of plants and/or micro-organisms to produce higher quality proteins and oils is likely to be possible, but restrictions on genetically modified products are a major constraint. All Norwegian fish feed producers have a policy that rejects the use of genetically modified organisms in fish feed;[23] the Scottish aquaculture industry has the same approach.

Non-marine sources of protein and oil

Microalgae and micro-organisms

6.32   Cultured microalgae is also a promising alternative for fishmeal and fish oil, as some microalgae have higher protein content and may be rich in long-chain n-3 PUFAs. Moreover, some species, such as *Spirulina, Chlorella* and *Scenedesmus*, include essential amino acids in the right proportions needed for fish. A WWF report noted that microalgae cultures are expected to have a significant role in fish feed production in 5 to 10 years.[24]

6.33   Fermentation of a bacterium of *Methylococcus capsulatus* with a supply of natural gas can also be used to provide a protein ingredient in fish feed. The use of such products in salmon feed production was approved by the European Union in 1995.[25]

Plants and other possible sources

6.34   Soybean has a protein content of 42-48% and is an important component of aquafeed. A number of other protein sources have been investigated, but some, such as meat, blood and bone meal, do not seem well suited to aquaculture.[26] Possible alternative protein sources include:

> poultry by-product meal; poultry feather meal; oilseed meals; cottonseed meal; peanut meal; sunflower meal; rapeseed meal and canola meal; peas and lupin; distillers' dried grains; brewers' dried grains; corn gluten meal; wheat gluten meal; plant protein concentrates and single-cell proteins.

ALTERNATIVE LIPID SOURCES

6.35   Lipids and their constituent fatty acids are important not only for fish health and growth, but also for human health (3.40 and appendix G). The replacement of marine lipid sources with plant-based alternatives therefore has important implications, both in terms of effects on the farmed fish themselves and on subsequent marketing and consumption. As well as the potential benefits of eating farmed fish, there are concerns about the potential contamination of farmed fish by heavy metals and organic compounds which have been transferred up the food chain from the feed (3.46–3.51).

6.36   Fish oil, produced during the manufacture of fishmeal, has been the traditional source of lipids in fish feed for a number of reasons. It contains all the dietary components necessary for fish health and it increases the amount of PUFAs in the flesh of fish. As we have seen (6.18), sources of fish oil are likely to become a limiting factor for the aquaculture industry before supplies of fishmeal. As fish oil prices rise, alternative oils will become more economically competitive.

6.37   The Norwegian Institute of Marine Research reported that world production of plant oils in 1999 exceeded 100 million metric tonnes, whereas fish oil production typically ranges from 1.2-1.4 million metric tonnes and decreases to around 800,000 metric tonnes in El Niño periods.[27] Soybean, corn, sunflower, oilseed rape and palm oils are the most abundant plant oils; other sources such as lupin are less common.

6.38   The main limitation for the use of plant oils is their fatty acid profile. Plant oils are rich sources of long-chain n-6 PUFAs, and have lower levels of long-chain n-3 PUFAs, while fish oils are high in long-chain n-3 PUFAs, and low in long-chain n-6 PUFAs. The ratio between the two types of fatty acid is important; with the ratio found in fish oil being more beneficial for human health. The use of rendered animal fat products also raises similar concerns, but may be another strategy for reducing dependency on fishmeal.

FARMING OF ALTERNATIVE SPECIES

6.39   At a global level, many herbivorous and omnivorous species are grown in fertilised ponds with little or no artificial feeding. Most of these can also be reared intensively using feeds with much lower fishmeal and oil content than that required for carnivorous species (typically 25% fishmeal or less compared with 30-70% for intensive carnivorous fish).[28] These species and systems are likely to become more competitive as the cost of fishmeal and oil rises.

6.40 Apart from some carp and catfish species, few such species can be raised in northern Europe; although they may be expected increasingly to enter international trade as the price of fishmeal and oil rises.

6.41 These species are however lower in long-chain n-3 PUFAs, and so, less desirable from this perspective. It may therefore be appropriate to consider salmonids as a vehicle to deliver fatty acids, and species such as carp and tilapia largely as the vehicle to deliver protein.

6.42 Data from Scottish businesses[29] indicate that in 2001 a total of 97 staff were employed in farming the following species:

- Arctic char, 7 companies at 10 sites producing 8 tonnes;

- Brown/sea trout, 19 companies at 26 sites producing 127 tonnes;

- Cod, 6 companies at 7 sites producing 207 tonnes; and

- Halibut, 7 companies at 12 sites producing 257 tonnes.

6.43 The British Marine Finfish Association estimates that within 10 years fish farms in the UK (principally Scotland) will be producing up to 30,000 tonnes of cod, 8-10,000 tonnes of halibut and 5,000 tonnes of haddock.

6.44 As well as the introduction of new species, it may be possible to improve the breeding stock of traditionally farmed fish to increase efficiency. This could be done through conventional breeding methods or through genetic manipulation. However, we are of the view that the latter option is not appropriate in the UK for the foreseeable future (6.81).

6.45 Farming of many of these alternative species will raise similar concerns to those already raised by the farming of Atlantic salmon, with respect to the impact on wild populations of the same species through interbreeding, and transfer of diseases and parasites. **We recommend that appropriate controls should be put in place at the start of farming of new species**, given that the aquaculture industry has caused substantial damage to wild populations with farmed salmon. There is also a requirement for research into these aspects for other marine species in the same way as for salmon. Lessons should be gained from the salmon experience where the industry developed for many years with inadequate controls before concerns were taken seriously.

LIFE CYCLE ANALYSIS AND VIABILITY

6.46 By comparison with meat production, fish farming is a relatively efficient means of providing protein for the human diet. This is mainly because fish are cold-blooded and have low metabolic rates. For example, 100 kg of feed mix (protein, carbohydrates, oil) produces an edible meat quantity of:[30]

- 65 kg of salmon;

- 20 kg of chicken; or

- 13 kg of pork.

6.47 For farmed salmon, approximately 18.5 units of energy are required to produce each unit of energy contained in the final edible product, while for mussels, this is reduced to a ratio of 3.2 to 1. Again, salmon compares reasonably well to some other sources of protein (table 6.3).

Table 6.3

Comparative energy usage for different aquaculture and livestock systems[31]

Product	Production system	Energy usage (MJ/kg protein)
Mussels	Intensive, long lines	116
Salmon	Intensive, cages	688
Grouper/sea bass	Intensive, cages	1,311
Tilapia	Semi-intensive, ponds	0-199
Catfish	Intensive, ponds	582
Catfish	Intensive, raceway	3,780
Carp	Intensive, recirculated	3,090
Beef	Rangeland	170
Beef	Feedlot closed	513
Beef	Feedlot open	1,350-3,360
Pork	Intensive	595-718
Poultry	Broilers	370

6.48 The EU-funded SEAfeeds project considered that it is difficult to make comparisons between the life cycles of farmed and capture fish.[32] The project observed that wild cod or salmon require more prey to grow than farmed fish, and that the 'natural' conversion rate of prey to predator fish is typically closer to 10:1. Catching, processing, distributing and feeding of forage fish to salmonids, however, may require more energy or transportation than capturing carnivorous fish from a well-managed stock. It also considered that a capture fishery was very unlikely to deliver as much product as aquaculture, or to introduce as much fish oil, to the human food chain, although the efficiency conversion between capture fisheries and farmed fish depends on the different weightings given to the various areas of sustainability, and the fish species used in comparison.

MAKING SUPPLY OF FORAGE FISH MORE SUSTAINABLE

6.49 The use of fishmeal and fish oil raises major questions of sustainability. On present trends, the amount of fishmeal and fish oil used in the livestock industry is predicted to continue to decline while the proportion used in aquaculture is set to increase. On current trends, it seems likely that there will be a shortage of fish oil within eight to ten years, given the present rate of growth of the aquaculture industry. **We recommend that the UK government and the Scottish Executive should promote a strategy to improve the sustainability of fishmeal and fish oil supplies. This should include steps to:**

- **increase the efficiency with which fishmeal and fish oil are used within the aquaculture industry;**

- **encourage the trend away from the use of fishmeal and fish oil in the livestock industry, so that the aquaculture industry is given preference of supply; and**

- **accelerate the development and use of viable alternatives for use in aquaculture. This should include research into the feasibility of substituting fishmeal and fish oil with alternatives, the farming of non-carnivorous fish and consideration of a tax or other economic instrument on the use of fishmeal and fish oil.**

INTERACTIONS WITH WILD FISH

6.50 There are three main types of interaction between farmed fish and their wild counterparts, raising three different areas of concern:

- abiotic interactions (e.g. habitat damage);

- biotic interactions (e.g. increased competition and predation); and

- genetic introgression (i.e. exchange) between farmed and wild stocks.

6.51 The risk of abiotic and biotic damage is perhaps greatest if the cultured species are exotic to the farm location. Globally, about 200 different species are used in aquaculture, but in practice much of the activity focuses on a few species with well-known requirements and assured markets. As a result, these species have been widely translocated around the world, and inevitably some have escaped into the local environment.

6.52 This raises the question of the genetic impact of farmed species on wild populations. Genetic variation is the foundation of biological diversity. Maintaining adequate levels of genetic variation, within and between populations, as well as between species that can hybridise, is essential for long-term sustainability and to ensure the evolutionary potential of aquatic organisms in the environment. It is important to protect genetic diversity *in situ* because wild populations harbour co-evolved gene complexes capable of continually responding to changing evolutionary forces in natural environments.[33]

6.53 A review of the environmental effects of mariculture carried out under the auspices of the UN Convention on Biological Diversity[34] noted that the genetic effects of mariculture are varied and highly significant for biodiversity. Understanding its genetic impacts demands a high level of understanding of the genetic structure of farmed and wild populations; something we do not have for any species.

6.54 Because much of the world's aquaculture relies on species cultured outside their native range, which may easily survive and reproduce outside their natural habitats, escapees are a constant concern.

6.55 Wild Atlantic salmon are a prime example of this problem. They can be adversely affected by interactions with their farmed relatives at several stages during their life cycle, both in freshwater and in the sea (figure 6-IV).

Figure 6-IV

Many wild Atlantic salmon migrate to distant feeding grounds, with the seas near Greenland and Faroe Islands being especially important. Wild salmon rely on adaptations developed over thousands of years. Genetic or other interactions with farmed salmon can reduce the survival of wild populations[35]

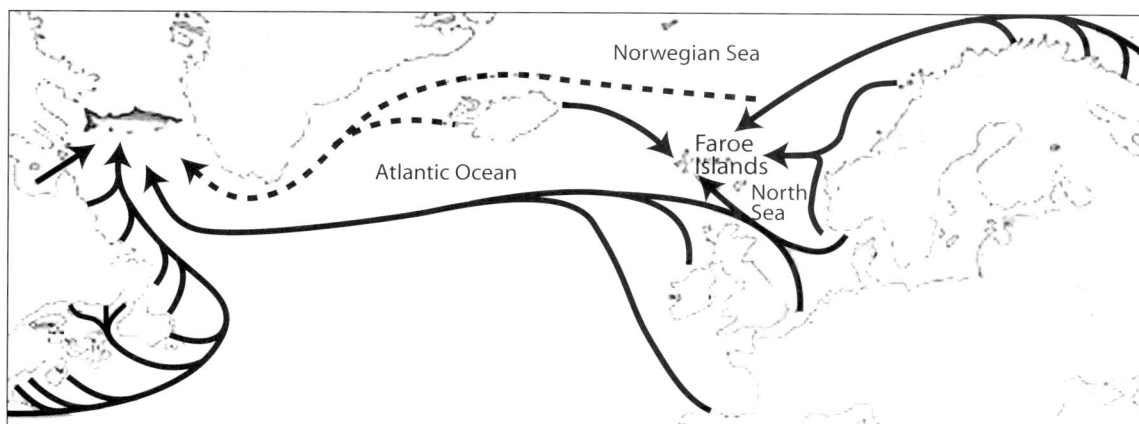

6.56 A 2003 study[36] estimated that some two million salmon escape every year in the North Atlantic, which is equivalent to about 50% of the total pre-fishery abundance of wild salmon in the area. Farmed Atlantic salmon are genetically distinct from wild populations as a result of their genetic origin (most farmed strains originate from Norway), founder effects (the use of a small number of parents that represent only part of the genetic variability in the original source population), directional selection (for favourable farming characteristics especially faster growth), inadvertent selection and genetic drift during domestication. Although previous studies have shown their breeding performance to be inferior to that of wild salmon, escaped farmed salmon do breed successfully and hybridise with wild fishes, thereby changing the genetic profile and recruitment capacity (fitness) of wild populations.

6.57 In evidence, the Association of Salmon Fishery Boards[37] reported that the total wild salmon catch in the salmon farming areas of the west coast of Scotland in 2001 was under 3 tonnes.[38] By comparison, the total farmed salmon production at that time was approximately 139,000 tonnes.[39] Reported escapes have been running at between 200,000 to 400,000 fish a year; up to four times the entire Scottish commercial and rod fishery wild salmon catch.

6.58 Evidence from the Environment Agency[40] noted that although no significant marine salmon farming occurs around the coasts of England and Wales, escaped fish have been found in these areas. In 2001, following a farm escape in Northern Ireland, up to 6% of the spawning escapement of north-west rivers in Ireland comprised salmon of farmed origin. The Environment Agency also considered that climate change had the potential to increase the viability of escaped non-native fish species and recommended that a precautionary approach should be adopted.

EFFECT OF FARMED FISH ESCAPES

6.59 Escapes matter because farmed salmon and hybrids between farmed and wild salmon have a much lower lifetime success than wild fish (perhaps as little as 2% for farm fish and 30% for hybrids compared with wild fish). Inevitably the lower lifetime success of these hybrids will reduce the overall 'fitness' of wild populations. The introduction of different forms of genes from farm to wild salmon (a process known as introgression) will also reduce the genetic heterogeneity within and among wild populations, because farmed fish have less genetic variation and only a few farm strains are in widespread use.

6.60   There is evidence that localised genetically distinct populations of Atlantic salmon exist. Loss of this genetic heterogeneity will reduce the adaptive potential of the species. Introgression will reduce the genetic variability in wild populations leading to inbreeding depression. In addition to these direct effects, offspring of escaped farm fish will reduce wild smolt (juveniles migrating to sea) output by competing for resources; wild parr and salmon will be displaced by the more aggressive and faster growing farmed salmon. As the hybrid and farmed salmon show reduced marine survival rates and a reduced lifetime success, the reduction in wild smolt production will reduce the fitness (i.e. adult return and subsequent juvenile recruitment) of the wild population irrespective of any direct genetic changes, to the population. This will further reduce the effective population size, resulting in lowered genetic variability (hence there is potential for a double effect). Parr of hybrid or farm origin show reduced levels of maturity and this may further reduce effective population size.

6.61   The overall extent of reduction in fitness in the wild population, as a result of both interbreeding and competition, will depend on many factors, including the availability of unoccupied juvenile habitat, relative numbers of wild, farm and hybrid salmon and mating success. As farm escapes are often repetitive, such reductions in fitness can be cumulative, which could lead to an extinction vortex for some populations with marginal population growth levels.

ACTION TO PROTECT WILD SALMON

6.62   At the European level, Atlantic salmon in freshwater has been named as a species of community interest in Annex II of the EC Habitats Directive. As a result, Scottish Natural Heritage has identified a number of rivers that could be considered as possible Special Areas of Conservation for wild Atlantic salmon.

6.63   Concern over the interaction between farmed and wild fish has also led to the development of a series of resolutions by the North Atlantic Salmon Conservation Organization (NASCO). These were consolidated into the Williamsburg Resolution in 2003.[41] The Williamsburg Resolution, to which the European Community is a party, requires signatories to minimise:

- escapes of farmed salmon to a level that is as close as practicable to zero through the development and implementation of action plans as envisaged under the Guidelines on Containment of Farm Salmon;

- impacts of ranched salmon by utilising local stocks and developing and applying appropriate release and harvest strategies;

- the adverse genetic and other biological interactions from salmon enhancement activities, including introductions and transfers;

- the risk of transmission to wild salmon stocks of diseases and parasites from all aquaculture activities and from introductions and transfers.

6.64 The Resolution states that the movement of reproductively viable Atlantic salmon or their gametes into the NASCO area from the outside should not be permitted. However, NASCO does not have the powers to prohibit activities, only to seek agreement among parties to the adoption of certain practices. The agreements have led to a commitment to annual reporting and the development of a series of indicators. EU Member States are putting in place measures to address the concerns covered by the resolution. A recent review by WWF and the Atlantic Salmon Federation, however, considered that progress towards the objectives of the Williamsburg Resolution had been poor.[42] While Scotland was among the countries that fared best in the evaluation (along with Norway), its overall performance was judged to be severely lacking in a number of areas. Here, we examine two specific aspects: the lack of a mandatory code on escapes; and exclusion zones around salmon rivers.

6.65 Within the UK, legislation requiring the notification of farmed fish escapes has been in place since May 2002, and the Scottish Executive has produced guidance for fish farmers on the action to take in the event of a notifiable escape. This advice, plus further measures to prevent escape, is reflected in a voluntary Code of Containment developed by Scottish Quality Salmon and the Shetland Salmon Farmers Association. The Code suggests that equipment should be sufficiently robust to withstand the range of expected weather conditions (equipment failure, especially during bad weather, is a major reason for escapes), that contingency plans should be in place should a failure occur and that there should be a target of zero escapes. The British Trout Association also has a code of practice for freshwater trout farms.

6.66 The Strategic Framework for Scottish Aquaculture also has a number of actions relating to containment and escapes. A containment working group, comprising industry, regulators and wild fish interests, has worked up a more robust version of the Code of Containment that will be included in a new industry-wide code of practice, to be issued for consultation by the end of 2004.

6.67 WWF and the Atlantic Salmon Federation have strongly criticised the existing Code for not being mandatory and the current Code is also very brief. While the Code does not have legal force, development consents require new fish farms to comply with the Code, and leases or licences will not be issued unless there is evidence that plans for compliance are in place. Nevertheless, given the threat that escaped fish pose to wild populations, and the possibility that escapes could be reduced through relatively simple measures, we are of the view that these voluntary arrangements should be strengthened.

6.68 NASCO's Williamsburg Resolution requires signatories to introduce exclusion zones to protect wild salmon stocks in a minimum set of rivers most essential to the survival of such populations. The Resolution only requires the zones to be introduced on a trial basis, and few countries have established even minimal zones. In evidence to the Commission, it has been suggested that minimum distances should be introduced between cage farms and salmon rivers. Figures of between 5 km and 30 km have been suggested,[43,44] but we recognise that this could have significant impacts on the aquaculture industry in Scotland, as well as practical limitations.

6.69   To address these issues **we recommend that:**

- **the UK government and Scottish Executive should publish an action plan describing how they will meet their obligations under the North Atlantic Salmon Conservation Organization's Williamsburg Resolution;**

- **the Scottish Executive and the Scottish Environment Protection Agency should fund research into the design of protection zones to separate cage farms from salmon rivers, including cage location based away from migratory routes of wild salmon, and apply the findings;**

- **the Scottish Executive should continue to work with the fish farming industry to strengthen its Code of Containment and to make the Code mandatory. In addition, the Guidelines on the Containment of Farm Salmon developed by the North Atlantic Salmon Conservation Organization should be reflected in the minimum standard for the construction and operation of fish farms;**

- **the Scottish Environment Protection Agency and the fish farming industry should collaborate to carry out further research to improve technical and operational standards on fish farms so as to reduce escapes. The findings of this research should be reflected in the Code of Containment; and**

- **the Scottish Executive should introduce regulations to prevent the outflow from smolt rearing units flowing into salmon rivers** (as is already the case in Norway).

STERILISATION

6.70   One way in which the ecological risks associated with fish escapes could be reduced is through sterilisation of farmed fish. This approach is already widely in use in certain branches of aquaculture. The technique, known as induced triploidy involves subjecting newly fertilised eggs to heat, pressure or chemical shock so that cells carry three sets of chromosomes instead of the normal two, rendering the fish sterile.

6.71   The US Pew Initiative on Food and Biotechnology[45] noted a number of drawbacks to induced triploidy. It reported that the effectiveness of triploidy induction varies greatly, depending on the species, the methods used and egg quality. Success rates range from 10-95%.[46] It is possible to screen out fish that fail to become triploid, but this relies on manually checking each fish.

6.72   A review for the UN Convention on Biological Diversity found that sterile fish can compete with wild populations fish for food, spread disease and disturb wild nesting sites.[47] Although sterile, induced triploid fish may carry enough sex hormones to enter into courtship behaviour and thereby disrupt reproductive processes and affect spawning success in wild populations. This particular aspect can be reduced by hormonal manipulation that results in all the farmed fish being female.

6.73   Compassion in World Farming has reported that triploid fish have been found with higher levels of spinal deformities, respiratory difficulties, low blood haemoglobin levels, a lowered ability to cope with stressful situations and higher rates of mortality.[48] The majority of rainbow trout reared by aquaculture are triploids. In evidence it was suggested there is a case for further expanding triploidy to other farmed species, but given the wider concerns we are not convinced that there is a strong case for increasing the use of triploidy.

GENETICALLY MODIFIED FISH

6.74 In the case of transgenic fish, recombinant DNA techniques are used to insert genetic material from one species into another. Such techniques are used for example to insert growth-promoting genes. These have resulted in both cold- and warm-water aquaculture species with growth rates enhanced by factors of between two and eleven. Part of this effect is due to increased food conversion efficiency, which also has the effect of reducing food waste in the water column. In addition to growth promotion, disease resistance is another important trait for which modification may be carried out.

6.75 Genetically modified (GM) salmon are not currently available for sale to UK consumers but government agencies in the US, Canada, Cuba and China are known to be reviewing procedures for the authorisation of commercialisation of transgenic aquaculture products.[49]

6.76 During 2003, the US Pew Commission on Food and Biotechnology produced a review of the use of transgenic fish in aquaculture.[50] This included an examination of environmental effects. The report noted that the use of transgenic fish would have a beneficial influence on issues, such as the potential for reduced feed wastage and reduced antibiotic and therapeutant use as a result of enhanced disease resistance. These factors would depend on the ability to introduce resistance to certain fish diseases. Increased feed conversion efficiency could also reduce pressures on wild fish used in aquaculture feed if adopted to a sufficient extent.

6.77 One particular concern is the mixing of populations of genetically modified farmed fish with wild fish populations. This depends on the probability of escape, the net fitness of escaped fish and the management measures in place to reduce escapes. The Pew Report notes that if novel genes do spread into wild populations, the possible environmental consequences are extensive, but difficult to determine, as they will depend on the most likely gene flow scenario. and the ecological characteristics of the transgenic fish and the fish community it might affect. One way in which this might be combated is through the use of sterilised farmed fish (but see 6.70-6.73).

6.78 The Pew Report suggests that the introduction of transgenic fish might reduce production costs to the extent that closed systems become economically viable. The environmental improvements would also need to be considered in the light of increased regulatory compliance costs and public acceptance to create the necessary markets.

6.79 The report also notes the possibility that genetic modification may accidentally induce increased tolerance within the farmed organism for substances toxic to humans (such as mercury) thereby raising food safety concerns. The same is also true for increased human allergens in genetically modified farmed species. The report calls for careful screening of products before they are put on the market.

6.80 In a separate report, the US Pew Oceans Commission[51] concluded that until an adequate regulatory review process is established, the US government should place a moratorium on the use of genetically engineered marine or anadromous species.

6.81   UK regulatory bodies have taken a strict line on commercialisation of genetically modified fish. The Department for Environment, Food and Rural Affairs (Defra) website notes that "it is difficult to envisage any circumstances in which the release of a GM fish would be permitted".[52] Similarly, the Scottish Executive notes that the Scottish salmon industry does not see a future for GM fish or fishmeal while consumer concerns are so high, although it might consider genetic modification if public attitudes changed.[53] The Agriculture and Environment Biotechnology Commission considered that commercialisation of GM fish raised significant environmental concerns because of the possibility of the fish escaping from the aquatic net pens used in offshore fish farms. It also recognised uncertainty with regards to the environmental consequences of GM fish escapes.[54] Given these widespread concerns, **we recommend that genetically modified fish should not be released or used in commercial aquaculture in the UK for the foreseeable future.**

LOCAL ENVIRONMENTAL IMPACTS

ENVIRONMENTAL CAPACITY

6.82   The concept of environmental capacity assumes that all environments have a finite ability to accommodate exploitation or contamination without unacceptable consequences. It can be described as "the ability of the environment to accommodate a particular activity or rate of activity without unacceptable impact".[55]

6.83   This concept can be used to assess the burden that aquaculture places on the environment and the potential for further expansion of the industry. Since aquaculture has a wide range of impacts, the Scottish Parliament's 2002 inquiry into aquaculture[56] recommended that distinctions be drawn between carrying capacity (the maximum production in relation to the available food resources in the area), assimilative capacity (the ability of an area to process wastes, etc.) and environmental capacity (the overall ability of the environment to accommodate an activity).

Figure 6-V
Main pollution pathways associated with salmonid cage culture[59]

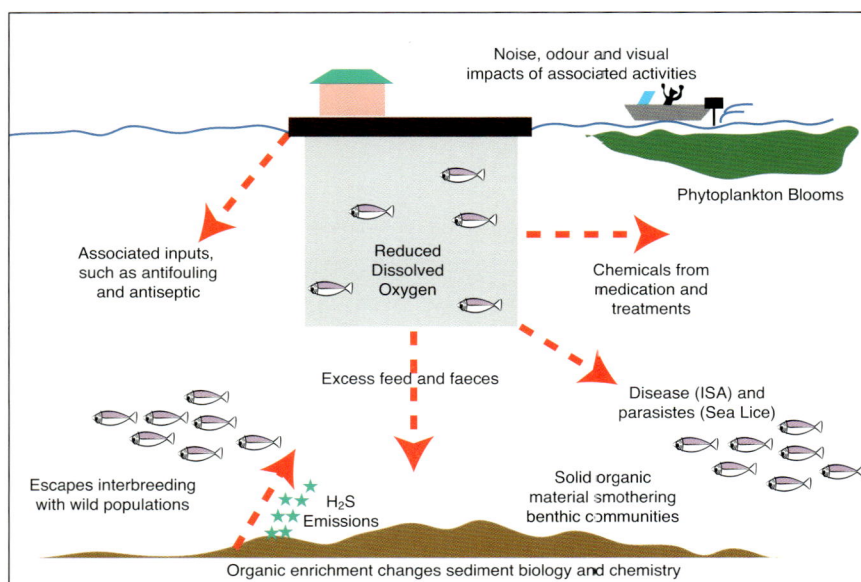

6.84 The issue of assimilative capacity is particularly important for aquaculture because many of its operations are open to the environment. For example, the marine finfish farming industry relies almost exclusively on rearing fish in floating cages.[57] This means that wastes are not treated before being discharged into the surrounding water. Instead, the industry relies on the environment to assimilate wastes through natural physical, chemical and biological processes. Since this capacity is finite, the Scottish Environment Protection Agency (SEPA) considers that nutrient enrichment will eventually become the main factor limiting production at any particular site.

6.85 Wider aspects of environmental capacity are also important, including the effect of farmed fish on wild populations, chemical pollution from sea-lice treatments and broader issues such as landscape. The 2002 inquiry by the Scottish Parliament recommended that these should receive more attention.[58] The range of local environmental impacts associated with salmon cage farming is illustrated in figure 6-V.

DISEASE AND CHEMICAL POLLUTION

6.86 Parasites and diseases are part of the natural biology and functioning of ecosystems, but if fish are raised under crowded and stressful conditions they can be more prone to disease. Disease can move in both directions between farmed and wild fish. Management methods can help reduce this problem but may not be able to eliminate it. A wide variety of techniques, including chemical treatments and vaccines, has been developed to combat disease and there have been substantial technical advances in this area.

6.87 Chemicals are also used to keep fish cages clean and to maintain water throughput and flushing. Since much finfish farming takes place in open net pens, chemicals can be dispersed into the surrounding water by a variety of routes including through treatment, spills, waste food and faecal matter. This section examines the diseases and parasites of farmed fish, and the release of chemical pollutants from fish farming, their effects and attempts to address them.

Parasites and disease

6.88 Sea lice have always affected wild salmon, but intensive farming has increased the size of the problem. It is now one of the biggest issues for salmon aquaculture in many areas of Scotland (the other major problem being infectious pancreatic necrosis in Shetland and the Orkneys). Sea lice are small, parasitic crustaceans that feed on their host causing fish to lose skin and scales (figure 6-VI). Louse damage around the head can be so severe that the bone of the living fishes' skulls can be exposed – a condition known as the 'death crown'. Sea louse infestation is a serious problem which if untreated can kill affected fish. Less common diseases, and the categories of concern for notifying outbreaks of diseases are given in box 6A.

Figure 6-VI
Sea lice on Salmon[60]

141

6.89 Sea lice have also had a major impact on populations of sea-going brown trout. As sea trout feed largely in coastal areas, as opposed to the high seas feeding of wild salmon, they have been even more affected by sea lice. In parts of western Scotland, the Orkneys and the Shetlands, sea trout are more important for recreational fishing, and thus for the tourist industry, than salmon.

BOX 6A   LESS COMMON DISEASES AND CATEGORIES OF CONCERN FOR NOTIFICATION OF DISEASES IN THE EU

Bacterial diseases
- Furunculosis
- *Vibrio salmonicida*
- Enteric redmouth (ERM)
- Bacterial kidney disease (BKD)

Viral diseases
- Infectious pancreatic necrosis (IPN)
- Infectious hematopoietic necrosis (IHN)
- Infectious salmon anaemia (ISA)
- Viral haemorrhagic septicaemia (VHS)

Parasitic diseases
- *Gyrodactylus salaris* (a skin parasite on parr)
- Proliferative kidney disease (PKD)

Notification of the outbreak of certain of these diseases is required under a EC Directive.[61] There are eight notifiable diseases, falling into three categories of concern:[62]

List I: Diseases exotic to the EU at the time of listing that pose a serious economic threat to aquaculture and for which treatment or vaccination is not available. Member States are required to take immediate action to eradicate the disease should outbreaks occur; infectious salmon anaemia falls into this category.

List II: Diseases that are established in parts of the EU and pose a serious economic threat to aquaculture and for which treatment or vaccination is not available. The EU is zoned by water catchment areas into Approved Zones and Farms (free of the disease(s)) and areas that are not approved. Movements into Approved Zones or Farms can only take place from those areas of equivalent or higher health status. The regime also provides for Member States to take action to eradicate these diseases in order to establish Approved Zones and Farms. There are two List II diseases at present: IHN and VHS. The Farm Animal Welfare Council report that IHN has never been found in Great Britain and VHS has been found only once, in farmed turbot in 1994.[63]

List III: Diseases that are a serious problem in some Member States and for which treatment or vaccination is not available or possible. With EU agreement national programmes can be established to contain or prevent the introduction of these diseases. Five List III diseases are notifiable in the UK: BKD, IPN in salmon, furunculosis in salmon, spring viraemia in carp and gyrodactylosis. The last of these has never been found in Britain.[64]

6.90 Although sea-lice are a major problem for the Scottish industry, heavy infestations are not notifiable. Some argue that voluntary management of the problem is working well in some areas, but there are no sanctions for poorly performing farms.[65] It may be possible to improve the control of sea-lice by requiring farms to report information on outbreaks to the Scottish Fish Health Inspectorate, and compelling farms to take corrective action.

Antibiotic use

6.91 Antimicrobial compounds such as oxytetracycline, sulphadiazine and amoxycillin are added to aquaculture feeds to treat bacterial infections. Typically, antibiotics are used intermittently for short periods (5 to 14 days) to control disease outbreaks. Their use has been declining for a number of years since vaccines were introduced. As a result, the amount of antibiotics used in aquaculture is now relatively low, especially compared with use in agriculture. In 2002, around 1 tonne of active antimicrobial ingredients was used in UK aquaculture compared with over 400 tonnes in livestock production.[66]

6.92 A number of concerns have been raised over antibiotic use in aquaculture.[67] These are:

- development of drug resistance in fish pathogens;

- spread of drug-resistant plasmids to human pathogens;

- transfer of resistant pathogens from fish farming to humans;

- presence of antibiotics in wild fish; and

- impact of antibiotics in sediments on rates of microbial processes, composition of bacterial populations, and relative size of resistant sub-populations.

6.93 However, a review of antibiotic use in Norwegian aquaculture[68] concluded that it was unlikely to pose significant environmental or other problems at the low levels used in developed countries. There remains concern over the higher levels of use in developing countries such as Chile, and in newer enterprises such as cod farming.

Sea lice treatments

6.94 A review of the availability of sea lice control products identified 11 compounds representing five pesticide types that were in international use on commercial salmon farms in the period 1997 to 1998.[69]

6.95 Of these, four compounds are presently in use in the UK (azamethiphos, cypermethrin, hydrogen peroxide and emamectin). Cypermethrin and emamectin are the most widely used and considered to present the greatest environmental risk, although there are many uncertainties (appendix I). Hydrogen peroxide, which degrades rapidly to water and oxygen, is not considered to be a hazard to marine life, but is falling out of use because of difficulty of handling and limited effectiveness.

Alternatives to chemical treatments for sea lice

6.96 Biological control of sea lice is possible by introducing a second fish species, wrasse, into salmon cages. Wrasse eat the lice and provide continuous control that cannot be achieved with chemicals except through repeated doses. The cost of using wrasse is about the same as conventional treatments, although there can be problems of welfare and disease transfer if the wrasse do not thrive under fish farming conditions.[70] It may also be difficult to source or breed sufficient quantities of wrasse and this approach is not widely used in Scotland. Other alternatives include new vaccines, novel products to increase salmon resistance and the use of lights to encourage salmon into deeper water where lice infestations are smaller.[71]

Anti-foulants

6.97 The growth of seaweed, barnacles, mussels, tunicates, etc., on marine structures is a significant management problem. They increase the weight of floating structures, block the mesh of nets reducing water exchange, and reduce the diameter of pipes so decreasing flows and increasing pumping costs.

6.98 In the 1980s, anti-fouling coatings were based on tributyl tin (TBT) with additional herbicides. TBT was subsequently found to have endocrine-disrupting effects on dog whelks, oysters and other molluscs at very low concentrations. The British Ecological Society reported that, although most TBT problems in aquaculture revolved around contamination of farms by nearby harbours and marinas, there have been cases of mussel farms being adversely affected by TBT treatment used at upstream salmon farms.[72]

6.99 Copper has replaced TBT treatment and 19 of the 24 anti-foulant products registered for use in Scottish aquaculture include copper, copper oxide or copper sulphate.[73] When such coatings are used on nets, copper can slowly leach out into the water. In addition, cleaning the nets can cause copper to accumulate in the sediment below the cleaning facility.[74] SEPA measured copper and zinc concentrations at fish farms within its West Region during 1996 and 1997.[75] Sediments directly beneath the cages and within 30 m of the farms were severely contaminated by copper and zinc at seven of the ten farms surveyed, with 'probable' adverse effects predicted on the benthic invertebrate community at these sites.

6.100 High levels of copper and other heavy metals in seawater are toxic to marine organisms.[76] The long-term ecological implications of high metal concentrations in fish farm sediment are unknown. Sediment biogeochemistry and physical characteristics influence the accumulation, availability and toxicity of sediment contaminants, such as trace metals, to benthic invertebrates. Even when metal concentrations in sediments substantially exceed background levels, metal bioavailability may be minimal and adverse impacts may not occur. To date there is little evidence that copper is transmitted through the food chain, as it is present in an organic form that is not directly toxic.[77]

6.101 New anti-foulant materials are under investigation and some are at a marketable stage.[78] New products tend either to use natural compounds, such as capsicum and enzyme-based derivatives, or slow fouling materials, such as polyurethane polymers or Teflon coatings. Early results in large-scale field trials of many of these anti-foulant mechanisms have been encouraging. Their use is likely to be extensive in the aquaculture industry in the future, potentially replacing toxic copper-based substances. Little is known however about the environmental impacts of some of the alternatives since the only data available are from manufacturers' own marketing information.

Conclusions on diseases and chemical pollution

6.102 Fish farms often have a detectable impact on the seabed in their immediate vicinity, but our knowledge of the fate and effect of common pollutants is far from adequate.

6.103 The Commission's Twenty-fourth Report recommended an improved scheme of 'reconnaissance monitoring' to detect adverse effects of chemical products in the environment, supported by improved environmental epidemiology and increased reporting of adverse effects. The Report also advocated greater substitution of hazardous chemical products, alongside measures to reduce the quantity of such products in use. Given the many uncertainties that still remain **we recommend that the Scottish Executive and the Scottish Environment Protection Agency should commission further research and monitoring into the long-term environmental effects of using therapeutants and copper anti-foulants in aquaculture, and into alternatives to such compounds.**

SEDIMENT AND WATER QUALITY

6.104 Most of the particulate effluent from cage farms consists of faecal material and uneaten fish feed, which is eventually broken down in the wider environment. The amount of faeces and feed depends on the type of food, temperature and disease status. There have been attempts to design bags to sit within cages and collect this waste material so that it does not enter open water, but these methods have not yet proved successful in Scotland and are associated with higher fuel and running costs.

6.105 Another way to reduce waste per unit of production is to ensure that diets are easily assimilated and to give good feed conversion ratios (product produced per unit feed). In this way, feed losses have been reduced to less than 5% in well-run farms.[79] This is important since fish feed is extremely energy-rich, and causes much greater organic enrichment than faeces on a weight-for-weight basis.

Effects on sediments

6.106 The solids from cage farms consist of a range of particle sizes and densities, with a range of settling velocities. The particles are affected by water currents that may vary with depth. The resulting dispersion may cause settlement well away from the farm, but usually the highest deposition rates are in the immediate vicinity. The eventual site of deposition will depend on local bathymetry, water movement and the way finer particles clump together to form larger, more rapidly settling particles. Bacteria may break down slow settling particles, leading to the release of nutrients into solution. On reaching the seabed, these particles may become incorporated into the sediment or may be re-suspended by near-bed currents, thus further dispersing them away from the cages.

6.107 Addition of organic wastes to sediments immediately causes an oxygen drain as bacteria degrade them. When the oxygen demand caused by the input of organic matter exceeds the oxygen diffusion rate from overlying waters, sediments become anoxic and anaerobic processes dominate. This creates 'dead-zones' where marine life fails to thrive.

6.108 The rate at which sedimentary ecosystems recover following the removal of cages or the cessation of farming is of considerable interest, particularly as the fallowing of sites and rotation of cages has become recommended practice in many areas. In a Scottish study of benthic recovery, communities adjacent to cages returned to near-normal 21-24 months after farming ceased, but to date no study has looked at recovery processes over a sufficiently long period to be certain about recovery times.[80]

6.109 Although the major effects of fish farming on sediments are relatively well understood, more research is needed into the dynamics of waste input, interactions between microbial and macrobiological processes in the sediment, how these influence the chemistry of the sediments, the physical processes of oxygen supply and sediment resuspension, and mixing by water currents. These interactions take place against a background of seasonal changes and a two-year farming cycle that results in great variation in the supply of organic materials to sediments.

6.110 In addition, inter-annual variability in biological factors, such as the supply of invertebrate larvae, could have effects that are not yet well understood. These aspects are important as they affect our understanding of: the assimilative capacity of sediments with respect to fish farm wastes; how chemical contaminants in sediments are redistributed to the wider environment; and how sediments consume oxygen and release dissolved nutrients into the water column.

6.111 In evidence, the British Geological Survey (BGS)[81] suggested that a set of indicators should be developed to assess the impact of the organic load placed on the seafloor by fish farms. BGS proposed that the variation in geochemical status over time should be examined by analysing sediment cores taken at proposed fish farm sites. This would help assess whether the environment could process nutrient and chemical wastes from aquaculture. Existing sites could also be assessed for nutrient cycling by examining transport pathways through sediment for the chemicals and organic matter released into the environment through fish farming. Seabed sediment and habitat mapping would assist in predicting the consequences of locating fish farms in a given area.

Biological oxygen demand

6.112 Extensive aquaculture relies on large volumes of water operating at low flow rates. In contrast, intensive systems require lower volumes to support the stock but higher flow rates. Water throughput rates of between 25,000 m³/tonne and 250,000 m³/tonne are commonly reported. In many systems, especially marine systems, there is no apparent consumption of this water; it is the alteration in water quality that is important.

6.113 A higher throughput of water results in a more dilute effluent from the facility. Aquaculture effluents typically have a biological oxygen demand (BOD_{14}) of 2-12 mg/l, which compares well with treated sewage effluent (20-60 mg/l), dairy effluent (1,000-2,000 mg/l) or silage effluent (30,000-80,000 mg/l). However, based on BOD per unit of output, aquaculture compares unfavourably, producing 200-1,300 kg BOD/litre output compared with 2-4 kg BOD/litre output for a dairy enterprise.[82]

6.114 The significance of this effluent depends on the degree to which it is dispersed in the environment and the ability of the environment to cope with it. Effluent is generally controlled through discharge consents, which limit aquaculture production at many marine sites. On land, businesses would be charged for discharge of wastewater. This is not the case for fish farms using seawater. Any system of charging for waste output however could have a major impact on the financial viability of the aquaculture industry as it is presently structured.

Nutrient enrichment

6.115 Figure 6-VII shows aquaculture's contribution to nitrogen and phosphorus discharges relative to total nutrient discharges. Although other sources make up of most of the nutrient input, aquaculture's contribution is significant in some areas. In Norway, for example, it has been estimated that the nutrient enrichment due to a fish farm is equivalent to that from a small town of up to 7,000 people.[83]

Figure 6-VII

Contribution of marine and brackish water finfish culture to total anthropogenic coastal discharges in selected countries[84]

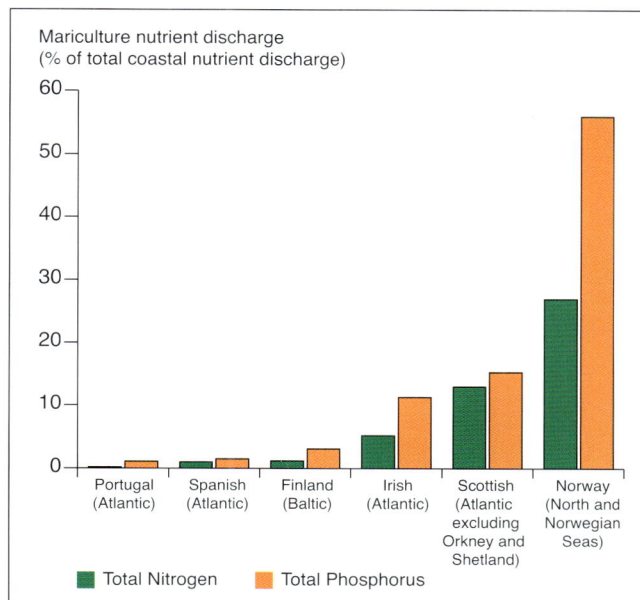

Aquaculture's contribution may also become relatively more important in future if other sources are reduced as a result of clean-up measures.

Toxic algal blooms

6.116 Much scientific literature has been published on algal blooms but it is still unclear why blooms suddenly appear, why they are sometimes toxic and lead to shellfish poisoning, or what the relationship is between fish farms and algal blooms. Blooms are thought to occur naturally, but also to be linked to nutrient pollution in some cases.

6.117 A review has listed six possible types of phyto-plankton causing different kinds of adverse effects:[85]

- plankton that produce a species-specific toxin that is concentrated by intermediate consumers in the food chain, resulting in a named human disease. These are often harmful even at low concentrations. An example is *Alexandrium tamarense*, which can cause Paralytic Shellfish Poisoning;

- mildly toxic plankton, such as *Gyrodinium aureolum*, which cause direct harm at high concentrations, for example, by killing fish;

- plankton that cause mechanical damage, for example, by clogging power plants inlets;

- plankton that when in bloom create 'red tides', 'white waters' or foaming, causing nuisance rather than harm – e.g. *Phaeocystis*;

- plankton that produce dimethyl sulphide; and

- extreme blooms that produce oxygen depletion on decomposition.

6.118 Harmful algal blooms have led to fishery closures in affected areas and bans on shellfish consumption. In the UK there have been outbreaks of Diarrhetic Shellfish Poisoning (DSP) and Paralytic Shellfish Poisoning (PSP). Figure 6-VIII shows the presence of shellfish toxins across Europe from 1991-2000. The maps indicate that most reported incidents are in the open sea, away from polluted or eutrophic coastal waters. Scottish waters appear to be affected by all groups of toxins. The Irish Sea and the southern North Sea, which have more nutrient enriched waters, feature comparatively few outbreaks. Like Scotland, western Spain, Portugal, Ireland and west Norway experience the highest frequency of blooms.

Figure 6-VIII
Presence of shellfish toxins across Europe, 1991-2000[86]

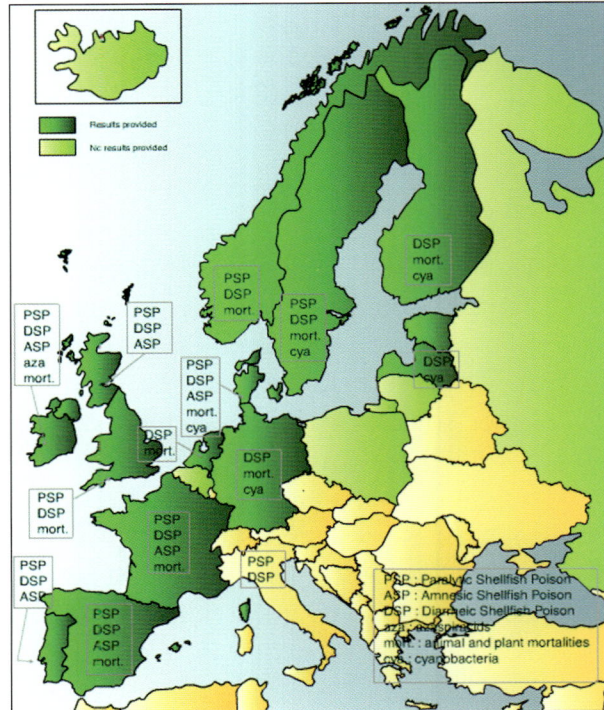

6.119 During 2003, consultants appointed by the Scottish Executive reported on the interaction between fish farming and algal communities in Scottish waters.[87] The report noted that:

> "In our opinion, it is very unlikely that fish farming should have a large-scale impact on the occurrence of harmful algal blooms, particularly on toxic algae, which are related to shellfish poisoning. Occurrence of such blooms, in general, appears to indicate that they are more common in pristine than in enriched waters and that they appear independently of fish farming activities."

6.120 The report nevertheless made it clear that fish farming was not a harmless activity and that excess nutrient loads were coupled to increased plankton production. While localised sedimentation and recovery effects underneath individual cages have been studied, there have been few long-term studies of changing nutrient and oxygen fluxes.

Conclusions on nutrient and water quality issues

6.121 Aquaculture has a wide range of impacts on the marine environment, including on sediments, nutrient cycling and water quality. **We recommend that the Scottish Executive and the Scottish Environment Protection Agency should develop a set of indicators to describe the pollutant load from fish farms, and that the performance of fish farms against these indicators should be monitored and published.** The indicators could include aspects such as organic load on the seafloor and the capacity for nutrient processing.

INTERACTIONS WITH WILDLIFE

6.122 Predators and scavengers cause considerable problems for aquaculture, including killing or wounding of farmed fish, increased stress and disease transfer. Predators include squid, birds and aquatic mammals. Predators are present at most farms due to the ready supply of food or, in the case of marine cage aquaculture, to wild populations of fish attracted by uneaten food waste. Cages may also serve as a roost or observation site for opportunistic scavengers. Although there have been several studies on predation in aquaculture, there are few reliable figures on the economic impacts. It has however been estimated that predator-related losses for the Scottish salmon industry in 1987 were £1.4-1.8 million.[88]

6.123 The main approaches to predator management (which focus on minimising the economic impact of predators) are:

- exclusion (predator netting and other physical barriers);

- harassment (acoustic deterrent devices, scaring devices and guarding); and

- removal (licensed shooting, trapping).

6.124 Removal methods are rarely considered appropriate, and are not a key target for technology development. Most outdoor farms deploy perimeter fences to protect against terrestrial predators and wires or netting over fish tanks, ponds and cages to protect against predatory birds. For water-based farms, underwater netting (on sides and occasionally bottoms) may also be used, for instance to protect mussel farms from eider ducks, or fish farms from diving birds and sea mammals. They can be reasonably effective if correctly sized and installed, but can be destructive if predators are caught in the netting. Scaring devices (usually acoustic) can be used against dolphins, seals, otters and bird predators. Acoustic deterrent devices (ADDs) are reportedly effective for up to two years, though the effect appears to diminish with time. This is especially so with seals who tend to learn through previous hunger and successes that these intense signals can be withstood.

6.125 Long-term impacts of ADDs on marine mammals are not conclusively known; however, seals and sea lions that are not deterred by the devices may experience hearing damage at close range. These sounds may also interfere with communication signals between animals and with passive listening abilities. The devices have also been linked to declines of baleen and killer whales, leading to a ban on their use in British Columbia. More complex, modern systems provide a 'ramp up' from a low level to reduce the chance of hearing loss in mammalian predators. Acoustic scarers for birds mostly involve sudden loud noises, with similar problems of habituation and greater issues of sound pollution. A laser rifle that scares away birds rather than killing or wounding them is also available and may be an avenue for further technology development.

FISH WELFARE

6.126 The welfare of farmed fish is important in its own right, but also has knock-on effects for the environment, as stressed fish can be more prone to disease and therefore need more treatment with therapeutants and antibiotics. In recent years there have been at least three reports addressing these issues. The first was the 1996 Farm Animal Welfare Council report[89] that made recommendations concerning the welfare of farmed salmon and trout; these are summarised in box 6B. The report signalled concerns about staff training, prompt response to problems, and the handling and transport of fish.

6.127 In 2002, Compassion in World Farming issued its *In Too Deep* report on the welfare of farmed fish.[91] This included recommendations on maximum stocking densities, lice treatments, the use of genetic manipulation (triploidy) and genetic modification of fish, and slaughter methods.

6.128 Defra's Science Directorate also held a workshop on farmed fish welfare during 2002.[92] Discussion focused on three key issues: ways to record welfare; methods of slaughter; and stocking density. The meeting examined the use of upper biomass limits, as stocking densities can be increased on farms with better water quality.

BOX 6B	SUMMARY OF WELFARE CONSIDERATIONS IN AQUACULTURE[90]

There should be adequate training for people running and working in aquaculture.

Fish should be conditioned to the proximity of staff to reduce fear responses.

Handling of fish should be kept to a minimum.

There should be a prompt response to welfare problems.

Dead and moribund fish must be removed daily where possible.

Water quality should be assessed frequently.

Fish farmers must record live fish movements on or off the site, fish mortality and medicine use and keep details of feeding and management regimes.

Fish populations should not be graded more often than is absolutely necessary, since most kinds of grading are likely to be stressful.

The welfare of fish in transit by road should be checked at least every four and a half hours.

Fish in transit should be kept in conditions that will allow them to survive a journey at least 50% longer than the anticipated duration.

Oxygen levels must be constantly monitored during transit and excessive changes in temperature and pH in transport tanks must be avoided.

Salmon producers should follow the Salmon Farming and Predatory Wildlife Code of Practice produced by the Scottish Salmon Growers Association (now known as Scottish Quality Salmon). The industry should consider developing codes that cover transport issues.

The welfare of farmed fish should not be adversely affected by limiting the availability of vaccines or medicines which are effective and do not pose food safety or environmental hazards. Ways must be found to achieve rapid access to medicines and vaccines in emergencies. There needs to be an increase in the range of vaccines and medicines approved for the treatment of fish. Well-tried and efficient medicines must not be lost unless adequate alternatives are available.

Mutilations which involve removal of sensitive tissue should not be carried out on farmed fish.

If a fish is to be stunned, the stun must cause immediate loss of consciousness that lasts until death which should happen without delay.

FISH RANCHING

6.129 Free-range aquaculture has developed from sea ranching of commercially important fish or shellfish species. Hatchery-reared stocks are released into marine or brackish waters where they can propagate or grow on natural foods, until they reach harvestable size, when they are captured using traditional fishing techniques. Twenty-seven countries now employ ranching as an alternative to aquaculture. Japan leads the world with a total of 80 ranched species including Pacific salmon, cod, blue crab and grouper. Salmonids are the most widely ranched group of fish. Approximately 400,000 tonnes of salmon are ranched in Japan, Russia and North America annually.[93]

6.130 Ranching has advantages and disadvantages over traditional pen aquaculture, primarily in terms of initial cost benefits. Feeding, the greatest part of growing costs (6.22), is only necessary during juvenile period when fish are grown in tanks. Since these tanks are only required through these early stages, they need only be of limited capacity. Benefits are also shown by the energy efficiency ratios, which are 13% and 25% for tank production and ranching of salmon, respectively. The great disadvantage however is the limited survival or return of fish of only 1-15%, compared with cage culture, in which about 90% of fish grown are harvested.

6.131 Several enhancements to ranching techniques have been applied, including the use of barriers to fish migration or artificial reefs for provision of habitat. Under the 'Free Fish Farming at Sea' system, hatchery-reared fish are conditioned in either tanks or cages to specific acoustic signals and subsequently released into the open sea. A number of additional facilities can be installed, such as artificial reefs to provide stimulus and conditions for the fish to stay in the general locality. These may also provide additional habitat for prey and protection from predators. The success and wider adoption of ranching depend on introducing additional controls to restrict harvesting by non-target fisheries, the generation of waste from feeding stations, and competition and genetic mixing between ranched and wild stocks.[94]

FARMED SHELLFISH

6.132 Table 6.4 shows production of farmed shellfish in the UK in 2001. A review carried out for the Scottish Executive,[95] estimated the first-sale value of the industry in Scotland to be around £5 million, not including the revenue from managed wild stocks. This review also noted the trend for increased overall production, an increase in the total number of operational businesses, but a slight decrease in the number of operational sites. In its evidence to the Commission, the Environment Agency pointed to the growing number of applications for shellfish culture, particularly for mussels along the east coast of England from the north-east to Kent.[96] The Agency considered that these could increase significantly over the next decade, although others have suggested that water quality might be a limiting factor.[97]

Table 6.4

Farmed shellfish production in tonnes for the UK in 2001[98]

	Scotland	England and Wales	Northern Ireland	UK Total
Pacific oyster	247	225	386	1,333
Native (flat) oyster	4	127	208	
Oysters (total)	251	352	386	1,541
Scallops	39	-0	–	39
Queens	58	–	–	58
Mussels	2,003	13,367	1,095	14,322
Clams	–	31	–	176
Cockles	–	105	–	147
				16,283

6.133 Shellfish species cultured in the UK are mainly filter feeders, extracting their food from the water column. Juveniles are supplied from hatcheries, in the case of oysters, or collected from wild populations. In the case of mussels, spat is collected in the wild. Shellfish receive no additional feed or medication in the grow-out phase. There have been problems with the collection/harvesting of mussel seed for relaying and on-growing where this occurs on a large scale. For example, inter-tidal mussel beds in the Wadden Sea almost disappeared during the late 1980s due to a combination of exploitation and low spatfall. This had a negative impact on bird populations for which the small mussels are a source of food, leading to increased mortality in eider duck and reduced breeding success for oystercatchers.[99] There have also been problems in some other areas of Europe. The carrying capacity has been exceeded in some areas of France, for example Marennes-Oleron where Pacific oysters are cultivated, and in some of the Spanish rias in Galicia as a result of intensive suspended mussel production.[100]

6.134 There are three principal environmental considerations in relation to shellfish farming: water quality, algal toxins and, in some areas outside of the UK, carrying capacity.

WATER QUALITY

6.135 All bivalve mollusc production areas are classified by UK regulations.[101] Areas are classified A, B or C depending on the number of faecal coliforms present. The industry is, therefore, highly dependent on the maintenance of good water quality. In evidence to the Commission, the Environment Agency[102] considered that there was a lack of robust microbiological water quality criteria to facilitate the design of remedial schemes to improve shellfish harvesting areas and ensure that hygiene requirements for shellfish products can be met. Other agencies[103,104] have also highlighted the need for action to improve water quality in shellfish growing areas.

ALGAL TOXINS

6.136 The second major constraint on many businesses is the prevalence and duration of closures on harvesting caused by the presence of algal toxins (6.118). Most notably, with respect to mussel growers, prolonged closures caused by the presence of algal blooms that cause Diarrhetic Shellfish Poisoning have threatened to close companies in north-west Scotland over two recent growing seasons. There have also been seasonal closures caused by the presence of blooms that lead to Paralytic Shellfish Poisoning. The scallop cultivation industry has been similarly affected by prolonged and widespread closures since 1999, because of the detection of the toxin that can cause Amnesic Shellfish Poisoning.

CARRYING CAPACITY

6.137 The significance of the depletion in phytoplankton as a result of the filter feeding activity of cultivated molluscs is dependent on their biomass and the prevailing hydrographic conditions. Studies on the reduction in phytoplankton biomass by mussel populations have demonstrated depletion in the range of 10% to 74%. In evidence the Joint Nature Conservation Committee[105] noted that mussel cultivation already takes place in protected areas (e.g. Natura 2000 sites) and that there are proposals for further expansion in these areas, some directly on protected benthic habitats (i.e. features for which the Natura 2000 sites were designated, e.g. Pembrokeshire Marine candidate Special Area of Conservation), thereby changing the benthic communities. Mussel biodeposits (i.e. faeces and pseudo-faeces) may contribute significantly to the total suspended load in estuarine and coastal environments and increased sedimentation could lead to anoxia and a change in the infaunal community.

6.138 As with finfish farming, the appropriate siting of shellfish installations is an important consideration in terms of limiting potential impacts. This could be assisted by a better understanding of the carrying and assimilative capacities of coastal waters for shellfish farming, and a research project has begun to design possible criteria.[106] Although such detailed information is not yet available, there are some controls on shellfish farming. New facilities require licences or leases from the Crown Estate or other relevant authorities. They are, however, exempt from the Environmental Impact Assessment (Fish Farming in Marine Waters) Regulations 1999, and are not considered in detail by the locational guidelines for fish farming (6.139).

LEGAL FRAMEWORK FOR AQUACULTURE

6.139 The siting of new fish farms is governed by the *Locational Guidelines for the Authorisation of Marine Fish Farms in Scottish Waters* (2003)[107] which divide potential sites into three categories, where:

- further increases in production are not recommended; or

- there is limited scope for further increases in production; or

- increased production could be considered without due concern.

6.140 There is a presumption against further development of aquaculture on the north and east coasts of Scotland. The majority of Scotland's wild salmon are found in these areas, and the guidelines are intended to reduce the potential for interaction between wild populations and escapes of farmed fish. However, while numerically the largest numbers of fish exist on the east coast, there are as many populations on the west coast.[108] These are important for the overall genetic diversity of the species as they are more varied in their life history.

6.141 The legislative framework for aquaculture is described in chapter 4. A key element missing from existing process is the requirement for every applicant to carry out an environmental impact assessment before any application for a new or extended fish farm is granted. Marine fish farming falls under schedule 2 of the Environmental Impact Assessment (EIA) Directive. The relevant authority is required to consider whether a project is likely to have a significant effect on the environment before deciding if an EIA should be carried out by the applicant. If and when an EIA is carried out, it involves comparing recent monitoring data from the site against statutory environmental quality standards (EQSs) and background concentrations. An application may not be granted if concentrations of dissolved oxygen, ammonium, nitrogen dioxide or nitrate do not comply with the EQS. Existing farms are also regularly monitored to check nutrient concentrations in winter, and chlorophyll-a levels in summer where the biomass exceeds 1,000 tonnes.

6.142 The EC Directive on Environmental Impact Assessment (EIA) applies to larger fish farms and aims to ensure that the environmental implications of development are fully considered. There is concern however that this regulation is not functioning well.[109,110] For example, only a few Environmental Statements that describe a farm's likely impact have been submitted, and those that have been produced are of variable quality. In addition, the EIA process does not appear to take into account all possible disease, parasite and genetic impacts. As a result, the Strategic Framework for Scottish Aquaculture (2003) proposed an independent review of application of the EIA Directive and further work with the industry to improve standards. **We recommend that an environmental impact assessment should be carried out for every application for a new or significantly modified fish farm.**

6.143 There has been some concern that part of the legal framework relies on the Crown Estate to grant a seabed lease in order to control fish and shellfish farms (or a licence from local councils in Orkney and Shetland). This has attracted criticism in terms of public accountability and possible conflict of interest, since the Crown Estate also has a duty to maximise its revenue.

6.144 The Water Environment and Water Services (Scotland) Act 2003 includes provisions to facilitate the transfer of planning responsibility for aquaculture developments to the planning authorities. This will bring to bear the established principles and practice of land use planning in an open and accountable way, separating development control and landlord functions in doing so. Local Authorities will need to develop expertise in controlling marine development, and there is a need for clear guidance to foster a consistent approach, as far as possible, between different authorities.[111]

6.145 The EC Water Framework Directive will have a major influence, requiring significant changes to the present regulatory framework with the objective of achieving good status for controlled waters through a system of River Basin Planning. There is still uncertainty though as to how the Water Framework Directive and Water Environment and Water Services Act will apply in practice. Existing industry codes of practice have gone a considerable way to improving the way in which finfish aquaculture is managed in Scotland; however these codes rely heavily on voluntary compliance and there are at present no obvious ways of requiring fish farm companies to operate in the best interests of themselves and their neighbours. There should therefore be a requirement in the regulation of finfish aquaculture (both freshwater and marine) to make either operational or locational consents conditional on adopting certain accepted management practices enshrined in industry or regulator codes of practice. It is hoped that secondary legislation that might emerge from the Water Environment and Water Services Act would address this issue.[112]

6.146 Further problems and opportunities may also be encountered with the growing interest in organic production and in the production of other finfish species (cod, halibut, haddock). A strategic approach will therefore be essential for planning how the coastal zone is going to be used by all forms of finfish aquaculture and how this in turn relates to the general quality of the environment, other aquaculture forms (shellfish, etc.), other commercial users (capture fisheries), other stakeholders (recreational interests) and indeed less easily defined concepts such as aesthetics and 'wild land'. These ideas fall under the concept of Integrated Coastal Zone Management (10.10), and in Scotland the best practical examples of this have been early attempts by the Highland Council to establish Aquaculture Framework Plans.

6.147 The European Commission has also drawn up a strategy for the sustainable development of European aquaculture. This outlines action across a range of areas over the next ten years to develop a sustainable and stable aquaculture industry, to guarantee long-term secure employment and development in rural areas, and to ensure the availability of healthy and safe products.[113]

CONCLUSIONS

6.148 As we have discussed, aquaculture can cause several localised environmental impacts. In the OSPAR region many of these are subject to regulation and most may be manageable. The exceptions appear to be disease and the rather newer concern about the genetic impact of farmed fish on wild populations which is less well understood and does not have an obvious parallel with other systems or types of farming.

6.149 The central issue, however, is the extent to which aquaculture could or should compensate for the decline in global fish stocks. The real potential for growth in aquaculture appears to lie not so much in marine fish farming in developed countries (although there may be room for some growth), but in freshwater farming of a wider variety of species, particularly in developing countries.

6.150 Evidence suggests (table 6.2) that some of the pelagic populations exploited for fishmeal are in a relatively healthy state at the moment, although others clearly are not. However, there a number of serious questions to which we do not have complete answers:

- Are the fisheries that are exploited for fishmeal largely independent of those that could be used for human consumption or is there significant overlap in the food chain?

- Are the populations of forage fish truly sustainable in the medium and long term given the pressures from aquaculture and the uncertainties related to climate change and other factors?

- Even if fish populations are healthy, what are the environmental impacts of removing forage fish from the ecosystem? For example, the sandeels of the North Sea are important in sustaining bird populations in the North Sea but have been heavily exploited by the Danish industrial fishing fleet.

- To what extent is it really necessary for aquaculture to use carnivorous fish fed on fishmeal? This is the only way at present to provide carnivorous farmed fish with the high level of protein they require and to ensure that the final product contains the long-chain n-3 PUFAs that are beneficial for human health. We discuss elsewhere the prospects for finding alternative sources of these compounds (chapter 3).

6.151 With wild fisheries declining due to over-exploitation, aquaculture is considered by some to be the best and most logical solution for maintaining and even enhancing supplies of seafood products. As aquaculture is a relatively new industry, without a well-established tradition in rural and coastal areas, it has naturally provoked a number of conflicts of interest and concerns over environmental issues as the scale of the industry has increased. Many of these issues are under investigation or are the subject of regulatory controls. At a time of increasing awareness of the health benefits of eating fish and with capture fisheries under increasing scrutiny with regards to sustainability, further expansion of the aquaculture industry seems inevitable. Our analysis suggests that there are three major issues that remain to be dealt with:

- the use of fishmeal and oil in aquaculture feed;

- interactions between farmed and wild fish populations; and

- disease (in particular sea lice).

6.152 We are convinced that sustainable management and technical innovation, such as the search for feed substitutes, are essential to ensure that the growth in aquaculture does not cause unacceptable environmental impacts. We have offered recommendations in respect of each of these three key areas to encourage the UK aquaculture industry to become more sustainable.

Chapter 7

UNDERSTANDING THE IMPACTS OF FISHING ON THE MARINE ENVIRONMENT

How can scientific knowledge help us understand the impacts of fishing on the marine environment? How should fishing be managed to avoid irreversible changes to marine ecosystems and loss of biodiversity?

INTRODUCTION

7.1 The seas have been treated as a source of food and income to be exploited as fully as possible. This has led to a reluctance to fish within sustainable limits and problems of over-exploitation. The harmful consequences of this approach for the fishing industry and the marine environment have been outlined in chapters 3 and 5. As the damage has become increasingly clear, scientists, policymakers and others have begun to recognise the need to review the way we think about the oceans. As we have seen in chapter 2, the seas are a set of interconnected ecosystems whose protection is essential to preserve a wide variety of environmental and other functions. The emphasis should now be on managing fisheries in the context of the whole marine environment.

7.2 Describing the marine environment in a way that can be used to help fisheries management remains a major challenge. Models of ecosystems are complex and contain many uncertainties. In order to deliver responsible governance of fisheries, the scientific community needs to produce advice that is relevant, responsive, respected and right,[1] but it must also provide a better means of dealing with these uncertainties about the marine environment that can be incorporated into the fisheries management process.

7.3 In this chapter we discuss the scientific advice given at present to fisheries managers and what the future prospects are for a holistic management regime that considers marine ecosystems, rather than just the status of commercial fish populations. We then describe the changes that should be made to the management framework to use the available scientific advice and minimise the environmental impact of fisheries. In chapters 8, 9 and 10 we detail how the components of this framework could be implemented.

FISHERIES MODELLING AND SCIENTIFIC ADVICE

Current situation

7.4 At the present time, the main role of government fisheries scientists is to provide data for stock modelling to inform the process of deriving total allowable catches; as such they can be thought of as 'fish accountants'.[2] Conventional approaches to fisheries management are deeply rooted in the models and methods of traditional population biology. Fisheries are reduced into their component parts with populations assessed species-by-species. Population data (derived from catch data) are entered into models from which the future status of the fishery and its yields are estimated.[3] Research has highlighted the problems associated with this over-simplification in modelling of marine organisms. For example, there is no explicit representation of other trophic levels, of the physical and chemical state of the sea, or of the spatial domain. There are also difficulties in making assumptions about how changes in the natural environment will affect natural mortality and recruitment of organisms. These difficulties increase where higher organisms, such as fish, are involved, since they have more complex life histories and behaviour than more simple organisms (box 7A).

BOX 7A   **FISH POPULATION MODELS**

The International Council for the Exploration of the Seas (ICES) uses fishery landing data and independent survey information from fisheries scientists on the number, size and age class of fish in order to estimate the proportion of fish populations that can be caught each year, consistent with what statistical models suggest would be sustainable stocks.

Fisheries biologists have been engaged in scientific analysis of fish populations since the middle of the nineteenth century. In the mid-twentieth century, a set of single-species models was created to define the total maximum sustainable yield for a given fish population. Fisheries management has made extensive use of these single-species models, which deal in detail with the age structure and recruitment of individual fish stocks. The core methodology used by ICES is virtual population analysis (VPA), developed in the 1960s. This modelling method assumes that there is a link between landings and the level of fishing mortality. The assumption allows catch predictions to be produced by fisheries scientists, who attempt to forecast fishable biomass from hind-casts of mortality estimates. The data are used to estimate the proportion of fish in each year class that survive from one year to the next; and the estimates are summed across year classes within each year to give an estimate of spawning stock biomass by year. Values for mortality are divided into natural mortality and that due to fishing.[4] However, the accuracy of single-species stock models is limited as they do not take account of interactions with other populations, species or the wider ecosystem, and accurate estimates of natural mortality are extremely difficult to make.

As a result of some of these shortcomings, fisheries management for most species is now based multi-species virtual population analysis (MSVPA), which models the structures of several interacting fish stocks. This has been used to reconstruct age-size and time-dependent estimates of trophic flows and mortality rate components, using the assumption that historical abundances can be inferred by back-calculating how many organisms must have been present in order to account for measured and estimated removals of those organisms over time.[5]

ICES also provides medium-term projections (5 to 10 years) of the state of various fish stocks, which are presented as probability distributions of future spawning stock biomass for different exploitative scenarios.[6] Uncertainties, however, arise in such projections because of limitations in the data, in the process representations in the model, and in its parameterisations.[7] In the case of medium-term forecasts, ICES estimates of the spawning stock biomass of various species have generally failed to characterise adequately the real uncertainty in future stock sizes.[8]

7.5 Models used in fisheries management (box 7A) address only a subset of the underlying factors that regulate ecosystem processes. They assume that populations are in equilibrium and functional groups are invariably aggregated over different species or age groups. When considering competition between a target fish and a top predator, for example, a simple model would predict that with the removal of the top predator, prey release would lead to an increase in the target fish stocks, thus benefiting the fishery. If, however, there exists a secondary fish predator that competes for the target species, but is also eaten by the top predator, a cull of the top predator would lead to a lag increase in the secondary fish predator, and ultimately a further decline in the target fish stock. This example illustrates the danger of omitting an important interaction when trying to make meaningful predictions about complex ecosystems.[9] Fisheries models also focus on commercially important species and often ignore important ecosystem components of limited direct commercial value.

7.6 Beyond the limitations in the basic assumptions underlying the statistical models, illegal landings and unrecorded discards are not included. There is also often a time-lag (of up to two years) between data collection and its use for the management of fish stocks. Thus, the year-class of fish used as the basis for setting quotas may already have been captured. The modelling process used to provide scientific advice for the setting of total allowable catch quotas therefore has major limitations and can, at best, only be described as having had marginal success for most fish species (7.7). The collapse of key fisheries such as the north-west Atlantic cod (box 5B) highlights the underlying weakness of the existing regulatory system. The EU has attempted to incorporate a more precautionary approach to setting quota levels for fisheries management. The need for fisheries to define precaution in an operational sense, has led to an urge to quantify uncertainty and identify the limits of knowledge,[10] but attempts to provide 'hard' predictive systems through population assessments and modelling have not been particularly successful. Fish population models can only ever be diagnostic of a limited number of factors rather than truly predictive.

7.7 An oft-quoted example of the success of single species management relates to some pelagic fish populations, such as North Sea herring and the Norwegian spring-spawning herring populations. These species have biological attributes of early maturation, short life-span and rapid rates of reproduction that lend themselves well to recovery from fishing pressure, unlike many demersal fish species. Both populations were heavily overfished and plummeted to critical levels in the 1970s. But international management plans were agreed on the basis of ICES scientific advice and, helped by some successful recruitment years, both stocks have now recovered. In both cases, the crucial aspect for recovery was that the management plan gave the fish a chance to spawn and reproduce through the application of complete bans on fishing and other technical measures,[11] that are easier to enforce in single species pelagic fisheries, and which were not undermined by by-catch of the species in other fisheries. The removal of species that are predators of herring from the ecosystems is also likely to have played a role in the recovery of these herring populations. The current fishing pressure on these stocks is considerably less than has been the case historically and now only a precautionary level of 15% of stock is taken in any given year. There are few species with these characteristics in the Common Fisheries Policy area, and the herring successes have been the exception rather than the rule. For example, the population of North Sea mackerel, fished to equally low levels in the 1970s has failed to recover. As described in chapter 5 most other fish populations managed under the Common Fisheries Policy are in a parlous state despite measures aimed at stock recovery.

7.8   Although the failure to apply scientific advice has contributed to the collapse of fisheries, it is also the case that the complexity and sensitivity of the ocean environment makes the forecasting of fish abundance an uncertain process[12] given the present state of scientific knowledge.[13] In addition, the process of making annual stock assessments of individual exploited fish species does not provide information on the changes in composition and nature of marine ecosystems caused by fishing. Nor can it tell us how fishing may continue without causing irreversible damage. The models used to assess population sizes do not take account of the evolutionary and ecological effects of fishing pressure on the marine organisms in the affected ecosystems, nor their basic biological responses such as habitat preference or mating behaviour.

7.9   Fisheries science, as presently used for regulatory purposes, may provide some limited information about the effects of fishing on fished species, but it is unable to provide information on the changes in the composition and nature of marine ecosystems caused by fishing.[14] One of the main causes of the environmental crisis outlined in chapter 5 has been the under-estimation of the complexity of marine ecosystems and institutional failure to deal with uncertainty about these ecosystems appropriately.[15,16] John Farnell (Director of Conservation, DG Fisheries, European Commission) has stated that "Scientists must be big enough to admit that their traditional approach to providing advice has to be abandoned in favour of a more comprehensive and robust approach ... managers must be big enough to accept that not everything can be done at once, proper management of scarce scientific resources requires clear choices about priorities".

Modelling marine ecosystems

7.10   As described above, management decisions for fisheries have always been made in the context of models of the system and are usually expressed in mathematical terms. To complement the move to a holistic approach, it would be desirable to have a better theoretical underpinning based on models that can more adequately reflect the interactions between different parts of the ecosystem rather than simply the population of single or multiple fish species. Ecosystem models may eventually develop sufficiently to help support management decisions and identify new options.

7.11   Marine ecosystems involve a range of trophic levels from top predators to phytoplankton, the chemical state of the ocean, including nutrient and oxygen levels, and the physical state, as reflected by the temperature, current, tidal mixing, bottom structure, etc. Given the complexity of such systems and the relative scarcity of data to develop, initialise and evaluate any model, great simplifications are necessary. Decisions have to be made about which features to represent explicitly, which to represent implicitly with representations of their feedbacks to explicit features ('parameterisation') and which features to ignore. The extent of any spatial or temporal resolution also has to be decided.

7.12   Models can be used to assimilate past data on aspects of an ecosystem and so obtain a view of the state of the whole ecosystem to the extent that it is presented in the models.

They can then be run forward in time to produce predictions of the future state of the ecosystem (box 7B). Models need to be refined and expanded to account better for uncertainties about ecosystems, evaluate the ecosystem level consequences of proposed actions and the setting of reference points for ecosystem indicators (7.18).[17]

7.13 As an example, macroecological modelling studies have suggested that in the North Sea the current biomass of fishes larger than 4 kg is only about 2.5% of its pre-trawling level, and the total biomass of fishes is nearly 40% lower than it would have been had trawling never taken place (5.36). Such modelling approaches are likely to prove useful in comparing the impacts of fishing across different ecosystems and fish communities.[18] A range of such models already exist or are under development. They are often highly simplified in their biology and usually do not explicitly represent the effects of important physical, chemical and spatial variation. Despite this, they are now useful as one input to the decision-making process and they should become increasingly so with time.

BOX 7B   **ECOSYSTEM MODELLING**

In an ecosystem model there should be some representation of the trophic levels from phytoplankton to fish and preferably also higher predators such as cetaceans, seals and birds. The specification in such a model of the interaction between the different species or biomass pools must be parameterised, but the experimental data used to determine the individual parameterisations directly are usually very limited or impossible to obtain. Therefore the parameterisations in such ecosystem models usually have to be tuned by comparison of the results for the whole model with past data.

Because of their temporal biological development, spatial movement and ability to make decisions, fish species are represented in some models in terms of a number of representative individuals, the state of each of which is predicted.[19] At the other end of the trophic scale, in an attempt to use only data from laboratory experiments, one model explicitly represents a very large number of individual phytoplankton and their interactions.[20] Interesting questions about the robustness and accuracy of the species-based models are being raised by such individual-based studies.

The nutrients that biological systems require can be specified and held constant in such models. Nutrient cycling and nutrient sources and sinks are, however, part of a more complete ecosystem model and their explicit representation has been found to be important for modelling in coastal regions.[21] Prediction of dissolved oxygen levels also forms part of some recent models. The representation of vertical and horizontal spatial dimensions allows a better description of the habitat and interaction of different species and trophic levels, from the benthos to the surface, at differing depths, temperature and other physical conditions. It also enables the representation of spatial heterogeneity in populations, diurnal movement of phytoplankton, and other spatial movements of fish, including seasonal migration. As an example, in the case of the ECOSPACE model, the ECOSIM model has been applied to each cell in a map with a representation of the communication between the cells, and equilibrium solutions have been obtained.[22] These and other marine ecosystem models are further described in appendix J.

7.14 The primary physical variable for a marine ecosystem is the water. Currents move nutrients and organisms around. In shallow seas, the movement backward and forward of tides leads to stirring of the water column and the suspension of nutrients. Such stirring is also provided by surface wind. Computer models of the physical system based on Newton's equations of motion are able to represent such features with increasing accuracy. There is also a feedback from the biology to the physics of the ocean as the penetration of solar heating in the ocean depends on the level of phytoplankton. Physical and biological models are starting to be coupled. In state of the art climate predictions a crude marine ecosystem model is included so as to represent its impact on the carbon cycle.[23] An example which is more relevant to the state of the marine ecosystem itself is the coupling of the physical model of the coastal seas developed by the Proudman Oceanographic Institution with the European Regional Seas Ecosystem Model (ERSEM, appendix J) embedded in it.[24]

7.15 Climate variability and change also affect marine ecosystems. In hind-cast mode the climate could be specified. Inclusion of the best possible information on climate variability and change on time-scales from years to many decades (possibly four dimensional oceanographic data from a climate model) in some predictions of the marine ecosystem will be important.

7.16 At present, however, single marine ecosystem model predictions are usually given as if the system were deterministic. Uncertainty in prediction arises due to limitations in understanding, in the representation of processes in numerical models, in the omission of aspects of the problem and in the data used to specify the initial state of the system. Some uncertainty is also inherent to stochastic variability in the marine environment, and outcomes can be highly sensitive to initial conditions. As a support for management, it is vital that the uncertainty be estimated in as realistic manner as possible. In order to understand the management implications of uncertainty we need to know both how great the uncertainty is, and how sensitive the system is to this kind of uncertainty.[25] One method for doing this is to perform an ensemble of predictions with initial conditions and parameters sampled from the ranges thought to be realistic.

7.17 Despite these problems, the creation of more complete marine ecosystem models will assist the integration of the management of all human activities that impact on ecosystems, making these models a desirable objective for the future. In addition to fisheries and human-induced climate change, the impact of nutrients and other pollution from rivers, shipping, oil and gas industries, and atmospheric deposition can in principle be represented. It is likely, however, that considerable effort and resources would need to be devoted to this area before models with useful quantification of uncertainty will be available to inform management decisions.

Use of indicators for fisheries management

7.18 A working group incorporating ICES, the Intergovernmental Oceanographic Commission (IOC) and the Scientific Committee on Oceanic Research (SCOR) is advising on the development of ecosystem indicators for fisheries management purposes. In the UK, the recent Review of Marine Nature Conservation has also recommended that the UK government should develop and agree indicators and procedures to monitor the state of marine biodiversity and the impacts of human activities. It is envisaged that such indicators would describe ecosystem state, activity specific ecosystem properties or impacts. The development of such indicators will be reliant on the available monitoring information and developments in ecosystem modelling.

7.19 An indicator provides a formal measure of performance that can be used to judge the success of a management strategy.[26] Indicators are matched to objectives, and progress towards these objectives is measured by use of reference points. These are values associated with specific states in relation to the indicator, known as limits and targets.[27] For example, for target fish populations, limit reference points may be set to ensure that society obtains the ecological, social or economic benefits that fisheries provide without compromising sustainability. ICES has defined a limit reference point as a value of a property of a resource that if violated is taken as *prima facie* evidence of conservation concern, such that serious or irreversible harm is likely to occur to the resource.[28] When the limit values of the indicator are breached this would act as a trigger for management action.

7.20 By contrast, target values are associated with achievement of the objective and provide a positive management goal.[30] Target and limit values may themselves be established through studies of the properties of the indicators and altered through adaptive management (7.47) as more knowledge becomes available. In addition to targets and limits, threshold values may also be established. Threshold values act in a precautionary context as an early warning reference

Figure 7-I

The relationship between reference points and reference directions for an indicator of fishing impact[29]

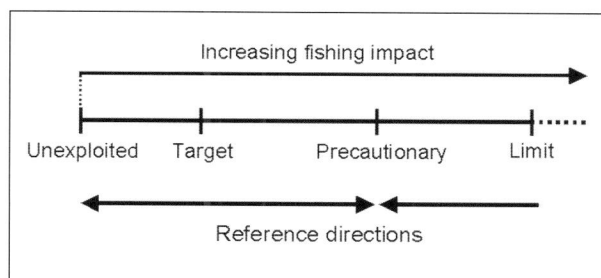

point, to reduce the probability of the limit value being breached and a negative outcome incurred (figure 7-I).[31] The indicator system is analogous to a traffic light system[32] – the limit values represent red, the threshold values orange and the target values green; and red would halt fishing.

7.21 Reference points are already used for the management of commercially exploited fish populations, although the majority of reference values relate to thresholds and limits rather than including targets as well. In the management of exploited species, indicators of the status of the harvested species, particularly biomass of the mature population and mortality rate, are core tools.[33] Examples of reference points that are already in use are given in table 7.1. There is still considerable debate about the framework in which such a system should operate and on how to deal with the uncertainties associated with reference points.[34]

Developing ecosystem indicators

7.22 Indicators should provide a readily understandable tool for describing the state of the ecosystem and for assessing trends. Beyond this general description they will ideally need to be:[35]

- quantifiable – Any property that cannot be measured on an agreed scale is of little use for management purposes. An estimate of quantity should also have explicitly stated degrees of uncertainty attached.

- simple – An indicator needs to be relatively easy to understand and interpret to make it useful to managers and stakeholders. It is more likely in such cases to get agreement on taking action.

- relevant – An indicator needs to directly relate to a management objective to be accepted by managers and stakeholders. The indicator should also clearly pertain to one or more identified assessment questions and clearly relate to ecological components or processes deemed important to an ecological condition.

- tractable – An indicator should relate to a property of a species population or ecosystem that can be affected by management of human activities. Choosing a performance measure over which you have little influence is clearly of limited use.

- faithful – An indicator should consistently convey reliable and accurate information to be of any use.

- comparable – A performance measure should be comparable from year to year to illustrate emerging patterns and changes in status in relation to changes in management over the period.

- cost-effective – The performance measure will need to be cost-effective in relation to its application, particularly in terms of any associated monitoring programme.

7.23 OSPAR has led the way in developing Ecological Quality Objectives (EcoQOs) as indicators of a healthy marine environment (appendix K). Their purpose is to enhance communication, transparency, effectiveness and accountability of management of highly complex natural systems. ICES has, however, expressed a number of reservations about the use of EcoQOs. Indicators can send conflicting messages; for example, positive action to reduce discarding in order to restore fish populations could have a negative impact on birds and mammals that scavenge discards. ICES has been also concerned that the tools which have formed part of management systems for many years may be dismissed because they fail the strict criteria for EcoQOs. It advised caution about the rapid introduction of indicators that are not based on proper understanding of ecosystems, and warned that monitoring a large number of EcoQOs could be costly.[36]

7.24 **Single-species indicators:** Assessing the impact of fishing on different components of an ecosystem is an important part of recent attempts to introduce ecosystem consideration into fisheries management.[38] Although indicator reference points have only been applied to target stocks, it would be relatively straightforward to extend such reference points to stocks of non-target species. ICES has suggested indicators could be applied to non-target species caught as by-catch (e.g. small cetaceans) or damaged by fishing gear (e.g. seabed fauna), ecologically-dependent species (e.g. predators dependent on harvested species) and species affected by scavengers (e.g. kittiwakes affected by skuas).[39] Various single-species indicators have been proposed for measuring the direct and indirect impacts of fishing on fish and benthic communities[40] and these could potentially be used within an appropriate management framework.

Table 7.1

Common reference points used in fisheries management[37]

Reference points based on yield per recruit

F_{max}	The fishing mortality rate that corresponds to the maximum yield per recruit. This value is often difficult to estimate because of the flat-topped shape of the yield per recruit curve, and it ignores future generations (limit). $F_{0.1}$ is often preferred
$F_{0.1}$	The fishing mortality rate at which the slope of the yield per recruit curve as a function of fishing mortality is 10% of its value near the origin (target). $F_{0.1}$ is always lower than F_{max}

Reference points based on biomass

B_{msy}	The biomass corresponding to a maximum sustainable yield as estimated from a production model (limit)
$B_{90\% \text{ R}', 90\% \text{ Surv}}$	The level of spawning stock corresponding to the intersection of the 90th percentile of observed survival rate (R/S) and the 90th percentile of the recruitment observations (limit)
B_{pa}	The threshold that ensures that there is a high probability of avoiding reducing the stock to a point, B_{lim}, below which recruitment is impaired or the dynamics are unknown (precautionary limit). B_{pa} is the biomass below which the stock would be regarded as potentially depleted or overfished
B_{lim}	The limit spawning stock biomass, below which recruitment is impaired or the dynamics of the stock are unknown (limit)
B_{loss}	Lowest observed spawning stock size (limit)
$B_{50\%}$	Biomass at which recruitment is one-half of its maximum level (limit)
$B_{20\%}$	Biomass corresponding to 20% of biomass when fishing mortality is zero (limit)
F_{msy}	The fishing mortality rate that corresponds to the maximum sustainable yield as estimated by a production model. MSY is the yield that should result if that exploitation level were maintained until equilibrium is reached (limit)
Z_{MBP}	The total mortality rate for maximum biological production (target)
F_{crash}	Represents fishing mortality axis as estimated by an *F*-based production model (limit)
F_{τ}	The threshold fishing mortality rate associated with extinction of the stock (limit)

Reference points based on spawning stock biomass per recruit

F_{low}	A level of fishing mortality at which recruitment has been sufficient to balance the mortality in about 9 out of 10 years. The likelihood of a decline in the stock at this level of exploitation is therefore low (target)
F_{med}	Corresponds to the level of fishing mortality where recruitment in half of the years has been sufficient to balance the mortality. There should therefore be a good chance of sustainability at the F_{med} fishing level. Long-term considerations have shown that little is gained by increasing *F* above F_{med} (limit)
F_{high}	A level of fishing mortality where recruitment has not been sufficient to balance the mortality in about 9 of 10 years. Exploitation at this level is therefore likely to result in a decrease in the stock (limit)
$F_{40\%}$	Fishing mortality for 40% of spawning biomass per recruit when fishing mortality is zero (limit)
F_{loss}	The level of spawning stock at which average recruitment is one-half of the maximum of the underlying stock-recruitment relationships
M_{BAL}	The level of spawning stock below which the probability of poor recruitment increases as spawning stock size decreases. The implications of a stock going below M_{BAL} are potentially serious because any sustained decrease in recruitment can lead to a progressive decrease in the stock, with the possibility of eventual collapse (roughly defined as a decrease to 5% of the unexploited stock size)
F_{lim}	The limit fishing mortality that should be avoided with high probability because it is associated with unknown population dynamics or stock collapses (limit). There are very few stocks for which F_{lim} is accurately known

7.25 **Indicators of ecosystem properties:** Defining ecosystem indicators beyond single-species reference points to imply 'emergent' properties of ecosystems requires a leap in the magnitude of complexity. This class of indicator of community status moves beyond aggregating data to reflect some underlying properties of the ecosystem. Such indicators require the implicit or explicit use of some form of ecosystem model, and so will be dependent on how well such a model represents the ecosystem.[41] As discussed (7.10-7.17), the assumptions underpinning such models remain an area of debate and further research is required to develop models of marine ecosystems. ICES has concluded that it is not possible to develop reference points for ecosystems because of a deficient knowledge of ecosystem dynamics and lack of a theoretical basis for identifying threshold and limit values. Ecosystem reference points and indicators would need to be flexible enough to allow account to be taken of the many complex variables in the marine environment, such as seasonal, annual and longer-term fluctuations in the demography of marine organisms, climate, human impacts and so on, all of which would entail a great deal of uncertainty.[42]

7.26 The use of indicators as management tools for an entire marine ecosystem is likely to be difficult unless an incremental approach is adopted within an agreed operational framework. Although the UK is funding much of the work on indicators, including piloting indicators (ECOQs) as required by the 5th North Sea Conference, there is still an obvious gap between theory and practice and the concept has yet to be applied in detailed policy-making. In the longer term, it will be necessary to know the minimum number of indicators that will ensure ecosystems continue to provide a specified range of ecosystem goods and services. Thousands of indicators and potential indicators could be developed and used. Many of the widely advocated indicators tell us little about what activity needs to be managed or how. For example an increase in fish diversity could be interpreted variously as an enrichment of biological diversity, or a disturbance in the ecosystem (mild disturbance usually increases diversity); it could also be caused by a natural disturbance that is part of a long-term trend.

FURTHER RESEARCH TO DEVELOP SCIENTIFIC ADVICE ON ENVIRONMENTAL IMPACTS OF MARINE CAPTURE FISHERIES

7.27 Scientific information is vital for providing evidence about the current state of the seas and informing the public debate. The UK invests heavily in marine research. Annual public sector research on marine topics was worth around £270 million in 1999/2000, while industry funded a further £220 million. Seventeen UK universities have large research teams working in this area, and at least another ten centres of expertise exist outside the higher education sector. The UK also has access to thirty-three research vessels. All of this activity could provide a wealth of information that could be more creatively disseminated.

7.28 While this represents a good base on which to draw, knowledge concerning the relationship between fisheries and the environment is often incomplete and methods to apply such knowledge to fisheries management are often lacking.[43] The ecological effects of fisheries are diverse in terms of scales, processes and biota. Many ecosystem properties and components that are affected directly or indirectly by fishing also show substantial variability caused by environmental factors. Thus, it will often be difficult to establish direct relationships between specific fisheries activities and ecological effects. There are also gaps in knowledge about the ecology of marine species from larval recruitment through to food web interactions.

7.29 Better scientific data, modelling, monitoring and reporting are therefore needed to improve our understanding of the impact of fishing on ecosystems. Few elements of the marine environment have been monitored or studied in a coherent and widespread manner other than commercial fish species. Most possible ecosystem indicators and their associated reference points have been developed in isolation. To facilitate further indicator development there needs to be:

- better understanding of the relative effects of fishing and other environmental variables on the indicators;

- greater statistical power of indicators and monitoring programmes to detect change in response to management action; and

- adequate monitoring programmes for estimating the value of indicators on spatial and temporal scales that are relevant to managers.

7.30 In addition to the development of indicators, the management of human activity in the marine environment, in particular fisheries, could be better supported by the use of well-tested marine ecosystem models, including techniques for estimating uncertainties. An important benefit of such research would be a better understanding of marine ecosystem states and processes under conditions of low human disturbance. Such research is needed to help to develop and refine ecosystem indicators and test model predictions against empirical evidence. Where fishing has significantly altered natural systems the risk remains that environmental services have been so eroded that features of value will be lost or that systems will not 'rebound' in their original form once pressures are removed and targets may never be reached.

7.31 The Natural Environment Research Council (NERC) is currently working with other bodies with interests in the exploitation and conservation of the marine environment to establish a joint venture to meet these research and management needs. This initiative will integrate the knowledge of fisheries assessment and management, fisheries ecology, marine ecosystems, oceanography, biology and economics held by UK academic, research council and government research organisations. Multidisciplinary research teams will be created under an organisational structure that cuts across traditional sectoral and institutional boundaries to provide science and management advice. We welcome this initiative to further develop understanding of the marine environment and **we recommend that the Natural Environment Research Council makes it a priority of the above initiative to fund research on the environmental impacts of marine capture fisheries and to ensure that this knowledge is transferred to policymakers, regulators, fisheries managers and others**.

7.32 The research sector also has its own specific needs. There are already organisations that provide researchers with access to marine information (for example, the British Oceanographic Data Centre). However, collating, synthesising and disseminating data sets, research results and other information are not trivial tasks, and people often need help to use these resources. **We recommend that Government should encourage universities, research councils and others to fund research on the marine**

environment and consider ways of improving the dissemination and use of marine data. The issue should also be considered by NERC as part of its review of marine science in 2004/05.

7.33 In addition to the research information provided by scientists, it would also seem sensible to look at the potential for fishers to collect data in a systematic way that would maximise benefit to managers with minimum disruption to fishing operations. The fish population assessments undertaken by UK and ICES scientists are not always trusted by the UK industry, despite the assessments being of equal quality and using the same methodology as those undertaken by other countries. This may be due to a lack of understanding of scientific methods by the fishing industry, the fishing industry's belief that it has useful knowledge that is ignored by scientists and fishers' perception that stocks are greater than estimated.[44]

7.34 Fishers' knowledge could make a valuable contribution to improving the quality of the underlying science. Studies show that many fishers have detailed local knowledge of fish ecology (for example, spawning and feeding areas), fish behaviour and of changes in fish distribution over time. These are the sorts of factors that are not accounted for by present fish population models used for regulation. With this knowledge fishers are able to concentrate their activities where fish populations are highest, and thus have a very different perception of population levels, particularly in species of fish, such as cod, which show herding behaviour at low population densities (box 5B). Fishers also have insights into species assemblages and the impact of fishing on the seabed. This wealth of information could be used to help improve conservation measures and stock assessments that would benefit the fisheries sector in the longer term. Methods are beginning to be developed that will allow fishers' knowledge to be included in fisheries assessment processes. In 2002, the North Sea Commission Fisheries Partnership used information from the industry on the state of stocks to feed into the ICES scientific process.

7.35 Fishers have also been involved in developing cod recovery plans at the UK and EU level and have contributed to knowledge of environmental change by reporting rare fish captured in their nets. Bodies like the Fisheries Conservation Group (of fishers, scientists and officials) have brought their experience to bear in developing more selective fishing gears, and in considering the effect of gear modifications. There are clearly difficulties with using fishers' catch and discard data if they are not complete, or gathered and processed correctly. Much may depend on setting the right context: fishers are less likely to report accurate information if it is against their self-interest. An atmosphere of co-operation needs to be fostered where the purpose of data collection and the status of the results are made clear from the beginning. Without agreement between fishers and scientists on the state of the marine environment, it will be difficult to achieve effective conservation measures, as successful enforcement and compliance with regulations is in part reliant on a shared understanding. In order to provide further data for understanding, modelling and management, monitoring schemes should utilise data provided by fishers. To this end, **we recommend that fisheries subsidies should support research and monitoring schemes that use information provided by fishers in order to supply data for modelling and management.**

7.36 Scientific advice and research play a crucial, sometimes controversial role, in the management of the marine environment. The current uncertainties argue for more science of the highest possible calibre, not less. We encourage government and other institutions to take the opportunity to help fund new and exciting research to help us understand the marine environment, and eventually to help us develop better, more sensitive ways of managing our impact on it. To make progress, such research will require continued gathering, monitoring and synthesising of data, activities that are equally deserving of adequate funding.

BETTER MANAGEMENT OF HUMAN ACTIVITIES IN THE MARINE ENVIRONMENT

THE ECOSYSTEM APPROACH TO FISHERIES

7.37 In 2001, the UN Food and Agriculture Organization (FAO) conference in Reykjavik, reached agreement that "in an effort to reinforce responsible and sustainable fisheries in the marine ecosystem, we will individually and collectively work on incorporating ecosystem considerations into management". The conference also agreed to "identify and describe the structure, components and functioning of relevant marine ecosystems, diet composition and food webs, species interactions and predator-prey relationships, the role of habitat and the biological, physical and oceanographic factors affecting ecosystem resilience".[45] The FAO has since produced technical guidelines for best practice with regard to introducing ecosystem considerations into fisheries management. As we have seen in chapter 4, management of fisheries and the marine environment must now operate in the context of an 'ecosystem approach' and a 'precautionary approach' through, for example, Annex V of the OSPAR Convention and the Bergen Declaration.

7.38 This more holistic view is the essence of the ecosystem approach to fisheries (EAF).[46] It stands in sharp contrast to previous philosophies for managing the marine environment that were fragmented, sectoral and focused on short-term economic gain. Within fisheries management, this change in mind-set is illustrated by attempts to move away from managing individual fish populations for maximum economic gain towards more precautionary controls on fishing that recognise the interdependence of predator and prey species within the food chain. Realising this concept will require managers to reconcile a range of issues such as genetic and species diversity, species rarity, habitats, food web properties and the ecology of marine mammals in a balanced and credible way when managing the marine environment.[47] This will involve incorporating a wider range of scientific advice into the management framework.

7.39 The 'ecosystem approach' is a term that has come to be used in a wide variety of ways (box 7C). The core concept lies in integrating the range of demands placed on the environment, such that it can support these demands without deterioration.[48] The various definitions share a common recognition that management strategies need to take account of interactions between different parts of an ecosystem. Managing a single aspect or exploited species is unlikely to deliver long-term sustainability. Many of the interpretations also stress the need to assess physical and chemical factors as well as biological ones. Management of human activities in the marine environment on this basis would need to take account of all important external drivers on the ecosystem such as climate change, pollution, dredging for sand and gravel and the impact of exotic species, as well as fishing impacts.

BOX 7C	DEFINING THE ECOSYSTEM APPROACH

The UN *Convention on Biological Diversity*[49] defines the ecosystem approach as:

"a strategy for the integrated management of land, water and living resources that promotes conservation and sustainable use in an equitable way."

The UN Convention on Biological Diversity has also adopted 12 principles associated with the ecosystem approach in 2000 (appendix E) and operational guidance for implementation,[50] which have been elaborated through detailed guidelines.

OSPAR has refined the concept for the marine environment as:

"the comprehensive integrated management of human activities based on the best available scientific knowledge about the ecosystem and its dynamics, in order to identify and take action on influences which are critical to the health of marine ecosystems, thereby achieving sustainable use of ecosystem goods and services and maintenance of ecosystem integrity."

The *Intermediate Ministerial Meeting of the North Sea Conference on the Integration of Fisheries and Environmental Issues* gave the following definition, since adopted by Greenpeace and WWF:

"An ecosystem-based approach involves considering all the physical, chemical and biological variables within an ecosystem. In the management of living resources this means decisions are based on the best available scientific knowledge of the functions of the ecosystem, including the interdependence of species and the interaction between species (food chains) and the abiotic environment, as well as knowledge of the temporal development of the ecosystem. It could therefore imply widening of the multi-species approach, currently used in fisheries, to encompass not only fish but other organisms which directly or indirectly depend on fish or on which fish depend, as well as other significant biotic and abiotic factors."

The *International Council for the Exploration of the Sea (ICES) Study Group on Ecosystem Assessment and Monitoring* defined it as:

"Integrated management of human activities based in knowledge of ecosystem dynamics to achieve sustainable use of ecosystem goods and services and the maintenance of ecosystem integrity."

The *US Ecosystem Principles Advisory Panel* gave the following definition:

"A comprehensive ecosystem-based fisheries management approach would require managers to consider all interactions that a target fish stock has with predators, competitors and prey species; the effects of weather and climate on fisheries biology and ecology; the complex interactions between fishes and their habitat; and the effects of fishing on fish stocks and their habitat."

7.40   The ecosystem approach therefore entails taking into consideration all elements that make up the ecosystem as well as the activities affecting it in order to ensure that the biodiversity, health and integrity of the marine environment is maintained. In evidence to us, ICES maintained that fishers were themselves part of the ecosystem,[51] a position similar to both that of the Convention on Biological Diversity, and the proposed EU Marine Thematic Strategy[52] that treat humans as part of the ecosystem. Although it is undeniable that fishers have huge effects on marine ecosystems, we do not believe it is helpful to achieving the objectives of the ecosystem approach that the modern industrial fishing fleet of the EU should be defined as part of the marine ecosystem.

7.41 The goals of the ecosystem approach to fisheries are frequently related to the idea of sustainability.[53] At present, it is clear that the present level of fishing activity in UK and EU waters is unsustainable environmentally, economically and socially. It is environmentally unsustainable because it both damages the marine environment and depletes fish stocks to the extent that they cannot sustain themselves even at today's levels, which are much below historic levels. Parts of the industry are economically unsustainable because the depleted fish stocks cannot provide an economic return on much of the capital, or adequate livelihoods for much of the labour, that are currently employed in the fishing industry. The falling level of employment in the fisheries sector and the associated changes in communities that have traditionally been involved in fishing activity means that it is not socially sustainable. Economic and socially sustainable fisheries are only possible if the fisheries are environmentally sustainable. Achieving environmental sustainability must therefore be the primary objective of the ecosystem approach to fisheries.

7.42 Defra interprets the ecosystem approach as a more strategic way of thinking that puts the emphasis on maintaining the health of ecosystems as well as human use of the environment, for present and future generations (rather than treating humans as a specific part of the marine ecosystem).[54] In practical terms, this means setting clear environmental objectives and basing management on the principles of sustainable development, robust science and precaution, as well as conserving biodiversity, involving stakeholders and developing integrated management plans that address the needs of different sectors. The process of deriving any actual objectives and making them operational remains somewhat vague. This is true of all the interpretations of the concept by international and national bodies. The complexity of ecosystems and the mass of data deriving from their analysis runs the risk of distracting attention from delivering practical actions to achieve the required environmental goals.[55]

A PRAGMATIC APPROACH TO IMPLEMENTING THE ECOSYSTEM APPROACH

7.43 While the definitions of the ecosystem approach vary in detail, one principle emerges clearly, the need to change the focus of fisheries management from fish stocks to conservation of ecosystems within the wider marine environment. However, the key factors that influence the stability of marine ecosystems are still imperfectly understood. The marine environment is dynamic and fluid; geographical connectivity makes boundaries difficult to define; and much remains unknown about marine biodiversity and ecosystems. As a result, there is some concern that international commitments have been made to adhere to the principles of the 'ecosystem approach' in the absence of clear means of implementing the concept.[56] These fears partly rest on the argument made by some fisheries scientists that an ecosystem-based approach can only begin with a thorough analysis of ecosystem attributes, boundaries, processes and interaction, but that it is unlikely that we can ever gain such a complete picture of the ecosystem.[57]

7.44 Many of the schemes devised by regulators and non-governmental organisations (NGOs) are complex and multi-disciplinary, and create considerable demands for additional information and monitoring,[58] but of itself the complexity engenders little change. The following important considerations need to borne in mind for any system attempting to limit the impact of human activities on the marine environment:

- our capacity to predict ecosystem behaviour is limited;

- ecosystems have thresholds and limits which, when exceeded, can result in major system restructuring;

- once thresholds and limits have been exceeded, some changes can be irreversible;

- diversity is important to ecosystem function and integrity;

- components of ecosystems are linked;

- ecosystem boundaries are open;

- ecosystems change with time.

7.45 In short, it is not possible to manage the marine environment, nor is it possible to predict the effects of any and every human activity on marine ecosystems. But it is possible to take pragmatic steps to protect the marine environment from excessive exploitation and damage.

7.46 All the definitions of the ecosystem approach given in box 7C are founded on the premise that the best guarantee for the long-term sustainability of commercial fisheries is a management system that maintains the structures and functions of the ecosystems of which commercial fisheries are a part. Such a management system is also a prerequisite to ensure protection of the whole marine environment from the impacts of fishing. Implementing such an approach in marine capture fisheries requires taking proper and careful account of the condition of ecosystems that may affect fish stocks and their productivity, but this does not necessarily require large amounts of additional information. This outcome can be met through measures designed to protect natural habitats, communities and species from the risk of degradation. Basic elements include adaptive management (7.47), co-management (7.49) and the precautionary approach (7.52).

Adaptive management

7.47 It has generally been envisaged that the ecosystem approach would be implemented on an incremental basis over an extended period, in part because an evolutionary rather than revolutionary move towards the ecosystem approach would be less likely to paralyse the fisheries management decision-making process and would maintain broad-based support.[59] It is also because both the changes of objectives for management and the knowledge base for management will need to be much more than incremental. New objectives will be added to the management agenda that will require decisions that cannot be based on existing scientific data.[60]

7.48 Limited knowledge should not, however, be used as an excuse for delayed implementation. The crisis in the marine environment means that such a delay will be highly detrimental. Processes that enable existing data to be used and more information to be added when it becomes available will need to be adopted. This process is known as adaptive management, or 'learning through doing',[61] which recognises the need to experiment with the management of the effects of human activities on complex systems in order to develop effective management processes over social and economic as well as ecological scales.[62] Adaptive management provides knowledge that can only be obtained by intervention. A basic model for adaptive management of the marine environment has been described and is being tested in a number of transboundary waters including the Black Sea.[63]

Co-management

7.49   Fisheries continue to have a disproportionate impact on the health of marine ecosystems compared with all the other economic sectors utilising marine resources. Applying simple and robust management measures to fishing such as drastically reducing the overall levels of fish mortality, using the least damaging fishing methods and reducing by-catch, together with setting aside some marine areas from any form of exploitation, are fundamental to sustaining marine ecosystems. Unsustainable fishing practices such as fishing for deep-water species (5.89, 9.44) will need to cease. Such broad-brush management measures will have to be introduced prior to the implementation of any co-management framework at a regional level, and we make detailed recommendations to this end in chapters 8, 9 and 10.

7.50   Beyond these broad-brush measures, local 'bottom up' initiatives will also be important for delivery of local environmental objectives. Projects such as 'Invest in Fish South West' are already attempting to manage the local marine environment through stakeholder groups of fishers, scientists, retailers and environmentalists.[64] Fishers in the south-west of England have undertaken innovative voluntary environmental measures, such as no-take fishing zones, and this may prove a successful model for managing inshore waters (out to 12 nm. The management regime will also have to reconcile French, Spanish and Irish interests that have traditional fishing rights in these areas. The project is part of a wider initiative led by WWF, the National Federation of Fisherman Organisations and Marks and Spencer, which aims to provide a long-term strategy for managing fishing fleets on a regional basis, taking important local factors into account.[65]

7.51   As part of the Sixth Environmental Action Programme Directorate-General Environment of the European Commission has drawn up the EU Marine Thematic Strategy, which includes guidance for the application of the ecosystem approach to management of human activities.[66] It contains a vision, principles and strategic goals and sets out the framework for delivering the ecosystem approach in Europe. The guidance suggests that objectives for the ecosystem approach are set by regional stakeholder groups and are applied at scales ranging from regional to local level (although some objectives would be the same in all areas). From these objectives, ecosystem indicators would be derived that would be used to measure the success of management procedures undertaken to meet those objective. There is a gap, however, between high-level policy development and implementation. Both the EU strategy and the UK Marine Stewardship Strategy have yet to elucidate various fundamental practical considerations of how they will be implemented, including what spatial scale they will operate at, the depth of understanding of the natural functional limits of ecosystems required and how the effects of environmental change will be dealt with. In chapter 10, we further consider the appropriate framework for managing the marine environment within UK and EU waters.

THE PRECAUTIONARY APPROACH

7.52 Our Twenty-first Report on Setting Environmental Standards, described the precautionary approach as a rational response to uncertainties in the scientific evidence relevant to environmental issues and uncertainties about the consequences of action and inaction. Even the best scientific assessment may not provide a clear basis for taking a decision on an environmental issue. The requirement for sound science as the basis for environmental policy should not be interpreted as a requirement for absolute knowledge.

7.53 The precautionary approach is not usually seen as requiring precise and complete scientific knowledge, but in fisheries science advice precaution has been defined in quantitative terms. The precautionary reference points have been accommodated within existing fish population assessment methods (7.6) and expressed using parameters such as mortality and biomass. Such precautionary assessments should in theory lead to lower precautionary quotas (total allowable catches) being set by fisheries managers. A quantification of risks in this way has been seen as the only way to avoid an interpretation of the precautionary principle that would require a policy of zero impact as a result of lack of knowledge.[67] ICES, however, is clear this is not a precautionary approach *per se:*[68]

> "In such instances good scientific judgement should be first used in order to determine whether better estimation techniques are possible. Results from such exercises should be clearly documented and should never be associated with the precautionary approach, since the precautionary approach applies to management and not scientific estimation."

7.54 Fisheries' managers have failed to apply the precautionary approach because of their interpretation of what kind of scientific knowledge is required to invoke its use.[69] They have burdened marine scientists with proving that a deleterious effect will occur before a decision is made to take protective management actions, as in the case of the Darwin Mounds, where clear scientific evidence of damage was required before a ban on trawling could be introduced. Such an approach does not account for the large scientific uncertainties, whether of fish population assessments or of the effects of fishing on the wider marine environment, and that they are unlikely to be entirely overcome, or even significantly reduced, in the near future. Despite such uncertainties, it is clear that existing levels of fishing effort are causing a crisis in the marine environment. **We recommend that human impacts on the marine environment should be managed in a fully precautionary manner. Fishing should only be permitted where it can be shown to be compatible with the framework of protection set out in this report.**

Reversing the burden of justification for fisheries

7.55 Reversing the burden of justification in capture fisheries is consistent with the precautionary approach as proposed in the 1995 UN Straddling and Highly Migratory Stocks Agreement. Article 6 (1) of the Agreement states that:

"States shall apply the precautionary approach widely to conservation, management and exploitation of straddling fish stocks and highly migratory fish stocks in order to protect the living marine resources and preserve the marine environment."

Article 6 (2) also states that:

"States shall be more cautious when information is uncertain, unreliable or inadequate. The absence of adequate scientific information shall not be used as a reason for postponing or failing to take conservation measures."

Similar international commitments have been undertaken under the 1995 FAO Code of Conduct on Responsible Fisheries. Despite these commitments, the precautionary approach has been incorporated into the fisheries discourse in a somewhat ambiguous manner and solely within the context of fisheries rather than being applied to the wider marine environment and all the environmental effects of fishing.[70]

7.56   A scientific assessment should present the range of possible interpretations of the available evidence, or the range of scientific possibilities and options concerning a particular course of action, accompanied by acknowledgement of the assumptions implicit in the assessment. This may produce results that conflict and point to a need for more research, rather than providing a definitive answer.

7.57   This has clearly not been the case for fisheries. There have been a number of problems in incorporating the concept of uncertainty into fisheries sciences, not least because fisheries science is responding to a regulatory process that calls for simplification and specification of problems. Although ICES is to be commended on its efforts to be transparent and move towards a system of comprehensive peer review, as with all regulatory science there is a continuous pressure to pare down specialist information (e.g. scientific data and interpretation) to simple essentials before it can be used in reaching management decisions. Uncertainties in data and the existence of alternative hypotheses are often not evident in the science documents upon which managers' base decisions.[71] As a result, the presentation of fisheries science has been replete with short-term, inadequate answers to complex problems, such as the use of total allowable catches.

7.58   The first step in overcoming the problems caused by scientific uncertainties is to provide an open appraisal of where they lie. In the case of fisheries, the level of uncertainty has been under-estimated substantially in the past, and there has been a failure to document legitimate differences in scientific opinion.[72]

7.59   The second step would be to apply the precautionary approach to the management of fisheries. **We recommend that the presumption in favour of fishing should be reversed. Applicants for fishing rights (or aquaculture operations in the marine environment) should have to demonstrate that the effects of their activity would not harm the seas' long-term environmental sustainability.** This could operate through the licensing system, marine planning, and we explain how this could function in chapters 9 and 10. There may also be areas that need to be entirely protected in order to fulfil the precautionary principle and achieve enhancement of ecosystems.

7.60   Such a procedure would have prevented some of the serious damage that has been caused by some new fisheries or aquaculture in the past. For example, deep-sea fisheries, where the activities of vessels that targeted seamounts or cold-water coral reefs, would have been prevented if there had been a responsibility on the industry to show that their fishing activities would cause no harm. These dangers are present today; such unacceptable effects are likely to be incurred following a decision by the European Commission to allow access to the Azores by the EU deep-sea long-lining fleet (box 7D).

7.61   Reversing the burden of justification turns on its head the way that the impact of fishing and aquaculture have been assessed in the past. Previously, the marine environment has been regulated on the basis of a presumption in favour of fishing. Unless harm to ecosystems or habitats could be demonstrated by the regulator, then it was acceptable for the activities to continue. This has not prevented marine ecosystems from being severely modified by fishing. Reversing the presumption would place the burden of justification on those seeking fishing rights. This process would promote greater attention to the biological state of the marine environment by both the industry and the regulator.

BOX 7D	THE AZORES DEEP-SEA FISHERIES

The seas of the Portuguese Azores Islands in the mid-Atlantic have been closed to other EU fishing vessels since Portugal joined the EU in 1986. This ended on 1 August 2004. In November 2003 the European Commission decided to open the 100-200 nm zone of Azorean waters to the EU deep-water fishing fleet. The decision was taken without ensuring limits on fishing activity or considering the impacts of increased fishing on deep-water species or habitats. In the Azores, a delicate balance between nature and commerce – which has been the hallmark of fishing there for centuries – could be lost overnight, with the opening up of waters to these foreign fishing fleets.

The Azores are a unique case as they do not have a continental shelf and the fisheries are all deep sea. Averaging 3,000 m in depth, the waters around the Azores contain extensive undersea mountain ranges (seamounts), deep-water coral reefs and volcanic hydrothermal vents that are rare in European waters. The region supports a diverse range of marine life, and is especially vulnerable to intensive fishing activities like trawling. The deep-water commercial fish species found around the Azores are long-lived and slow to reproduce and even modest fishing pressure could seriously deplete stocks (5.89).

Using small vessels and traditional fishing methods, including a ban on trawling in the deep-water fisheries and use of hand-drawn lines of hooks, the people of the Azores have fished this area for generations and have avoided damaging the environment. Local fishing also represents 5% of the Azores' gross domestic product. Azorean protection has maintained their deep-water resources in better condition than those elsewhere in the Atlantic that lie outside national jurisdiction. Although last minute negotiations prevented access by bottom trawlers,[73] the European Commission opened this pristine environment to the deep-water fleet without any environmental or socio-economic impact assessments or ensuring controls to prevent over-exploitation were in place. The entire area of the Azores deep-water fishing grounds could be fished out by the EU long-line fleet in as little as 18 days.[74] The decision was taken against the will of the European Parliament which voted to maintain access for Azorean vessels only.

As a result, the Autonomous Region of the Azores (part of Portugal) has brought a case against the European Commission in the European Court to suspend a Fisheries Council decision on EC Regulation 1954/2003, which allowed the opening up of their waters.

This is first time in European Court history that it has been argued that in implementing the Regulation, the European Commission had breached its own requirement to integrate environmental policies into its Common Fisheries Policy. The European Commission's defence has been to argue that the principle of equality of access takes precedence over the precautionary principle. This case could set an important environmental precedent that will determine much of the future course of EU fisheries policy. Temporary protection was granted to the Azores until the end of 2004 under emergency CFP measures. Discussions continue on whether this should be made permanent.

Strategic environmental assessment and environmental impact assessment

7.62   The first step in facilitating the reversal of justification would be to ensure all maritime users adopt appropriate environmental assessment procedures for their activities. This may involve strategic environmental assessment, environmental impact assessment or other appropriate assessment carried out under the EC Habitats Directive. The present way in which such assessments are carried out varies in quality, and accordingly such studies vary in how useful they are in delivering the ecosystem approach.[75]

7.63   At present, the EC Directive on Environmental Impact Assessment (EIA) does not apply to capture fisheries. Similarly, while the EC Directive on Strategic Environmental Assessment (SEA) applies to "projects and plans affecting the environment", but it is not currently applied to fisheries. Fisheries should be brought fully within this framework. This will start to address the internalisation of costs for environmental impacts, as already happens for the exploitation of oil, gas and aggregates.[76] Both EIA and SEA processes have proved effective in reducing the impact of other marine industries on the environment. The second step would be to define where fishing activities can and cannot be allowed through use of marine spatial planning. We make recommendations concerning the application of SEA and EIA to fisheries in chapters 9 and 10.

Spatial management

7.64   Human activities within the marine environment will need to be managed at the appropriate scale. One of the basic principles of the ecosystem approach is that management decisions should be taken in an ecosystem context. Given the correlation between the many types of impact and the complexities and costs involved in understanding, monitoring and regulating fisheries specifically in relation to many impacts on different types of ecosystem, it would seem most effective to regulate overall fishing pressure on marine ecosystems at the regional sea level, as defined by the Joint Nature Conservation Committee (2.55). This framework, based on a development of the 'large marine ecosystem' concept, defines appropriate boundaries on the basis of ecosystem properties, rather than on commercial interests alone.[77] The recent Defra-sponsored Review of Marine Nature Conservation has also recommended that the UK government should apply a conservation framework of wider seas, regional seas, marine landscapes, important marine areas, and priority features in UK waters and discuss with other countries in the north-east Atlantic biogeographic region the potential for extending the framework to their waters.

7.65 Management on a regional sea basis will mean developing institutional arrangements to bring together science and technology to inform the development of regional seas management plans. Delivering science at a regional scale will be necessary for the success of decentralised management. Integrated multidisciplinary research should be supported to synthesise and develop knowledge within spatially resolved systems, particularly with regard to ecosystem models, moving the agenda from what is there to what it does. We make further recommendations concerning the implementation of spatial management in chapter 10.

Precautionary use of indicators

7.66 If the precautionary approach is used appropriately within the fisheries management framework, it is likely that the role of indicators of ecosystem status will become increasingly important. As the management regime recommended by this report will be based on objectives to protect marine ecosystems, two classes of indicator are needed – states and pressures. Measures of the pressures on the ecosystem indicate when management measures are required and measures of the state of the environment indicate whether objectives of protecting ecosystems have been met. It should also be noted that indicators have been widely used in other areas of general environment reporting, such as pollution control and assessment of overall environmental status.

7.67 As outlined previously (7.22-7.26) there are many problems to be overcome in the development of new and more complex indicators. ICES believes, however, that with regard to the state of the ecosystems, indicators beyond those for single species are unnecessary, because if the individual components of the ecosystem are protected there is a high probability there would be minimal community and ecosystem effects. A suite of single-species indicators or single-factor indices for major target species, by-catch species, indicator species and other vulnerable species (for example, abundance indices of sensitive species, proportion of mature individuals in sensitive populations) would be a more pragmatic approach than attempting to develop ecosystem level indicators.[78] Such indicators would be relevant to a large group of issues simultaneously, and would synthesise the overall state of the ecosystem without aspiring to track specific interactions.

7.68 There is already a scientific basis for such indicators, which could be improved over time as knowledge of marine ecosystems grows, and management systems can be adapted accordingly.[79] To this end, there has been research into the use of such a suite of individual indicators in an aggregate collection to determine ecosystem status.[80] It is also likely that some system or multi-species level indicators will be developed such as those relating to trophic interactions between species and community size structure, which could be monitored through size compositions or average trophic level of catch, once a broader scientific understanding of these areas is achieved.[81] Indicators of pressures on the ecosystem, such as the amount of fishing activity occurring in a given area, could be developed from readily available data. As understanding of marine ecosystems increases, these more complex indicators can be adopted in line with an adaptive management approach.

7.69 In order to embed the use of indicators within an appropriate risk management framework it will be necessary to utilise thresholds as well as targets and limit reference points for indicators. It is imperative that indicators are not treated as another piece of information provided by scientists to the fisheries management debate. Thresholds are used routinely in risk management contexts to trigger commencement of moderate regulatory restrictions and are useful means of incorporating uncertainties about indicators.[82] Although it is likely that if indicators are not at the target then some form of management action will be needed. Formal standards of precautionary decision-making are required for governance that is both transparent and accepted by involved parties. Use of threshold points would allow managers to agree on specific management actions with stakeholders if thresholds are violated in advance of any crisis within a co-management framework. This is preferable to imposing more drastic measures from above at the point that reference limit points for indicators are breached. If levels remain above that of the threshold, managers would be in a position to negotiate with stakeholders the actions needed to meet the required target limit of an indicator.

7.70 The UK government is already obliged to develop ecosystem indicators in relation to marine trophic levels under the Convention on Biological Diversity. The EU has also proposed that the ecosystem approach management system underpinning the EU thematic strategy will operate using a comprehensive set of indicators of the pressures and states of marine ecosystems. To this end, **we recommend that the UK government should adopt a suite of indicators that reflect the state of marine ecosystems in UK territorial waters and measure progress in conserving the marine environment.**

CONCLUSIONS

7.71 Current approaches to the conservation, management and protection of marine biological diversity will need to be radically changed if they are to be sufficiently coherent and robust to support marine ecosystems.[83] In order to achieve this change the focus of scientific advice must shift from commercially exploited fish populations to a more holistic view of the marine environment. The Commission recognises that scientific advice can carry little weight when there are high short-term political, social and economic costs when moving towards sustainability and so it will require high-level government commitment to overcome these obstacles.[84] There is little doubt that allowing ecosystems to recover from their degraded state will inflict some economic hardship on fishers in the short term, as is further discussed in chapter 9. As a result of lack of political will, progress has been slow towards a new framework for fisheries management, despite the failings of the current system to manage the environmental impacts of fisheries and the legislative imperative to move to an ecosystem approach to fisheries management.

7.72 In addition to the lack of political will, there are a number of practical problems in making any such approach operational, as has been outlined in the preceding sections. This has led to suggestions that such an approach should be introduced incrementally as scientific knowledge of marine ecosystems increases. We strongly oppose such a cautious approach; we believe that there is an urgent need to protect the marine environment from over-exploitation by fisheries. The broad principles of the ecosystem approach can be realised now through simple precautionary and adaptive management measures.

7.73   The complexity of marine ecosystems defies extension of present management principles to ecosystem management, and questions remain as to whether it is possible, or desirable, to manipulate patterns of trophic interactions and ecosystem processes to achieve a desired effect.[85] The intention of the ecosystem approach is solely to manage human activities and their effects on ecosystems, not to manage the ecosystem. Although it is clear that more research is needed on the ecological effects of fisheries, this is not a rationale for delay in taking urgent action to protect marine environments. This is particularly so for high impact fishing methods such as bottom trawling, which is clearly very damaging to benthic fauna, or deep-sea fishing. Implementing a precautionary and adaptive management style could achieve the environmental protection goals of the ecosystem approach and at the same time take into account the complexities and dynamic nature of the ecosystems involved, even in the absence of extensive knowledge or understanding of how they function. In Alaska, for example, some ecosystem-based management measures have already been introduced, such as control of direct and incidental catches; a prohibition on fishing of forage species (on which other fish, seabirds and marine mammals depend); protection of habitat for fish, crabs and marine mammals; and temporal and spatial controls of fishing.[86]

7.74   Science, coupled with the precautionary approach, needs to provide a far stronger and firmer basis for making management decisions. It should consider effects on all aspects of the marine environment rather than fish stocks alone. In order to take a longer-term strategic view of the sustainability of fisheries impacts, performance targets will need to be set against a continuously moving baseline. Such an adaptive management approach would involve taking what we know about ecosystem function and trying to adapt our technology to conserve it. Targeted research and investment would also be required in those areas that are most likely to reduce the uncertainties in a cost-effective way. In the absence of scientific information on the impacts of fisheries on biodiversity, conservation measures should be adopted in line with the precautionary approach, particularly in cases where fishing activities are likely to result in serious or irreversible damage.

7.75   As our knowledge of marine ecosystems increases, it will be possible to introduce more sophisticated management of activities in the marine environment. This will become increasingly important if we are to manage activities in the marine environment during climate change. Modelling should take account of scenarios that include other natural and human drivers of the system, such as accounting for predictions for positive and negative North Atlantic Oscillation (NAO) years separately (2.12-2.13). However, models of such a complex system will never be able to predict the ecosystem effects of any and every action. The information from ecosystem models will contribute to the decision-making process but not replace it.

7.76   More immediately, management of fisheries could make use of much simpler, holistic models, and of environmental assessments and indicators as management tools. Managers of the marine environment need to have some confidence in how any given fishing activity will impact key aspects of the ecosystem. This will be assisted if the appropriate form of environmental assessment has been undertaken for such an activity. It is possible, even with existing knowledge, to design a set of indicators to assess fishing activity against ecosystem targets. Scientists should also attempt to establish limit reference points for all target and non-target species within an ecosystem and managers should determine best fishing practice to guarantee minimum levels of disturbance to the ecosystem. Reference points for indicators that relate to the virgin or unexploited situation may also be appropriate for assessing the overall impact of fishing, especially with regard to the setting of baselines. The management objective, however, may not be the unexploited state as society may deem some impacts acceptable. In the absence of precise scientific knowledge, estimates of habitat and species target levels based on modelling can be used as indicators of a healthy environment.

7.77　The priorities of fisheries managers must change to that of protecting the resource in the long term rather than protecting economic gain in the short term. The failure to maintain stocks above safe biological limits by controlled landings means that fishing pressure must be controlled through reductions in fishing capacity and effort directly and by setting areas aside from fishing,[87] as is further described in chapters 8 and 9. In addition to the management measures proposed thus far, there is also a need to evolve some form of tenure or ownership, such that fishers can reconcile their economic interest with long-term conservation at a local level. Further to this, there needs to be direct involvement of stakeholders in data collection and decision making with the spatial scale of management and data collection set at the appropriate spatial level (the regional sea framework). It will also be necessary to widen the group of stakeholders in the marine environment to include those without an extractive interest.

NEXT STEPS

7.78　The broad principles of the ecosystem approach can be effectively implemented as a package of straightforward precautionary management measures that reduce the overall environmental impact of fisheries. Such measures would lower the requirement for detailed understanding of marine ecosystems. Ecosystem states would be allowed to adjust without having to predict what form that adjustment will take. There would still be a requirement to monitor for responses to management measures, but success or failure would not then depend on whether specific ecosystem attributes had improved. Rather, the important measure would be the degree of success in overall reductions in mortality rates of marine species and reduced impacts on the ecosystem as a whole.

7.79　The implementation of these management measures will require a new framework, discussed in more detail in following chapters. The strength of these measures is that they are a pragmatic approach to implementing the ecosystem approach to fisheries. In chapter 8, we describe the requirement for better protection of critical sites, species and habitats in line with the EC Habitats Directive and OSPAR agreements. This would include extra measures to control fishing within Special Areas of Conservation and Special Protection Areas. In addition to protection of these areas, a much larger network of no-take marine reserves should be established with complementary management of adjoining waters.

7.80　In chapter 9, we outline the need for reduced fishing pressure. More precautionary levels of fishing effort should be established in line with the best scientific advice, to ensure that fishing is sustainable. The fishing pressure on target and non-target species should also be reduced through management measures to lower rates of by-catch and discarding. There should be incentives for use of more selective and less environmentally damaging fishing gears, combined with spatial planning and strategic environmental assessment of management plans to control where gear types are used.

7.81 In chapter 10, we describe how an improved institutional framework and guiding principles set out in a marine act would facilitate the introduction of marine spatial planning to formulate management plans that set the context for the protection and use of the sea.

Chapter 8

MARINE PROTECTED AREAS

The marine ecosystem as a whole needs protection, as well as individual species and habitats. Could marine protected areas (MPAs) and marine reserves play a role in providing such protection and lead to healthier seas in the OSPAR region?

INTRODUCTION

MAKING THE ECOSYSTEM APPROACH A REALITY

8.1   In this chapter, we examine a potentially important step towards the practical implementation of the ecosystem approach; the establishment of large-scale networks of marine protected areas (MPAs) in UK waters and across the OSPAR region. We consider the potential merits of this option in detail, which we envisage forming part of a broader package of measures to protect the environment and improve the management of fishing (7.79-7.81).

8.2   There is widespread concern about the damage being done to our seas, particularly by fishing. In chapters 5 and 7, we argued that one reason for the failure to respond effectively to the problems linked to fishing has been the over-reliance on management approaches that focus on individual commercial species, without taking account of the broader consequences for the environment.

8.3   This position is now changing, and a broad international consensus has emerged that seeks to protect ecosystems as well as particular features of the environment. This means protecting representative areas that reflect the diversity of habitats and process upon which all species depend, alongside special measures for particularly vulnerable aspects of marine communities. A move from agreement in principle to practical action is, however, long overdue. We now consider one element – MPAs – that could help deliver these objectives and make the move to an ecosystem approach a reality.

WHAT ARE MARINE PROTECTED AREAS?

8.4   The term marine protected area (MPA) is used in different ways, but put simply, it describes any marine area that is afforded some kind of special protection, usually to benefit conservation and/or fisheries. In this report, we have adopted the general definition put forward by the World Conservation Union (IUCN) in 1994[1] which we use to describe any type of protected area:

> **Marine protected area:** An area of land and/or sea dedicated to the protection and maintenance of biological diversity, and of natural and associated cultural resources, and managed through legal or other effective means.

8.5   Many different types of management regime fall under the broad heading of MPAs (appendix L). In this report, we have examined the effectiveness of a subset of more highly protected MPAs known as marine reserves.

> **Marine reserves:** areas in which the extractive use of any resources (living, fossil or mineral) is prohibited, along with any form of habitat destruction.[2]

8.6   We have focused on this type of protection because reserves are closed to fishing (i.e. they are **fishery no-take zones**). It is this feature that we are particularly concerned with since it is integral to the success of marine reserves, and because fishing is the major pressure on the UK's marine environment. We examine the evidence that reserves can help restore ecosystems and protect them for the future, while also helping fish populations to recover.

8.7   We recognise that the beneficial effects of reserves may also arise from the additional protection they receive that limits disturbance from other extractive industries. For example, measures in highly protected areas may include bans on all uses including dumping, dredging, construction and the extraction of all living and non-living resources. In a few extremely sensitive areas, there may even be bans on human access but these are rare. This wider variety of controls may be necessary in some places, but it is beyond the scope of this report to consider these issues in-depth.

8.8   There is considerable scientific consensus that the benefits of MPAs and reserves are much increased if areas are linked together into ecologically coherent **networks**. We explore whether this concept could be applied to the seas around the UK to help combat over-fishing and other threats to the environment.

WHY MARINE RESERVES?

8.9   The key reason for establishing marine reserves is that unlike most other management options they can protect the entire ecosystem, from spawning fish, to the creatures living in the ocean depths, to the seabed itself. Designed in the right way, they can protect commercial fish, non-commercial species and features of the seabed that might be damaged by trawling and dredging. This makes them one of the most simple and straightforward means for implementing the ecosystem approach.

8.10   Another advantage is that while reserves need to be properly designed, they do not require a comprehensive understanding of individual sites before they are designated, since their objective is to protect a representative spectrum of the ecosystem, rather than individual attributes. This flexibility makes reserves an ideal tool for the marine environment, where data collection can be expensive and slow, and the inter-relationships between species and with their environment are often not well understood.

8.11   There are usually two main driving forces behind the creation of reserves. The first is the desire to protect species, habitats and ecosystems, and the second is to benefit fisheries by relieving fishing pressure, particularly during critical parts of the life cycle such as spawning. The primary focus of this report is on the use of reserves to help marine ecosystems to recover from the effects of human activities and to preserve and improve them for the future. However, we also examine some of the important advantages that reserves can have for fisheries, research, education and tourism.

8.12 One of the first sea parks to be established anywhere in the world was created in the Bahamas in 1958[3] and formal marine reserves have existed for more than two decades. A recent analysis of the effects of protection within these no-take marine reserves shows that they have important benefits for both conservation and fisheries.[4] They are also the best way to protect resident species and provide protection to important habitats. In addition, they can provide a critical benchmark for evaluating impacts and threats to ocean communities.

8.13 Reserves offer distinct advantages in terms of ecosystem protection compared with existing fishery management tools. They can also achieve the same level of control over catches of target species as traditional effort control, in addition to reducing mortality and damage to other species.[5] They can be simpler to establish and enforce than some current fishery measures that have fewer conservation benefits, and may cost the same, or even less, to run (8.47).

INTERNATIONAL EXAMPLES OF MARINE PROTECTED AREAS AND RESERVES

8.14 International initiatives such as the UN Convention on Biological Diversity, the World Summit on Sustainable Development and OSPAR all set goals for establishing MPAs (4.54). In addition, the 2003 World Parks Congress called on the international community to establish a global system of marine protected area networks by 2012. It recommended that these networks should be extensive and include strictly protected areas (i.e. no take zones) amounting to at least 20 to 30% of each habitat.

8.15 There is, however, a marked gap between the level of ambition expressed in such commitments and the size of the area currently protected. Currently, less than 0.5% of the world's oceans are protected.[6] This is very different from the situation on land, where nearly 12% of the area is protected.[7] Protected areas at sea tend to take the form of small isolated sites averaging around 1-20km^2.[8] While these may offer some protection to specific locations, they are unlikely to help ecosystem functions to recover on anything other than a very small scale, or to conserve migratory species.

8.16 However there are welcome signs of change. As well as the multilateral commitments described above, several governments are taking national action to develop MPAs and reserves (box 8A). Perhaps the most ambitious of these was the decision of the Australian Government in 2004 to build on a long history of marine planning by creating the world's largest ever marine protection plan for the Great Barrier Reef Marine Park. Thirty-three percent of the park is now inside marine reserves which exclude all fishing (box 8A).

Box 8A EXAMPLES OF NATIONAL POLICIES TO ESTABLISH LARGE-SCALE MARINE PROTECTED AREAS AND RESERVES

Australia has recently increased the protection given to the Great Barrier Reef Marine Park. The park is the largest World Heritage Site protecting one of the richest, most complex and diverse ecosystems in the world. It covers an area larger than the UK, the Netherlands and Switzerland combined – equal to about 345,000 km^2.

The reef itself is formed from a maze of over 2900 reefs and 940 islands and coral islands known as cays. It is home to 1500 species of fish, over 5000 species of molluscs, 350 kinds of hard corals, 350 types of star-fish, sea cucumbers and sea urchins, 30 species of cetaceans and 22 species of sea-birds. It also has an estimated economic worth of more than £470 million a year, of which £270 million is linked to tourism, £98 million to commercial fishing and £105 million to recreational fishing and boating.[9]

To ensure the park is preserved for the future, an extensive modelling and consultation exercise was carried out to identify areas that needed greater protection. Seventy different habitats were identified within the park and these were used to help design eight different types of management zone (figure 8-I). The zoning process enables highly protected areas to exist within an integrated management plan that allows for other uses of the park, such as recreation and limited fishing activities. This makes the park more practical to manage, reduces conflicts between different types of use and helps preserve ecological functioning over a variety of temporal and spatial scales.[10]

Canada is also committed to a network of no-take marine reserves under its Oceans Act. Parts of the Grand Banks fishing grounds are also closed to various types of fishing.[11]

New Zealand has been a world leader in establishing a national network of reserves,

Figure 8-I

Section of revised zoning plan for the Great Barrier Marine Park. The eight types of management zone are described below and some are shown on the map. Pink = Preservation zone (a 'no-go' area for the public); Green = Marine National Park zone (a no-take zone, where fishing and other extractive uses are not allowed; Orange (not shown) = Scientific research (areas undisturbed by extractive activities); Olive green (not shown) = Buffer zone (a protected zone with most but not all fishing prohibited); Yellow = Conservation park zone (protected area with limited fishing); Blue = Habitat protection zone (reasonable use but trawling prohibited to prevent habitat damage); Light blue = General use zone.

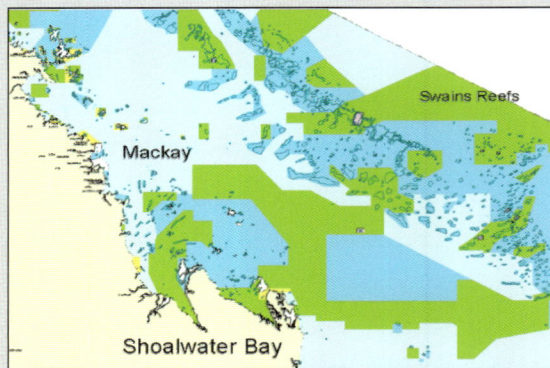

some of which date back 21 years. There are 16 no-take marine reserves, protecting 7% of New Zealand's territorial waters, with a target of 10% coverage of its marine environment by 2010.

The percentage of **South Africa's** coastline that is protected from fishing will rise to 19% in a network of 24 no-take marine reserves. The target is to protect 20% of the area of the exclusive economic zone.

In the **United States**, a Presidential Executive Order was issued in 2000 to strengthen and expand a comprehensive national system of MPAs.[12] It encourages government departments and agencies to work together to enhance existing sites and recommend or create new ones. It also led to the establishment of the National MPA Center[13] to assist the process and disseminate information on MPAs.

Examples of marine protected areas include the Florida Keys National Marine Sanctuary which covers nearly 3000 nm², of which 150 nm² is in no-take reserves. The Channel Island National Marine Sanctuary in California (ca. 1250 nm²) has a network of ten marine reserves and two conservation areas covering a total of 100 nm². In addition, California's Marine Life Protection Act (1999) requires the State to design and manage an improved network of marine protected areas, including highly protected areas within state waters. Less permanent, no-take fishery zones have been established in US Federal and State waters to allow the recovery for fish populations (8.25).

HAVE MARINE RESERVES BEEN SUCCESSFUL?

8.17　In a study of around 80 marine reserves, the biomass of organisms inside the reserves was on average nearly three times higher than in unprotected areas, while organism size and diversity was 20 to 30% greater.[14] Analyses of reserves from all over the world demonstrate that predictions of population recovery in reserves are robust and repeatedly observed.

8.18　Evidence from reserves show that substantial benefits can build up quickly. Improvements are usually seen within two to five years of reserves being established, although the benefits may take longer to materialise for animals that are long lived, slow growing and late to reproduce. Habitat complexity may take even longer to recover. This means that the benefits of reserves may grow for decades and be sustained over the longer term.

8.19　While this report focuses on the role that reserves can play in protecting ecosystems, it is important to consider their benefits for fish populations for two reasons. First, fish are themselves an important part of the ecosystem and an indicator of its health. Measures that benefit fish populations will help to improve the overall environment. Second, by sustaining fish populations, reserves can have important social and economic benefits for the fishing industry. Indeed, three National Research Council Committees in the US have argued that no-take marine reserves are essential for the delivery of sustainable fisheries.[15]

8.20　There is good evidence that reserves do benefit fish. Inside reserves, adult fish populations are larger, longer-lived and more fecund than those in fished areas.[16] Not only do larger and older fish produce more eggs, but the quality tends to better than for first time spawners. Many species benefit from protection, including molluscs, crustaceans as well as fish of a wide variety of sizes, life histories and mobilities. While the scale of the effect varies between reserves, there is a consistent increase in fish biomass in marine reserves across the world. Box 8B provides details on other beneficial aspects of reserves.

8.21　Reserves can also increase fish populations outside their boundaries through what are known as 'spillover' effects. First, as numbers of fish build up within reserves, adults and juveniles can migrate to fished areas outside the reserve. This often leads to higher catch levels immediately outside the reserve and fishers concentrating on fishing the boundary. Second, reserves can be important for protecting spawning and nursery sites. Eggs and larvae from protected species can then be exported outside the protected area on ocean currents to help build up populations elsewhere.

8.22 Reserves can also act as 'insurance' should other management measures go wrong, and since they are untouched by fishing they can provide important scientific information about what a more natural ecosystem might look like. Thus they can help provide information on natural rates of fish mortality that could help improve the accuracy of fisheries models, and data on ecosystem functioning that could be more widely applicable.[17]

Box 8B	FISHERIES BENEFITS OF MARINE RESERVES

Recent reviews have shown that average values of all biological measures are strikingly higher inside marine reserves compared with reference sites (either the same site before the reserve was created or equivalent sites outside the reserve).[18]

Spillover effects have been demonstrated in various species, including crabs in the Sea of Japan,[19] lobsters in Newfoundland[20] and New Zealand,[21] bream in New Zealand[22] and reef fish in Kenya.[23]

In the Scandola Nature reserve in Corsica, population densities of 11 commercial fish species were five times higher in the reserve than in fished sites after 13 years of protection.[24] Similarly, in the Columbretes Island Marine reserve in Spain, lobsters were 6 to 58 times more abundant than at fished sites.[25] In South Africa, four shorefish species were between 5-21 times more abundant within the Tsitsikamma National Park than outside,[26] and a sevenfold increase in larger predatory fish was seen in coral reefs over 11 years of protection at Apo Island in the Philippines.[27]

There is also evidence that reserves enable animals to grow bigger, for example, 35% of blue cod (*Parapercis colias*) in New Zealand's Long Island-Kokomohua reserve were bigger than 33cm after five years of protection, compared with less than 2% in fished areas.[28] Likewise, in the Tasmanian Maria Island Reserve, fish that were larger than 32 cm became six times more common after six years of protection.[29] In the Everglades National Park in Florida, US, the most common size range for grey snapper (*Lutjanus griseus*) was 25-30cm compared with 15-20cm in exploited areas.[30]

This increase in bigger animals translates into increased reproductive potential. In New Zealand reserves, egg production of the lobsters (*Jasus edwardsii*) at deep-water sites increased by over 9% per year of protection.[31] Egg production in snappers was 18 times higher in reserves than in fished areas.[32] After 20 years of protection in the Edmunds Underwater Park in Washington State USA, lingcod (*Ophiodon elongatus*) produced 20 times more eggs than in adjacent fished areas and copper rockfish (*Sebastes caurinus*) 100 times more.[33]

Other examples of successful marine reserves include Chumbe Island in Tanzania (Zanzibar), no-take zones in Belize, marine reserves in Chile, protection of clam fishing grounds in Fiji and the Soufriere Marine Reserve in St Lucia which has an effective community-based network of no-take zones. All these areas have seen significant rises in fish biomass, and where the fishery impinges on the seabed, they have had the secondary effect of protecting, enhancing and stabilising the habitat.[34]

CAN MARINE RESERVES WORK IN TEMPERATE WATERS?

8.23 Much of the evidence for the success of marine reserves and other protected areas comes from sites in tropical and warm waters. It is partly historical accident that reserves have been mainly associated with such sites. High profile areas such as the Great Barrier Reef are understandably quick to attract conservation attention. However, effective protection can be designed for virtually any habitat. Successful examples can be found in the deep sea as well as on continental shelves, and in diverse habitat types such as coral reefs, kelp forests, seagrass beds and gravel banks.[35]

8.24 Research shows that reserves can also work well in warm-temperate and temperate waters. For instance, successful reserves, fishery closures and protected areas have been established in the waters around New Zealand, South Africa, Chile, the US[36] and Europe (box 8B and 8.25). In 2004, an ICES Working Group reviewed six examples of temperate MPAs of which five were in Europe.[37] Four were judged to have met their objectives, with selected species showing signs of recovery, while two were less successful. In one case, the effects of a short-term area closure appear to have been counteracted by natural changes in temperature that caused the species of most interest, plaice, to move out the area. In the second case, seasonal closures intended to protect spawning cod stocks were introduced too late to have the desired effect. Enforcement, carried out mainly by vessel monitoring systems, was judged to be effective in the majority of cases, resulting in large reductions in effort within the closed areas.

Recovery of fish, shellfish and benthic organisms in temperate waters

8.25 One of the best documented examples of temperate reserves are the closed areas established on Georges Bank and in southern New England, off the north-east coast of the US. In 1994, NOAA's National Marine Fisheries Service closed a number of areas to mobile fishing gears. The resulting reserves now cover an area of about 20,000 km[2] and represent some of the largest closures anywhere in the world.[38] The action was part of a comprehensive management plan, which included major reductions in fishing effort, and which aimed to allow severely depleted stocks to recover.

8.26 As a result of the measures, there have been increases in demersal species such as haddock, yellowtail and witch flounders. One of the clearest effects has been on scallops. Their density has increased by up to a factor of 14 within five years, with a corresponding increase in larval export to fished areas (figure 8-II).[39] The closures on Georges Bank have brought the scallop fishery back from the verge of collapse and show that reserves can work at large scales and for commercial fisheries.[40]

Figure 8-II

Recovery of scallops as a result of closures on the Georges Bank

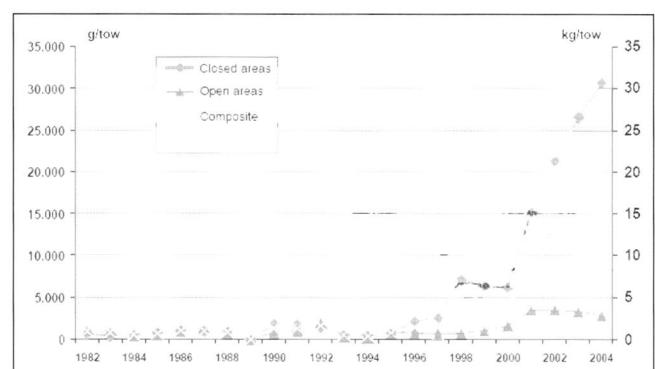

8.27 The results indicate that less mobile species benefited most from the closures and that it could take some time for the full benefits to be seen for long-lived temperate fish species. The researchers also concluded that closures need to be carefully designed to meet specific conservation and management objectives

MARINE PROTECTED AREAS AND RESERVES IN EUROPE

8.28 In our view, marine protected areas and marine reserves offer unique and important conservation benefits. Some successful reserves do exist in Europe, for example, in Corsica, Spain and elsewhere, but they could be much more widely used as a management tool. The ecosystems of northern waters are heavily depleted, yet they still contain many species of interest, as well as recently discovered landscape forms such as deep-water coral reefs, that are in urgent need of protection. While a range of fishery closures and other restrictions operate in European waters, the conservation benefits of these measures are limited (8.33 onwards) compared with those potentially on offer from large-scale MPAs and reserves.

8.29 Evidence from areas such as the north-east coast of the US shows that successful reserves can be established in relatively cold seas that have been heavily exploited by fishing for several centuries. This suggests that there is no scientific reason why many more marine reserves and other forms of MPAs should not be established within the OSPAR region to provide ecosystem protection and sustain fish populations. It is more likely that the reasons for their limited use to-date are related to the relative neglect of the marine environment, the commercial pressure from fishing and the political complexity of fishery regimes.

8.30 Although not a marine reserve, there is at least one example of large-scale integrated management in European waters – the Wadden Sea – and this management model could help inform the development of future initiatives in the OSPAR region. The Wadden Sea contains Europe's largest wetland and is an ecosystem of global importance, situated on the coast of Germany, the Netherlands and Denmark (figure 8-III). The region is co-operatively managed by the three neighbouring countries and is protected by numerous measures. Parts of it have been variously designated as national parks, nature and wildlife reserves, and planning guidelines have been developed to manage activities in the Dutch part of the Wadden Sea. It has also recently been designated by the International Maritime Organisation as a Particularly Sensitive Sea Area covering approximately 15,000 km².

8.31 In 2002, a trilateral Wadden Sea Forum was established to bring together all the environmental and economic stakeholders, as well as local and regional governments. Its task is to develop improved socio-economic proposals for the region that are compatible with protection of the Wadden Sea environment.

Figure 8-III
The Particularly Sensitive Sea Area of the Wadden Sea[41]

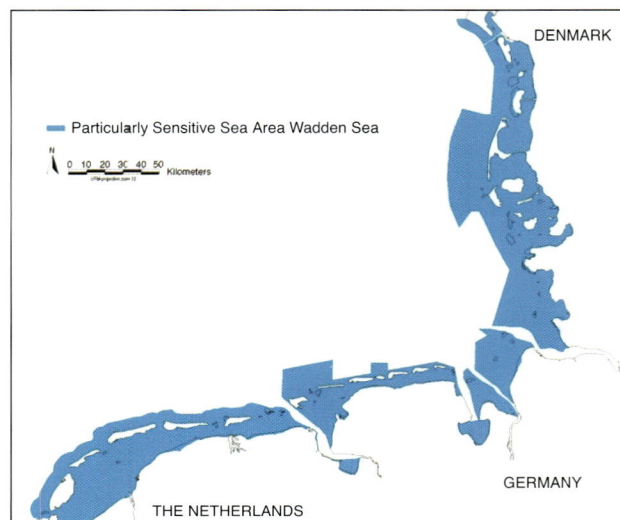

DENMARK

— Particularly Sensitive Sea Area Wadden Sea

0 10 20 30 40 50 Kilometers

GERMANY

THE NETHERLANDS

A proposal to protect the North and Baltic Seas

8.32   Greenpeace has also recently proposed that the Wadden Sea, along with areas of the North
        Sea such as the Dogger Bank, Cromarty, the Viking Bank (all off the coast of the UK), and
        parts of the Baltic Sea, should be designated as marine protected areas. Greenpeace used
        a mixture of expert opinion and knowledge of existing sites to develop proposals that
        would lead to the protection of over 40% of the North and Baltic Seas.[42]

THE CURRENT SITUATION IN THE UK

8.33   The UK already has a number of area-based measures that seek to protect parts of the
        environment or control fishing (chapter 4). If the UK were to decide to strengthen its
        protection of the sea, these areas could be considered for inclusion in a future system of
        marine protected areas. These measures include:

- Three statutory Marine Nature Reserves;

- Conservation areas such as the Natura 2000 sites designated under the EC Habitats and
  Birds Directives;

- Areas that are closed to fishing to allow a particular commercial species to recover or
  to protect its critical life stages;

- Other closed areas around ports and navigational hazards such as artillery ranges,
  wrecks, pipelines, oil and gas rigs, wind-farms, etc (10.38).

8.34   Over a thousand such sites exist in UK waters.[43] Many of these areas contain valuable
        wildlife or fishery assets where better protection could be very beneficial. These places
        could form part of a future network of marine protected areas in UK waters. However,
        they were not generally established with this purpose in mind and they have some
        important limitations in terms of design, size and degree of protection. For the most part,
        they have not been assessed for their ability to contribute to conservation objectives.

UK Marine Nature Reserves

8.35   At present, there are only three Marine Nature Reserves in the territorial waters of UK:
        Lundy, Skomer and Strangford Lough. Of these, only Lundy incorporates a no-take zone
        based on a byelaw introduced by the local Sea Fisheries Committee in 2003. This section
        accounts for just over 3 km^2. Thus, out of England's 48,000 km^2 of territorial waters, only
        0.006% is permanently set aside for wildlife to recover from fishing pressure. As well as
        being small, at least one reserve, Strangford Lough, has been unable to preserve the habitat
        for which it was designed (horse mussel beds) as a result of the failure to prohibit scallop
        dredging.

8.36   Another limitation is that the legislative framework underpinning Marine Nature Reserves
        is weak when it comes to its practical application. A single objection is enough to prevent
        a Marine Nature Reserve being created.[44] The legislation also fails to provide the means to
        prohibit fishing. Hence, Lundy's no-take zone was established by a fisheries byelaw, rather
        than by statutory means under the Marine Nature Reserve legislation.

Sites protected by the EC Habitats and Birds Directives

8.37 The EC Habitat and Birds Directives are two of the most important pieces of European conservation legislation (chapter 4). They have led to protected sites being established in estuaries and coastal waters across Europe. These are known as Special Areas of Conservation and Special Protection Areas, which along with Ramsar sites designated under a separate Convention, are collectively known as Natura 2000 sites. The UK has nominated a number of areas as candidate Special Areas of Conservation which, with one exception, are all within 12 nm of the coast (figure 8-IV). There are currently 66 marine Special Areas of Conservation covering around 5000 km². Only one wholly marine Special Protection Area has been designated to date, but work is under way to identify further sites.[45]

Figure 8-IV

Marine candidate Special Areas of Conservation in the UK

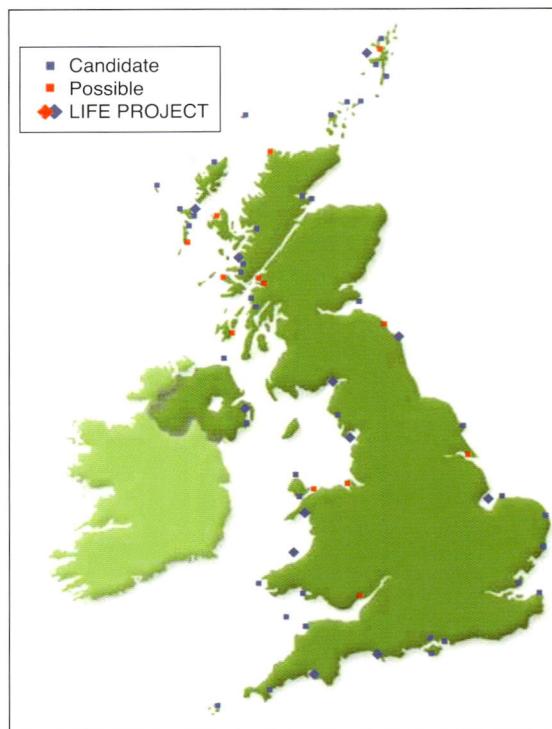

8.38 The Joint Nature Conservation Committee (JNCC) and the country nature conservation agencies have put considerable effort into identifying Special Areas of Conservation and Special Protection Areas. While it has taken some time to identify sites and develop management plans for them, the process has not been without success. For example, the Wash and North Norfolk Coast marine Special Area of Conservation has won international praise for the way in which fishers have been involved in its management. The governing group is led by the Eastern Sea Fisheries Joint Committee and includes 12 other organisations, mainly from local and regional government.

8.39 Until recently, it appeared that Special Areas of Conservation could only be proposed within the UK's 12 nm limit. However, a High Court ruling in 1999 established for the first time that the EC Habitats Directive applies to all seas over which the UK exercises jurisdiction, which effectively means out to the 200 nm limit. The Darwin Mounds is the first candidate Special Area of Conservation outside the 12 nm limit to be submitted by the UK. It is possible that further offshore Special Areas of Conservation will be designated, although this process may be made more difficult by a lack of scientific information that might be offset in part by wider use of proxy geophysical information (2.56). The Darwin Mounds also set another important precedent since it is protected from fishing by special Common Fisheries Policy measures.

8.40 The EC Habitats and Birds Directives were primarily designed to protect the terrestrial and coastal environment, and as a result they are not well tailored to marine situations. In evidence to us it has been argued that the Directives' coverage of marine habitats should be extended, for example, to seamounts and other deep-sea features[46,47] (appendix E).

A recent report[48] also noted the over-emphasis on specific habitats. The fact that sites fulfil certain functions, such as acting as spawning grounds, or coincide with oceanographic processes such as upwellings or eddies, has not been enough to secure their designation.

8.41 It has been suggested to us that the EC Habitats Directive provides for the possibility of designating linked habitats.[49] On land, this could include migration corridors, while at sea it could include linked habitats such as spawning and nursery grounds. However, Member States have made little use of this provision to-date. By comparison, the OSPAR initiative, while not comprehensive, covers considerably more aspects of the marine environment, in terms of linked habitats and individual habitats and species.

8.42 The EC Habitats Directive also fails to list a number of important marine animals in its Annexes of protected species. Among many others, these include the Atlantic salmon outside of fresh waters, basking shark, common skate, spotted ray, orange roughy, ocean quahog and flat oysters. The selective nature of the two Directives reduces their value for use in building representative networks of marine protected areas, or in providing protection to sites to enable re-colonisation and recovery of species that have been eliminated by fishing or other impacts.

8.43 Another major concern is that the sites designated under the EC Directives are often poorly protected from fishing. This is despite the fact that fishing is one of the most important causes of damage to the marine environment. Instead, there is a reactive, ad-hoc process by which protection has to be specifically invoked through local byelaws, Ministerial orders or measures under the Common Fisheries Policy. Action is thus usually only taken on the basis of proof of existing threats or damage, not in the interests of precaution. It also depends on Member States being willing to use their discretion to tackle the impacts of fishing. The conservation value of these areas could be considerably improved if fishing were integrated into the strategic management plan for such sites, or banned where it was causing significant damage. Indeed, it can be questioned whether such sites can meet their conservation objectives at all without proper protection from fishing.

8.44 In summary, the EC Habitats and Birds Directives afford some protection to parts of the marine environment. However, worthwhile protection will require much more stringent restrictions on fishing and other harmful uses than currently applied. The conservation outcome could be considerably enhanced if the Directives reflected the interconnected nature of the marine environment and included more species and habitats. This would also increase the chance that the Natura 2000 sites could help establishing the core of a network of UK MPAs in the short-term.[50] **We therefore recommend that the UK government should:**

- **amend legislation to allow UK Marine Nature Reserves to be designated even where there are objections;**

- **introduce measures to protect all designated sites (such as Natura 2000 sites) from the adverse effects of fishing. If such measures cannot be agreed under the Common Fisheries Policy, the UK should introduce unilateral measures to protect these sites;**

- **review the operation of the EC Habitats Directive and consider how the Directive's ability to protect the marine environment could be improved and extended to the wider environment as well as vulnerable areas;**

- **with the Joint Nature Conservation Committee, develop proposals to extend the Annexes of the EC Habitats Directive to provide adequate coverage of important marine species and habitats (see paragraphs 8.40-8.42); and**

- **use the findings of the above review to press the EU to amend the EC Habitats Directive.**

Areas closed to fishing

8.45   In the UK, a limited number of local initiatives have been set up to close areas to fishing. These have been implemented through voluntary agreements, fishery regulations and byelaws. The sites include experimental areas closed to scalloping off the Isle of Man,[51] the Inshore Potting Agreement around Start Bay in Devon,[52] the area closed to mobile gear in the St Abbs-Eyemouth Voluntary Reserve[53] and two areas of reef closed to scallop dredging in Lyme Bay in Devon.[54] Benefits have already been seen in some areas. For example, in the Irish Sea, densities of scallops increased nearly four-fold in ten years after an area around the island of Skomer was closed to scallop fishing.[55]

8.46   There is historical evidence that larger scale closures can play a role in stock recovery. For example, during both world wars, fish populations increased considerably in the North Sea because fishing was severely restricted.[56] The Common Fisheries Policy has taken advantage of this fact and begun to impose area closures to protect the breeding and spawning grounds of several fish species. These areas include the 'Plaice Box', which was intended to protect juvenile plaice and sole in the North Sea (although there is evidence that it is not proving effective for plaice – 8.24).[57] A 'Mackerel Box' has been established in the western Channel and there are closed spawning grounds for herring and cod (figure 4-II). An assessment of the seasonal closure of the 'Cod Box' in 2001, suggested that it was poorly timed, had only short-lived benefits and led to increased effort outside the closed area.[58]

8.47   The Common Fisheries Policy closures cover relatively large areas, but are designed to help individual fish populations recover from the effects of over-fishing and to prevent premature capture of young fish. Fishing for other species continues unabated in the area. As a result, the closures appear to offer little protection to other species and parts of the ecosystem, and their full environmental effect has not been evaluated. The cost of implementing fishing closures, in terms of selection, demarcation and enforcement, may be similar or less than those of establishing fully protected reserves that could achieve a far broader set of conservation objectives, and might well provide more effective protection for commercially important species.

Conclusions on existing measures

8.48   We already have numerous policies for protecting the sea. However, it is clear that many of the current conservation measures are operating on a tiny scale. The degree of protection is often weak, with fishing continuing unabated even in areas of international conservation importance. High standards of proof are needed to identify sites and the need for action to prevent damage. This is inconsistent with our view that a precautionary approach should be taken to protect the marine environment. In addition, the legislation and processes are complex and unable to protect the wider ecosystem.

8.49　In our view, the level of protection needs to be considerably strengthened to allow a degree of recovery from the effects of human activity, and to provide a more sustainable future for the marine environment and fishing. Creating significant areas of MPAs and marine reserves could offer much greater protection from fishing, one of the main ecosystem-altering activities, and represent a real step towards implementing the ecosystem approach.

THE CASE FOR NETWORKS OF MARINE PROTECTED AREAS (MPAs) AND RESERVES

8.50　Rather than establish isolated MPAs and reserves, the scientific evidence suggests that it is much more effective to link protected areas together in a network that is better able to accommodate the larger scales on which ecological processes operate.

8.51　A network of marine protected areas can be defined as a collection of individual sites that are connected in some way by ecological or other processes. For example, marine areas are connected by the movement of water, which also transports materials, animals and plants between them. Networks of MPAs are particularly appropriate for highly mobile populations. A single MPA may not be large enough to support self-sustaining populations of such a species, whereas a network of MPAs that link up different stages in its lifecycle may be more effective. Networks are also an important way of helping to protect habitats or populations on which there is only limited information on their requirements or vulnerability.

8.52　Five key principles for designing networks of marine protected areas have been suggested:[59]

1) Networks should represent the full spectrum of biological diversity, not just a subset of habitats or species of special interest by reason of commercial importance, rarity or endangerment;

2) Habitats should be replicated in separate protected areas;

3) Networks should be designed to ensure that protected areas are mutually supporting;

4) The total area protected and its distribution into different MPAs should meet the objective of sustaining species and habitats in perpetuity; and

5) The best scientific information, local and traditional knowledge should be used to guide choice of protected areas.

8.53　It is important to be clear about the objectives that a network or individual MPA is intended to achieve. A variety of organisations have set out possible criteria to assist their design, including OSPAR, IUCN, the U.S. National Center for Ecological Analysis and Synthesis and the Great Barrier Reef Marine Park Authority. Some examples of these criteria are described in appendix L.

8.54　Broadly speaking, networks targeted at conservation need to include areas that are most valuable for a particular aspect of biodiversity, and capture elements of the ecological processes that support that diversity. To help sustain fisheries, networks need to include

195

productive areas that are important for some aspect of the lifecycles of commercial species. However, there are many instances, including the examples in this report, that show that it is possible to combine these objectives and provide a positive outcome for both biodiversity and fisheries.

8.55 With the selection criteria in mind, there are four main ways to select and prioritise candidate sites: modelling; ad-hoc choice; relative scoring or ranking by expert opinion; or a combination of two or more of the approaches. Whatever method is chosen to identify potential sites, there will be important practical considerations. Establishing marine protected area networks is not a simple task, particularly where marine habitats are under joint or international jurisdiction. Careful consultation with a broad range of stakeholders during the design and implementations stages is also vital for success (8.91). However, we have enough scientific information to begin to identify the most vulnerable habitats and environments that could form the core of any future network. In addition, adding almost any piece of habitat to the system will be beneficial during the early stages of establishing MPAs.[60]

HOW BIG DO NETWORKS NEED TO BE?

8.56 There is much debate about how large an area should be devoted to MPAs reserves. The size of an individual site or network will depend on its objectives, the species concerned, the current population, and the effectiveness of the management regime in waters outside the reserve areas. In broad terms, however, the larger a reserve is, the more it is likely to benefit a range of species, habitats and processes.

Managing risk

8.57 Building a system of reserves may be regarded in a similar manner to building an investment portfolio, with a certain amount of spread betting to cover uncertainties. Spreading risk means including in the portfolio reserves from a wide variety of habitats and locations and therefore maintaining responsiveness to changing conditions. These include seasonal and annual fluctuations as well as the risks from longer-term factors such as climate change. There needs to be careful choice of the ecological and other criteria used to select the areas (appendix L), and the performance of established reserves must be monitored.

8.58 Managing risk was one of the factors taken into account in deciding how much of the Great Barrier Reef should be protected in the recent rezoning exercise. The Science Advisory Panel for the Great Barrier Reef Marine Park identified the following arguments for setting aside substantial amounts of the marine environment as no-take zones.[61] These general principles, which are not location-specific, are outlined in box 8C below.

Box 8C   HOW MUCH OF THE ECOSYSTEM NEEDS TO BE PROTECTED BY MARINE PROTECTED AREAS?

Risk minimisation – protecting a large proportion and replicate examples of a marine area – in total 20% or more – will reduce risks of over-exploitation of harvested resources and consequent effects on the ecosystem, whilst leaving reasonable opportunity for existing activities to continue in the remaining areas.

Connectivity – the life cycles of most marine organisms mean that offspring from one area often replenish populations in other areas (referred to as 'connectivity'). As more areas are closed to extractive activities, the benefits to the whole system through such connectivity (both among reserves and between reserves and non-reserves) is expected to increase, thereby offering greater security for conservation.

Resilience against natural and human catastrophes – for any one disturbance, much of the network of protected areas should remain intact so that affected areas can recover more quickly and completely through replenishment from other non-impacted no-take areas.

Harvested species – the protection of 20-40% of any fishing grounds in no-take areas can help bring about better management of fisheries, including through increasing fish populations, and permits no-take areas to maintain more natural communities as a whole.

Maintenance of ecological services and goods – in no-take areas, ecosystems can function in a more natural manner that contributes to maintenance of ecological processes. This leads to more sustainable delivery of ecological goods and services to both the environment and humans.

Is 20% protection enough?

8.59   Very few models show significant fisheries benefits if reserve coverage falls below 10%. It is more widely accepted that at least 20% of each habitat needs to be protected in order to provide at least some degree of support for fisheries and biodiversity. This figure has been supported by a U.S. National Research Council report on marine protected areas[62] and adopted by the Marine Conservation Biology Institute in an open letter signed by 1600 scientists.[63]

8.60   By preserving 20% of the habitat, it is assumed that 20% of the biomass of an individual species is protected, and that this may be enough to allow the population to sustain itself. However this is not a precise figure, and there are many uncertainties, including: determining the true size of fish populations, levels of fishing mortality, and the minimum population size that would allow the species to persist. For example, there is evidence that some sedentary species such as lobsters may be self-sustaining at 10% of their unfished biomass, but many more mobile species need to be maintained at 15 to 40%, or even more. Along with other experimental evidence this suggests that 20% protection may not be enough to maintain some species. In 2003, the World Parks Congress recommended that at least 20 to 30% of the sea should be in highly protected areas, where fisheries and all other extractive uses are excluded and other significant impacts minimised.

Achieving a fisheries benefit

8.61 Achieving a more substantial benefit would therefore require greater investment in reserves. The Scientific Advisory Committee for the Great Barrier Marine Park conducted a detailed study of the question of how large reserves should be. They concluded that a minimum of 10-40% coverage would be needed to conserve biodiversity, whereas a larger percentage, 20 to 50%, would be need to minimise fishery collapses and guarantee species persistence for 20 years or more.[64]

8.62 Figure 8-V summarises the findings of 39 studies worldwide that indicated that reserves covering between approximately 20 and 50% of the seas are needed to maximise catches. Reserves sizes of 40-50%, or even more, may be needed in cases where intervention is delayed and fish populations have become heavily depleted.[65]

Figure 8-V

Summary of studies examining the percentage of the oceans that should be protected in order to maximise the measure(s) of reserve performance considered[66]

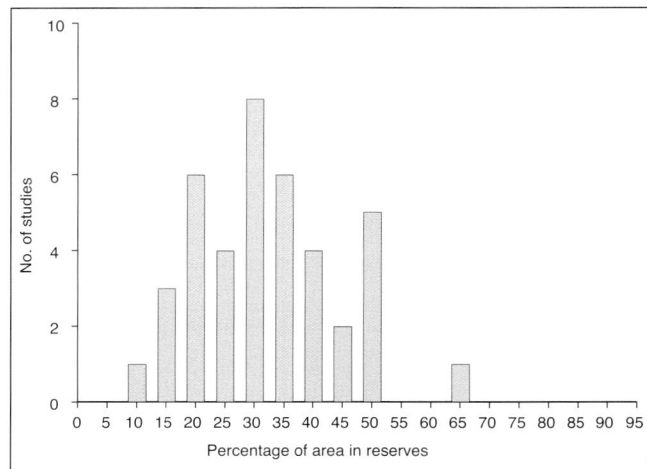

8.63 Given the level of uncertainty, the heavy exploitation of commercial species, plus the need to manage risk in the face of large-scale environmental perturbations such as climate change, we view a figure of 20% protection as very much a minimum level for the OSPAR region. Instead, our view is that **30% of UK waters should be no-take reserves in order to deliver the kind of recovery that is needed to protect the environment and make fish populations sustainable in the long run.**

THE FUTURE FOR MPA NETWORKS IN THE UK

8.64 Many of the current measures are aimed at protecting the 'crown jewels' of the marine environment – highly valued aspects of the environment, or unspoilt areas that approach a pristine state that was once commonplace. These areas undoubtedly need to be preserved and indeed many need greater protection. However, it is becoming increasingly apparent that establishing protected areas on an opportunistic basis can lead to unbalanced conservation portfolios.

8.65 The rationale for a network of marine protected areas is very different. MPA networks preserve ecosystem function by protecting a set of representative sites. They can be designed using systematic methods based on clear objectives. In the UK, a national network of MPAs and reserves could be built up from individual sites, each designed to deliver a particular sectoral goal or goals such as conservation or fishery benefits. Such a system could also deliver a set of national goals that would be determined in consultation with a wide range of relevant stakeholders.

8.66 An MPA network might well include existing designated areas, sites that have only a limited degree of protection or areas that are closed to fishing for other reasons. However to deliver significant improvements in ecosystem protection, a significant proportion of the area would have to be in marine reserves, where fishing would be excluded.

IRISH SEA PILOT PROJECT

8.67 Given the established substantial benefits of MPAs and reserves, a key issue is how to design them most effectively. Work is already underway to identify potential networks in UK waters, the most notable example being the Irish Sea Pilot. The objective of this JNCC study was to develop a strategy for marine conservation in all UK waters and the OSPAR region, using the Irish Sea as a case study.

8.68 A key part of the work was to map the physical and biological characteristics of the Irish Sea. However, biological information can often be difficult and expensive to collect, especially at sea. So the project tested the concept of 'marine landscapes' in which geophysical and hydrographic data are used as proxies for biological records (2.57). The study successfully used this approach to identify habitat types worthy of management attention.

8.69 The project went on to investigate the scope for a marine area network using a model known as *Marxan*. Although this was the first time the model had been tested in the UK, it is the most widely used decision-support system for establishing marine protected areas.[67] The model has been used in the rezoning of the Great Barrier Reef Marine Park in Australia[68] and the California Channel Islands National Marine Sanctuary in the US.[69]

8.70 The model reflects a range of habitat attributes. It then uses a mathematical procedure to identify near optimal protected area networks that meet conservation targets while minimising some measure of costs, usually the total area to be protected. Places known to be priorities for protection can be locked in, and undesirable areas locked out. *Marxan* can be run many times to provide alternative solutions to meeting any given set of targets. From these, a selection frequency or 'irreplaceability value' can be calculated for each planning unit, indicating its relative importance for meeting the given targets.

8.71 An example of the outputs from the model is shown in figure 8-VI. Since the Irish Sea Pilot was essentially a proof of concept, it did not recommend a specific network design. However, it did suggest that an ecologically coherent network of important marine areas should be a crucial element of future marine conservation, and that it could be developed at the regional sea level, based on principles and criteria used in the study.[70] Nonetheless, the

Figure 8-VI

A possible protected area network for the Irish Sea. Areas in purple represent Special Areas of Conservation and the best examples of estuarine habitat. Red areas indicate additional sites selected by modelling that could form part of a network of marine protected areas covering 20% of the Irish Sea

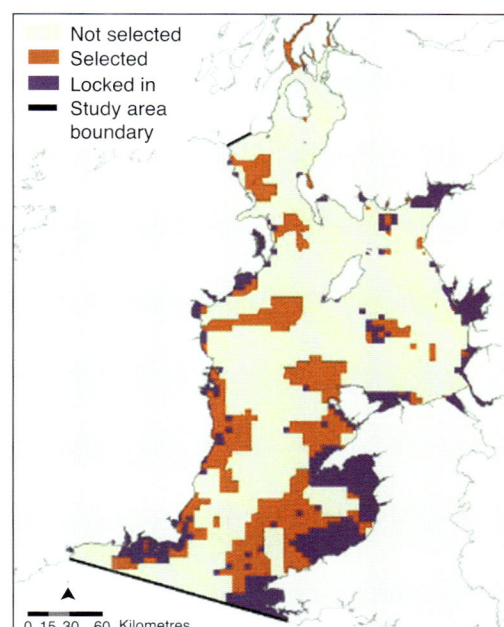

199

report noticeably falls short of recommending that these areas should be fully protected from the impacts of fishing. Moreover, sites that have been selected on a species-basis may not guarantee representative protection for habitats or ecosystems.

DEVELOPING RESERVE NETWORKS IN THE IRISH AND NORTH SEAS

8.72 This section describes more recent modelling work carried out for the Royal Commission which is intended to extend the work of the Irish Sea Project. More comprehensive protected networks were designed to cover around 30% of the area of both the North and Irish Seas. They were developed with the same model, *Marxan*, and habitat maps used in the Irish Sea Pilot Project, together with a set of design principles from the US National Center for Ecological Analysis and Synthesis Marine Reserves Working Group.[71] These examples are for illustration purposes only. They demonstrate some of the methods that can be used to build protected area networks, and the wide range of possible options available to managers.[72]

8.73 The model was used to examine two types of scenario. The first were *Biodiversity Only* scenarios whose primary objective was conservation. In the second set of *Fishery + Biodiversity* scenarios, the needs of commercially valuable mobile and migratory species were incorporated into the design process. This was done by identifying areas of great importance to them, i.e. spawning and nursery sites, and assigning these high priority for protection. In this way, fishery protection and rebuilding objectives were included alongside conservation goals in designing the networks (more detail is provided in appendix L). Such reserves could help improve catches of familiar species such as cod, whiting, haddock, hake, plaice and sole as well as scallops, lobsters and crabs.[73]

8.74 In modelling the Irish Sea, detailed information on marine seascapes was already available from the JNCC's Irish Sea Pilot Project.[74] This enabled options to be explored for the entire area, including the coast. In the North Sea, less information was readily available in coastal areas, but data on habitats and fisheries were used to design networks for offshore regions from 12 nm to the edge of the exclusive economic zones of countries bordering the North Sea. Results for two of the ten best model networks from each scenario are shown for the Irish Sea in figure 8-VII, and for the North Sea in figure 8-VIII. The results reveal the great flexibility in choice of sites for protection using decision support tools like *Marxan*.

8.75 Figures 8-VIIc and 8-VIIIc show the number of times an individual grid (known as a planning unit) was included in a network, based on the summed scores from 1000 runs of each scenario. It is clear from this figure that there are few sites that can be considered 'irreplaceable' in the *Biodiversity Only* networks (i.e. few occurred in large numbers of solutions), emphasising that conservation objectives can be met in a flexible way with many design options available to planners.

8.76 In the *Fisheries + Biodiversity* scenario (figures 8-VIIe and 8-VIIIe), areas of high spawning and nursery value were important in meeting network targets in the North Sea, showing up as planning units with a high frequency of inclusion in networks. In the Irish Sea, such areas were locked into the scenario and so are included in all networks. In this case, planning units adjoining spawning and nursery areas also appear in many network solutions as the program clumps planning units selected for protection to minimise the cost of meeting targets.

Figure 8-VII

Two of the ten best Irish Sea reserve network configurations. Based on 10,000 runs of Marxan for the *Biodiversity Only* scenario (a,b), and for the *Fisheries + Biodiversity* Scenario (d,e). Number of times each planning unit was chosen for inclusion in a reserve network from 1000 runs of Marxan for the *Biodiversity Only* (c) and *Fisheries + Biodiversity* (f) scenarios. Coverage = 33 to 38% (appendix L).

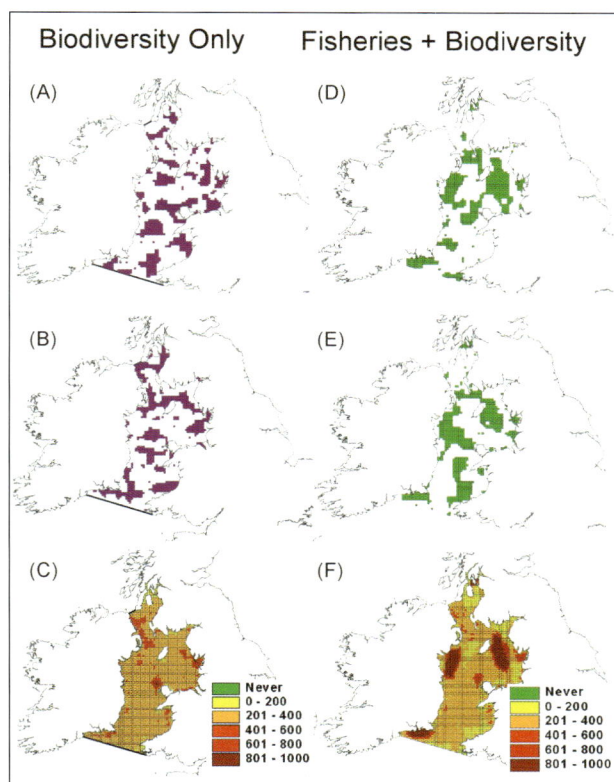

Figure 8-VIII

Two of the ten best North Sea reserve network configurations. Based on 10,000 runs of Marxan for the *Biodiversity Only* scenario (a,b), and for the *Fisheries + Biodiversity* Scenario (d,e). Number of times each planning unit was chosen for inclusion in a reserve network from 1000 runs of Marxan for the *Biodiversity Only* (c) and *Fisheries + Biodiversity* (f) scenarios. Coverage = 29 to 34% of the total area (appendix L).

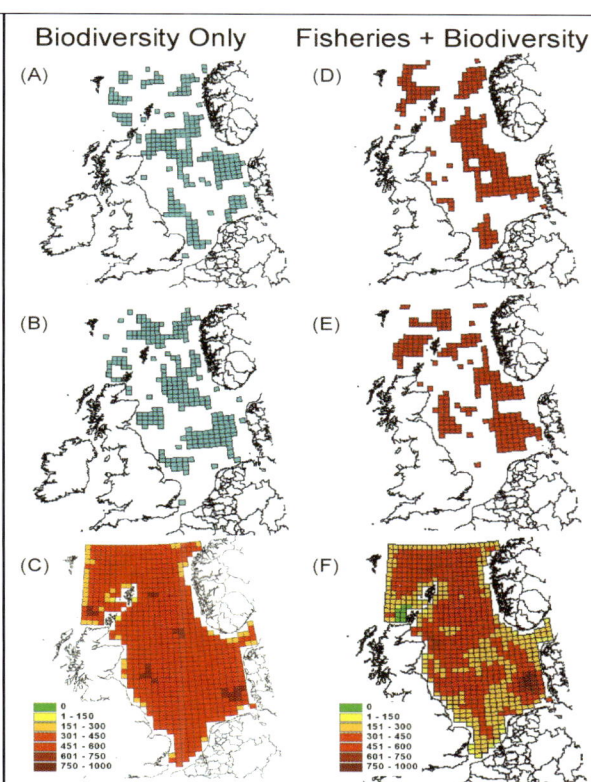

COST OF MPA AND RESERVE NETWORKS

8.77 A worldwide survey of the management costs of 83 MPAs found that costs decreased as distance from the shore increased and that small reserves cost more to run per unit area than large reserves.[75] Hence there are significant economies of scale for larger MPAs. The same authors found that the cost of running a global system of MPAs covering 30% of the oceans would be £6.5 to £7.5 billion a year. This is less than the present global spend on subsidies to commercial fisheries, which is in the region of £8 to 16 billion annually.

8.78 In the examples for the North and Irish Seas, running costs are estimated to range from around £4.7 to £7.5 million per reserve per year (appendix L). The costs are slightly lower for the *Fisheries + Biodiversity* networks, reflecting the greater efficiency of the model in identifying suitable networks at minimum cost in terms of protected area coverage.

8.79 Although the total area covered by the protected area network in the Irish Sea is only around 10% of that in the North Sea, estimated annual management costs are similar because the Irish Sea network is divided into smaller, more numerous protected areas than the North Sea and includes territorial seas from 0 to 12 nm from the coast, all of which tend to increase costs.

8.80   The estimated costs of running a national marine reserve network in the UK compare favourably with the cost of similar measures on land. For example, the National Parks in England and Wales cost around £35 million a year compared with a total of £9.4 to £15 million a year for protecting both the North and Irish Seas which cover a much larger area. The Royal Commission has also calculated that a marine network would be cheaper on an area basis costing an average of around £25-35/km² per year for the North Sea, and £240-370/km² per year for the Irish Sea. This compares with about £2450/km² for National Parks.

8.81   The costs are also small compared to the potential benefits. Fisheries in the Irish Sea were worth £60 million in 2002, and those in the North Sea were worth £226 million (UK fleet only). Theoretical work and evidence from field studies suggests that networks of MPAs will increase revenues from many fisheries. A 10% uplift in fishery productivity in the Irish Sea and a 2-3% uplift in the North Sea would pay for the running costs of the network. Such values lie well below increases in fishery landings seen in places with well-established protected area networks. Other benefits could include increased tourism and recreation opportunities[76] as well as the protection of wider ecosystem services.[77]

8.82   A UK network of protected areas and reserves would benefit marine ecosystems and make fish populations more sustainable, therefore the fishing industry should benefit from reserves in the medium and long-term. This is particularly evident when compared to the outlook for the industry if significant measures, such as reserves, are not taken to encourage recovery in fish populations. Without a 'large stock approach', the Prime Minister's Strategy Unit report[78] suggests that the threat of collapse could continue to hang over a number of commercial species for many years.

8.83   UK landings of fish have declined steeply in value from a high of nearly £800 million in the mid-1990s to about half that figure by 2003.[79] If the management regime remains as it is, then catch values can be expected to continue to decline on the present trajectory as a function of falling fish populations. However, if a marine reserve network were implemented in 2005 across 30% of UK waters, catches could be expected to dip initially before swinging upwards as populations of commercially important species rebuild and productivity recovers.[80]

8.84   Some commentators regard the measure of benefits from a marine reserve network as the difference between catches at the time of implementation and those measured at any given point thereafter. However, as figure 8-IX shows, another way of assessing benefits is to compare the difference between extrapolated catches without reserves and those at any given time following the onset of protection. This would lead to a much greater level of benefit being ascribed to reserves for the example shown in figure 8-IX.

8.85   Although the diagram is illustrative, the timescale over which reserve benefits accrue is based on experience in the similar environment of Georges Bank in the US.[82] The catch value with reserves would be expected to level off after 30 to 40 years of protection, as habitats and long-lived species recover.

Figure 8-IX

Potential benefits from a marine reserve network established in 2005 and covering 30% of UK waters
The solid line shows actual UK catch values from 1992-2003, converted to 2003 values.[81] The dashed line
extrapolates the present trend of falling catch value for a business as usual management scenario. The
dotted line shows catch values following implementation of the reserve network. Where there is no
reduction in fishing effort at the time of reserve creation, theory predicts an initial decline in catch per
unit effort due to reduction of fishing area, followed by an increase as reserves rebuild stocks. The solid
arrow shows the difference between catch values at the time of protected area implementation and those
in 2022 (shown by the horizontal line). The dashed arrow shows a second measure of benefit, which is
the difference between the extrapolated trend of declining catch value and catches in 2022.

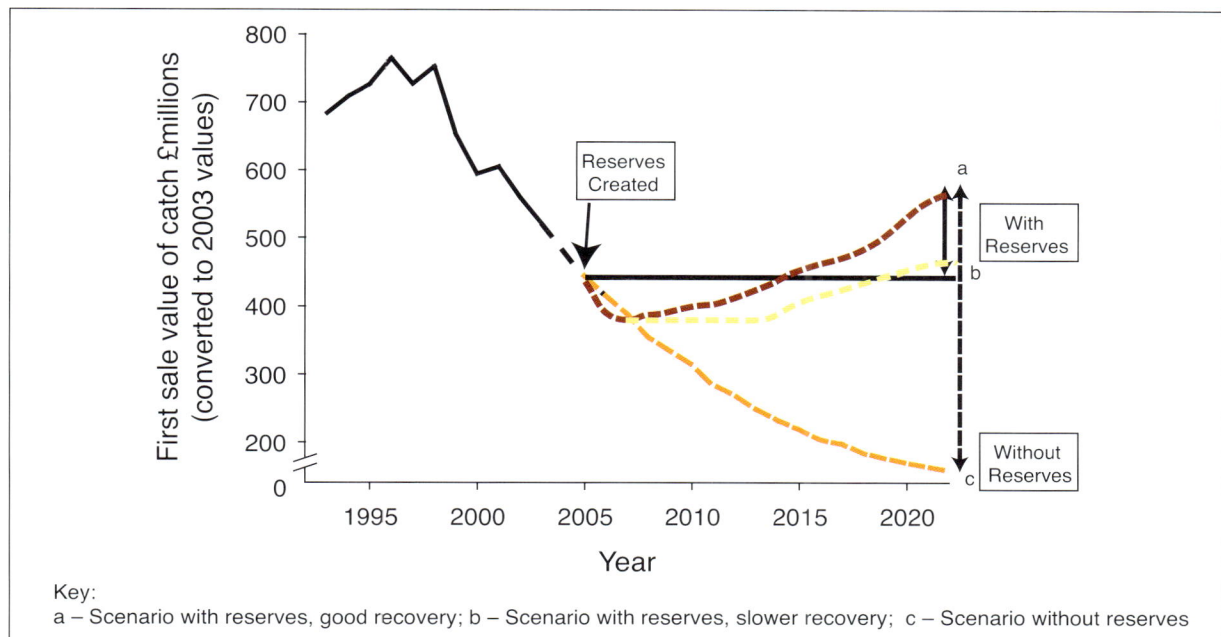

Key:
a – Scenario with reserves, good recovery; b – Scenario with reserves, slower recovery; c – Scenario without reserves

8.86   There is a justified interest in what the impact would be on the fishing industry during a
transitional period, between the implementation of a reserve network and the beginning
of recovery in the ecosystem. Closing areas to fishing will lead to a redistribution of fishing
effort and this will have two main effects; one on fishers and one on fish populations and
habitats. Both can be mitigated.

8.87   The first impact is socio-economic because some fishers will have to adjust their fishing.
Better catches in future could improve the financial security of individual fishers and the
industry in the longer-term, but in the short-term some may struggle. In a previous case,
a one-year payment was made to fishers to counteract falling catch rates after the
establishment of a reserve network, but was no longer necessary after five years when
catch rates were well above previous levels.[83] In Australia, a structural adjustment package
is on offer following the rezoning of the Great Barrier Reef Marine Park. This includes
restructuring grants, business advice and assistance to leave the fishing industry, as well as
aid to employees and communities.[84] **It may be appropriate to consider ways of
assisting the fishing industry through the transition to protected areas in the UK.
This could be done in tandem with measures to reduce overall capacity, which we
discuss in Chapter 9.**

8.88 The second impact of redirection of fishing effort is on fish populations and their habitats. Some argue that redistribution of fishing effort will be harmful because it could move effort to places where fish and their habitats are more vulnerable to depletion or damage or are presently little fished.[85,86] Such an effect is possible, but it can be avoided by using knowledge of the distribution of fish stocks and habitats to identify the most vulnerable areas and include them in protected areas. However, areas that are currently little exploited have almost certainly received considerable historical impact, since the seas around the UK have been fished over and over for centuries. While our seas continue to be productive and of conservation interest, they are far from being pristine habitats. Such areas may thus be relatively little affected by increased fishing. Measures to cut total fishing effort would also markedly reduce the size of this effect. Moreover, it has been suggested that the impact of displacement on the ecosystem would be less under a system of permanent closures rather than temporary ones,[87] since complete closures would lead to a one-off, permanent redistribution of disturbance.[88]

CONCLUSIONS ON MPA NETWORKS IN NORTH AND IRISH SEAS

8.89 To design practical reserve networks in these two areas would require further work employing a wider range of information on both biodiversity and fisheries. For example, in the North Sea, the analyses shown here did not incorporate oceanographic considerations such as persistent frontal features, primary productivity, plankton or current strength. Some of these water column characteristics were accommodated in the classification of seascape types in the Irish Sea analyses,[89] but other kinds of data would still be valuable in more comprehensive analyses. For instance, data on the distribution of gravel extraction, dredging activity and oil and gas licenses would help in designing networks around known sources of impact.

8.90 In summary, there is sufficient information to design comprehensive, representative and adequate networks of marine protected areas for UK waters. The seas around this country have been scrutinized in great detail since at least the mid-19th Century. There is certainly more information available for UK waters than for the Australian Great Barrier Reef where the *Marxan* model was used successfully to rezone the park and establish a practical management programme.

CONSULTATION

8.91 Whatever the method for selecting MPAs, it is essential that the development of a network should be supported by a well thought out consultation process, together with a careful explanation of the underlying principles. Consultation should be carried out at regular stages through the development of proposals and with a wide range of stakeholders. The objective should be to work with stakeholders to produce a design that provides optimal benefits while avoiding unnecessary restrictions and costs, and which allows the system to be as effective and easily enforceable as possible.

8.92 While proposals for MPAs can be controversial, there is a now wealth of experience on how to conduct such consultations. Examples include MPAs in the Gulf of Mexico,[90] the Florida Keys National Marine Sanctuary[91] and the extensive consultation carried out as part of the rezoning of Great Barrier Reef Marine Park (box 10.B).[92] Government, regulatory and other bodies also have a great deal of experience of conducting consultations on relevant aspects of conservation policy. For example, JNCC has recently completed an in-depth consultation to establish Natura 2000 sites. While consulting on a network of MPAs and reserves might initially require a greater level of funding and resources than past exercises, the benefits from such networks are likely to be substantially greater than those obtained by a piece-meal consultation process on small-scale sites.

THE POLICY RESPONSE

8.93 The UK and other European countries have committed themselves through OSPAR to establishing MPAs (although these would not necessarily be fully protected from fishing). However, practical proposals to establish such areas are embryonic and the legislative instruments needed to ensure that they receive adequate protection are missing.[93]

8.94 To-date, the European Commission has announced its intention to spend €5 million on research investigating the benefits and applications of MPAs over the next three years. MPAs are also mentioned as a possible policy tool as part of the European Commission's proposals for the EU Marine Thematic Strategy.

8.95 In the UK, the Prime Minister's Strategy Unit report recommended that marine protected areas should be established on an experimental basis, and their economic and biological impacts studied. This process would begin in areas of multiple use, but the degree of protection, size and timescale for implementation is not evident from the report. We note, moreover, that the UK government is not committed to implementing any of the recommendations in the report.

8.96 We view these responses as too tentative and too slow. Firm evidence exists that marine reserves can provide habitat protection and form part of an effective response to the effects of over-fishing.[94] Evidence of the latter is all too obvious in the OSPAR region. We also have sufficient information to identify some of the most vulnerable sites that could form the basis of future networks. There is a strong case for establishing large-scale protected areas as part of a suite of management measures. Therefore **we recommend that the UK government should:**

- **develop selection criteria for establishing a network of marine protected areas so that, within the next five years, a large-scale, ecologically-coherent network of marine protected areas is implemented within the UK. This should lead to 30% of the UK's exclusive economic zone being established as no-take reserves closed to commercial fishing; and**

- **develop these proposals in consultation with the public and stakeholders.**

A BALANCED PACKAGE OF MEASURES

8.97 The creation of MPAs and reserves could be an immensely important step in turning round the state of the marine environment, but it will not be enough on its own. There are some threats such as contamination from pollutants and climate change that MPAs cannot prevent; they can only limit their effects. Equally, some processes operate on a larger scale than it would be feasible for a network of MPAs to encompass. Protected areas will provide important benefits for many species, but other highly migratory species will need additional protection outside these areas.

8.98 Therefore **we recommend that the MPA network should be implemented as part of a balanced package of measures to improve the management of human impacts on the marine environment and reduce fishing effort.** MPAs should be part of a suite of measures, which must include reduction in fishing effort to sustainable levels, gear modification and other forms of marine protection. These tools need to be built into a comprehensive marine management framework, which is further discussed in chapters 9 and 10.

CONCLUSIONS

8.99 As outlined in previous chapters, over-capacity in fishing fleets combined with damaging fishing techniques, poor management and limited enforcement has resulted in a widespread degradation of species, habitats and food webs. As a result there is wide international consensus that management of the marine environment should be based on an ecosystem approach that is holistic and pro-active, rather than narrow and reactive. This is made all the more important by the cumulative and changing risks that the marine environment faces from pollutants, human activity and environmental change. We need a robust and flexible response to these threats.

8.100 UK legislation is generally not well suited to protecting the conservation value of the marine environment. Only very small areas have been protected, and the prospects for the future do not look much brighter if we continue to rely on traditional measures. However, a potentially more flexible and effective management option exists in the form of marine protected areas and reserves. These have been used extensively world-wide and shown to be successful. They are also perhaps the single most practical tool for safeguarding ecosystems, rather than selected species and habitats – something that is essential if we are to retain a robust and flourishing marine environment in the face of human activity and long-term environmental change.

8.101 We therefore view a network of MPAs and reserves as a key tool for bringing about a significant improvement in ecosystem protection. There are, however, some important barriers that need to be overcome to allow for their development. The OSPAR agreements and the EC Habitats and Birds Directives provide a useful starting point for building a coherent network of large scale MPAs that protect the marine environment. However, the legal framework needs to evolve to provide the underpinning for MPAs and reserves, and in particular, to provide a way of protecting them from fishing. **Changes are needed both at the UK and European level, where the UK government should be prepared to make the case for a new EC Directive for the designation of large-scale MPA networks protected from the effects of fishing.**

8.102 Second, implementing this new approach to the marine environment will require a co-operative management process across all sectors involved, as economic and social initiatives will be as much a part of a recovery process as conservation initiatives. In addition to 'top down' approaches to the regulation of activities impacting the marine environment, 'bottom up' approaches should also be built into regulatory structures that promote, support, facilitate and reward local, community based, NGO and private sector initiatives to improve the quality of the marine environment.

8.103 Third, MPAs, reserves and networks can, in the right circumstances, make an important contribution to sustaining the marine environment. Experience, such as that on the Georges Bank, has shown that they are most effective when developed as part of comprehensive package of measures to manage fishing effort, control the use of damaging gears, etc. There are have been examples, where the benefits of MPAs have been undermined by infringements, or by failing to control all the major fishery impacts.[95] In our view, MPAs should be part of a more integrated approach to our use of the sea – one based on spatial planning of all major activities in the sea, not just conservation and fishing. The following chapters make further proposals in these areas.

8.104 Stewardship of the marine environment requires joined up government from the regional to the international level, and the UK Government will need to apply pressure at the EU level to be able to implement many of the measures suggested in this report. UK seas extend over 867,000 km², the coastline stretches over a total length of approximately 12,000 km and its maritime environment contains many highly productive areas as well as valuable wildlife areas. At a time when international policy is moving towards greater protection of the marine environment, based on an ecosystem approach, **we recommend that the UK government should take this important opportunity to provide leadership by improving the marine environment for the benefit of this country, and the other nations with which we share it.** Setting the agenda in this way, would be consistent with the Government's objective of putting sustainable development at the heart of policymaking, and would improve the UK's ability to shape future European policy on marine issues. A better marine environment is also, without doubt, an essential part of securing a brighter future for the fishing industry.

Chapter 9

IMPROVING FISHERIES MANAGEMENT

What management measures are required to conserve the marine environment as a whole and ensure the future sustainability of commercially exploited fish populations?

INTRODUCTION

9.1   This chapter is concerned with reducing the environmental damage caused by fishing through the use of improved management measures, including restrictions on effort, fishing practices and equipment. Such measures aim to benefit the environment as a whole as well as populations of commercial fish species, and thus promote a more sustainable fishing industry.

9.2   The marine environment is being affected by intensive fishing on a global scale. Innovative, potent, and highly capitalised catching technology is being used to meet increasing demand at a time of stable or falling supplies of wild fish (chapter 3). Long-lived fish with low natural mortality have been removed from marine ecosystems and replaced by short-lived species with more volatile population dynamics. There have been changes in predator-prey relationships, and habitats have been degraded by fishing gear (chapter 5).

9.3   The previous chapter explored how protecting parts of the sea could help the natural environment recover from the effects of fishing and rebuild populations of commercial species. While the marine protected areas discussed in chapter 8 have great potential, to be most effective, they need to be combined with a range of other management measures in order to reduce the impacts of particularly damaging activities carried out by the industry.

9.4   The single most important step of this kind would be to ensure that the capacity of the fishing fleet is more closely matched to the level that can be sustained in the medium- to long term, and it is this element of an improved management system that we consider first.

VESSEL DECOMMISSIONING

9.5   The overcapacity in the European fishing fleet has been well documented and its effects are described in chapter 5. The European Commission Directorate-General Fisheries stated in 1994 that:[1]

> 'We know that fishing mortality rates are too high on almost all commercially important fish stocks. Scientific advice is that mortality rates should be reduced on average by about 40%. Current levels of fishing mortality are generated using only a fraction of the fishing effort that the fleet is capable of exerting. This would suggest that there is more than 40% overcapacity in the fleet.'

The European Commission's estimate was based on the premise that 30% of target fish populations were in danger of collapse and a further 20% were overfished.[2] This analysis is also reflected in the European Commission's 2002 Green Paper on the Reform of the Common Fisheries Policy (CFP).

9.6   Such overcapacity leads to unsustainable pressures on fish populations, making them more difficult to manage. It also reduces the profitability and stability of the industry, by leaving it highly exposed to short-term fluctuations in price. Since the commercial fish populations are inadequate to sustain the existing fleet at present levels of profitability, many boats will need to land more fish than their quota allows in order to survive financially.[3] Thus overcapacity creates strong incentives to fish illegally. It is estimated that 80% of UK boats may have landed some illegal fish,[4] and some have suggested that 50% of cod in Britain is landed illegally.[5] In addition, there have been continual improvements in the fishing efficiency of the fleet as a whole (a process known as 'technological creep') that can offset actions to reduce capacity such as decommissioning.[6]

9.7   Fishing capacity is currently defined in terms of tonnage and engine power, but there are many other factors that determine the fishing mortality generated by the fleet. Decommissioning fishing vessels can nevertheless be an effective way of reducing capacity if schemes are properly designed. Decommissioning schemes are most effective if carried out as quick one-off exercises. Repeated decommissioning programmes over a number of years can encourage excess capacity by reducing investment risks for fishers and acting as a financial safety net rather than a single opportunity to leave the industry.[7]

9.8   Care must also be taken to ensure that fishing effort does not shift to boats under 10 m in length, which are subject to a less strict regulatory regime. We recognise the practical reasons why fishing regulations have cut-off points that exempt smaller fishing vessels from regulation. We are, however, aware that smaller vessels with sizeable capacity have been built to benefit from cut off points. **We therefore recommend that the UK government should review the activities and environmental impact of smaller vessels that do not fall under the full set of fishing controls to ensure that the benefits from our recommendations are not reduced.**

9.9   A study by the Prime Minister's Strategy Unit[8] reported on the effectiveness of various rounds of decommissioning between 1993 and 1996 under the EC multi-annual guidance plan (MAGP) III. Around half of the fishers that took up the scheme would have continued fishing if it had not been for decommissioning, but around a third claimed they would have left the industry anyway. Of those taking part in decommissioning, around a quarter returned to sea and another quarter moved into the marine recreation sector. However, the following round of decommissioning under MAGP IV from 1997 to 2001 was ineffective in reducing the EU fishing fleet capacity as the objectives set were too weak.

9.10   Under the 2002 reform of the CFP, the reduction of fishing capacity needed to meet the fishing limits set in the context of multi-annual fisheries management plans is now the responsibility of member states. The last round of decommissioning by the UK was in 2003. The scheme cost £30 million in Scotland, where it aimed to reduce cod fishing by 15-20%. In England, £6 million was spent to reduce the whitefish fleet by 15-20%. There is a separate £5m scheme in operation for the Northern Irish fleet.[9]

9.11 The 2003 schemes did reduce the size of the industry, but in 2004 the Prime Minister's Strategy Unit suggested that a further round of decommissioning was needed in order to help the fleet become financially viable.[10] It recommended that UK fisheries departments should consider funding the removal of a minimum of 13% of the whitefish fleet (on top of the 2003 decommissioning scheme) as part of an overall package of management reforms. It estimated that this would require between £40 and £50 million of public money.

9.12 The report also suggested that the fishing industry would benefit from tying up a further 30% of the whitefish fleet for four years to accelerate fish population recovery, but that this should not be supported by public funds. Since the whitefish fleet operates on very tight financial margins it seems unlikely that many of these vessels would ever return to sea. Presently, it appears that little progress has been made in establishing an industry tie-up scheme, or to promote decommissioning beyond 13%. This is of concern, because the proposed public decommissioning scheme was based on an optimistic scenario of strong fish recovery, perfect compliance and no further warming of the north Atlantic. Under a more pessimistic scenario, reductions in fleet size of up to 52% would be required.[11] It therefore appears that stock recovery and industry profitability will continue to be at risk unless there are further rounds of decommissioning or other forms of intervention.

9.13 The recommendations from the Prime Minister's Strategy Unit and the European Commission were primarily made with the aim of improving the sustainability of the industry and commercial fish populations rather than improving the state of the marine environment. For this reason, along with those raised in 9.12, we are of the view that further action needs to be taken to reduce fishing pressure in EU waters. The result of this additional decommissioning and the regime of effort controls discussed in later in this chapter should be to decrease the overall fishing effort in the UK's exclusive economic zone to sustainable levels. The percentage of decommissioning required remains a matter of some debate and will vary between the various fisheries sector and areas, but is likely to be largest in the case of the whitefish demersal fleet.[12]

9.14 Introducing large-scale marine protected areas, including the 30% no-take reserves recommended in paragraph 8.96, is likely to mean some fishers will have to adjust their fishing. Better catches in future could improve the future security of the industry, but in the short-term some may struggle. It will also be necessary to reduce fishing effort to avoid large-scale displacement of activity from these areas (5.83).

9.15 **We therefore recommend that the UK government and fisheries departments should initiate a decommissioning scheme to reduce the capacity of the UK fishing fleet to an environmentally sustainable level, and ensure similar reductions are made in EU fleets that fish in UK waters.**

9.16 **We also recommend that funds should be made available to help the transition of the industry during the establishment of the UK network of marine protected areas and no-take reserves.** We agree with the Royal Society of Edinburgh[13] that assistance should be made available for those who wish to leave the industry and **recommend that the UK government should review arrangements for EU Structural Funds and other funds to promote economic diversification in fisheries dependent areas.**

REGULATING FISHING EFFORT

CATCH CONTROL

9.17   The existing system of European fisheries management relies on controlling the amount of fish landed. The system is based on limits set by the International Council for the Exploration of the Sea (ICES) for particular species known as total allowable catches (TACs) which are then allocated as quotas for individual vessels/fishers.

9.18   Quotas for vessels sometimes take the form of individual transferable quotas (ITQs), which can be bought and sold. They have been introduced in Iceland and have been considered for the UK. However, ITQ systems rely on similar science and assessment methods as TACs and other quota allocation methods.

9.19   ICES has adopted a precautionary approach to suggesting TACs for European waters, and catch controls have been used successfully to prevent some species being over fished. But the political process of translating those suggested limits into quotas for Member States by the Fisheries Council has meant that scientific advice and the numbers attached to estimates of sustainable yield have seldom been heeded (4.102).

9.20   Catch controls can encourage the race to fish since fishers have an incentive to fill their quota as fast as possible before stocks are depleted. Malpractices, such as overstating reported catch in order to retain historic quota, share have also occurred. Catch controls are also an ineffective way of addressing environmental impacts of fishing, such as the ecosystem effects described in chapter 7, because they have little influence on the types of fishing methods and gear employed. In addition to this, their value for managing fish populations is severely reduced where there are high levels of misreporting or illegal fishing, since this reduces the quality of the data on which the quotas are based. They also encourage practices such as high grading (throwing away low value fish to fill the vessel with higher value fish) and discarding of by-catch or over-quota species. Again this increases the uncertainty over the total catch so that in some cases actual levels of mortality may be well above precautionary limits, increasing pressure on ecosystems.

9.21   Another problem with catch controls is that they do not reflect the interaction between species that is crucial for long-term sustainability of commercial species and the marine environment. These problems are particularly severe for mixed demersal fisheries where the levels of by-catch of non-target species are the highest.

9.22   The reformed Common Fisheries Policy retains catch controls, but proposes modifications to the existing system. TACs will continue to be set annually on the basis of scientific advice. But under the reforms, fisheries management will be constrained by two types of multi-annual plan: recovery plans to help rebuild stocks that are in danger of collapse; and management plans to maintain other fish populations at safe biological levels.[14] These plans may cover single or mixed fisheries.

9.23   Despite these efforts at reform, we consider that TACs have disadvantages for all fisheries and we agree with the Royal Society of Edinburgh that they are an inappropriate management tool for mixed demersal fisheries. We also support the ICES view that TACs should not be used for deep-water species (9.51).

EFFORT CONTROL

9.24 In contrast to catch control, effort control restricts how much fishing activity takes place and is arguably more directly linked to the actual impact on the marine environment. Effort is a direct measure of fishing pressure, and so is a more effective method of control. Calculations of fishing effort are based on the number of fishing vessels, the average potential catching power of a vessel in the fleet, the average intensity of operation of a vessel per unit time at sea and the average of time at sea for a vessel. One factor that cannot be accounted for is the skill of the fishers; those who are more efficient will have higher catches and a greater overall impact for the same calculated effort.[15]

9.25 All the components of effort may be controlled, but in practice restrictions are especially applied to the amount of time that a fishing vessel and, by extension, the fleet as a whole, is allowed to fish, in order to keep fishing mortality within sustainable limits. A key advantage of the system is that it should be simpler to enforce because the emphasis is on monitoring vessel activity (for example, through satellite tracking) rather than the amount of fish caught. Effort control should therefore reduce the incentive for illegal landings, and could be combined with a requirement to land all fish caught, so as to reduce discarding. Effort control would also reduce the burden on fisheries science to produce the data to estimate TACs for each species. Assessments of the size of fish populations would still be necessary but would have a reduced regulatory role.

9.26 The European Commission and the European Council of Ministers have indicated a willingness to begin managing fish populations on the basis of effort control, and have already introduced such a system in EU Western Waters (box 9A). A simple system of effort control has also been instituted in the interim to demersal mixed fisheries in Skagerrak, Kattegat, the North Sea, West of Scotland and the Irish Sea to protect cod stocks and prevent overshooting of quotas. Vessels are allowed a limited number of days of fishing per month depending on gear type and area. These restrictions will remain in place until a replacement system of effort control is agreed. Although these regulations seek to limit effort rather than reduce it, they demonstrate that effort-based management could have a role in future. More sophisticated effort control regimes are already in use in the US and the Faroe Islands.

Making effort control work

9.27 We conclude that effort control is a simpler way of attaining environmentally sustainable fisheries than catch control. Effort control also has its problems that will need to be addressed carefully if it is to be effective. Effort levels are usually set for individual vessels. These take account of vessel engine size (kW), and specify the gears to be used and the areas to be fished within the given days at sea. Such an approach is, however, likely to be challenged by technological creep. This could allow fishing effort to increase without changing the technological specifications of a vessel by, for example, enabling longer tows, larger gears or longer fishing hours. There is also a likelihood that disproportionate effort will be directed at species of highest value, and individual vessels will attempt to build portfolios of effort entitlements using different gears and fishing different areas.[16]

<table>
<tr><td>**Box 9A**</td><td>**EFFORT CONTROL IN WESTERN WATERS**</td></tr>
</table>

The October 2003 Fisheries Council adopted an earlier European Commission proposal establishing a fishing effort based regime in 'Western Waters' (figure 9.I).[17]

The regime seeks to ensure that there are no increases in fishing effort, or major shifts in effort between areas, by keeping a register of effort deployed by vessels within the fishery (in kilowatt-days – the product of the power of the vessel in kilowatts and its time at sea expressed in days, e.g. a 500kW vessel at sea for two days would represent 1,000 kilowatt-days of fishing effort). Member States are required to monitor any changes to ensure that fishing effort does not increase. The effort limits apply to demersal fisheries as well as fisheries for scallops, edible crab and spider crab. Effort calculations will be based on average effort over the period 1998-2002 and a register of vessels over 15 m involved in these fisheries will be kept. These lists may be amended in future, so long as there is no increase in effort.

Figure 9-I
The extent of the 'Western Waters' within which the effort regime will apply[18]

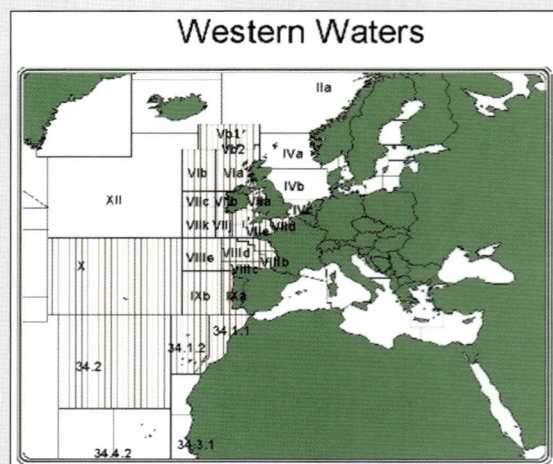

A biologically sensitive area to the south-west of Ireland has also been identified, where a specific effort regime will apply. Member States are required to allocate the identified level of fishing effort to relevant fisheries within this area. A strengthened control package has also been applied, requiring all vessels to report in and out of the biologically sensitive area.

This was the fourth time that the proposed regulation had been discussed by Ministers. The European Commission acknowledged the difficulty of securing agreement on this measure, partly because the regulations abolished previous measures preventing the entry of Spain and Portugal into this fishery.[19]

9.28 There are also likely to be problems in setting up any system of effort allocations, and the distribution of effort allocation between Member States is proving to be a major stumbling block in attempts to introduce effort controls into EU fisheries management (box 9A). It will be necessary to convert vessels' catch quotas as a percentage share of total allowable catch into effort allocations as a percentage of the total allowable fishing time.[20]

9.29 An important aspect of effort control is the ability to determine where fishing takes place and it is likely that specific areas will need to be excluded from fisheries, or allocated for use by specific fisheries, fishing gears or individual vessels (9.32). This could be achieved through a system of spatial plans (10.22), closed areas (8.96), licensing (4.138) and local management arrangements. **We recommend that the UK government should move towards managing fisheries on the basis of effort controls (in terms of kilowatt-hours at sea) within the next three to five years. We recommend that the UK**

government should take steps to ensure that appropriate effort controls are introduced throughout EU waters in the shortest possible time-frame.

Effort quotas

9.30   Effort controls limit the number of boats or fishers who work in the industry and the time the gear can be left in the water. They may also limit the power or size of the vessels and the periods when they can fish. As noted above, access rights to fisheries also may need to be limited, e.g. through establishment of a system of no-take marine reserves and licensing for fishing in specific areas. Other methods currently in use to control effort include individual transferable effort quotas (ITEQs), licences and the decommissioning schemes described earlier (chapter 4).

9.31   Individual transferable effort quotas (ITEQs) have not been as widely applied as individually transferable catch quotas (ITQs), but are an alternative management option. An evaluation for the Department for Environment, Food and Rural Affairs (Defra) of controls on fishing[21] considered that an ITEQ system in the UK would be less effective than either ITQ or existing management systems, as managers would still need to comply with EC regulations (i.e. collect landings data at the individual fisher and species level) in addition to bearing the costs of ITEQ monitoring.

TECHNICAL CONSERVATION MEASURES

GEAR RESTRICTIONS

9.32   Detailed measures to limit the 'how, when and where' of fishing can provide the spatial sensitivity and flexibility towards meeting specific biological/conservation goals for a given species or habitat. In particular, gear regulations in tandem with spatial planning provide the key to protecting sensitive habitat types.

9.33   The spatial aspect of gear restrictions can be realised through use of closed areas (8.33 onwards). These could include areas permanently closed to all fishing, specific fisheries closures, seasonal closures to safeguard spawning stocks or short-term, real-time closures to protect abnormally large populations of juvenile fish present in an area. It may also be necessary to selectively exclude certain types of gears from a particular area or zones. The licensing system (4.138) could be a key element in developing more regionally sensitive fisheries management.

9.34   **We therefore recommend that the UK government should make greater use of renewable fishing licences to regulate UK fisheries, by linking licensing to marine spatial plans, reductions in fishing effort, gear restrictions and improvements in vessel monitoring technology** (we discuss further how this framework could function in chapter 10).

9.35   Some gear types seem to be particularly destructive, either in terms of damage to benthic habitats or as a result of the high rate of by-catch in the fishery (chapters 3 and 5). The following section describes some studies and measures designed to reduce the impact of fishing gears.

9.36 The Shifting Gears study[22] in the US ranked the severity of the environmental impact of ten types of fishing gears and subsequently linked these to three categories of policy response. The gears ranked as having the highest impact were bottom trawling, bottom gillnets, scallop dredges and mid-water gillnets, those with medium impact were pots and traps, pelagic longlines and bottom longlines and lowest impact resulted from use of mid water trawls, purse seines and hook and line. The study suggested that gears in the highest impact category should be managed through the introduction of bans in ecologically sensitive areas. The medium impact category should be subject to mandatory modification of gears (such as the use of bird scaring lines in longline fisheries), with the low impact category requiring less stringent policies. The authors noted, however, that such a ranking should be both fishery and area specific. In addition to the ten techniques rated in this study, suction dredging for cockles and other shellfish has been cited as a particularly destructive technique.[23]

9.37 **We recommend that the UK government should rank the impacts of gear specific to UK fisheries in relation to their impact on habitats and press at EU level for the introduction of appropriate policy responses.** Such a review should take account of the marine landscape classification and seabed mapping exercises being carried out by the Joint Nature Conservation Committee. The progress of this review should not, however, delay action where specific gear types are already known to be highly damaging.

9.38 In UK territorial waters, some habitats will have been heavily modified by fishing gears over a long period and will have been converted into states that are resilient to the impacts of those gears. Therefore, it is important for such a gear review to be broad enough to consider historical modification of marine habitats by fishing, as well as present-day effects of running a trawl over a given piece of seabed. This would help ensure that unmodified habitat is preserved from the impacts of damaging gears, but allow fishing where habitat has been permanently degraded so long as sufficient areas of such habitat are protected elsewhere.

9.39 On land and in coastal areas above the low-water mark, a system is in place to protect sensitive and scientifically valuable sites (known as Sites of Special Scientific Interest). This is operated by English Nature and the Countryside Council for Wales, with statutory powers under the Wildlife and Countryside Act 1981.[24] Similarly, there are some marine habitats (such as maerl beds and coral reefs) that are so sensitive that they should receive total protection from damaging activities such as trawling and other destructive fishing techniques. **We recommend that the Joint Nature Conservation Committee should develop a list of potentially damaging operations, which should be avoided in all areas of marine conservation importance.**

9.40 On the other hand, areas of mobile seabed that have already been heavily trawled will be little affected even by scallop dredging and other destructive techniques. It would not be necessary to introduce a blanket ban on such methods, provided that sufficient areas of these habitats are protected from fishing to reflect a full spectrum of the marine habitats around the UK. In the absence of such a protected area network, **we recommend that the UK government should introduce plans to give complete protection to sensitive marine habitats from destructive fishing techniques in specific areas through a new process of marine planning and strategic environmental assessment (10.22–10.28) that would approve the use of gears only in those areas**

where they will not cause significant environmental harm. This is consistent with the position of reversing the burden of justification, so that fishing activity is only allowed where the effects are shown to be acceptable. It is also compatible with a marine spatial planning approach to management.

SELECTIVE AREA CLOSURES

9.41 A limited number of areas have already been selectively closed to certain types of gear. In the Sound of Arisaig inshore Special Area of Conservation, the use of trawls is banned in the most sensitive areas (such as maerl beds), but other less damaging types of activity such as creel fishing for lobsters, are permitted (table 9.1). Similarly, in Loch Torridon[25] an initial voluntary agreement between different groups of fishers led to the demarcation of separate areas for creel fishing and a mixed mobile gear fishery. The second phase of the Torridon initiative was the development of a management plan for the creel-only fishing area linked to accreditation by the Marine Stewardship Council. Box 9B describes the experience of the razor shell fishery in the Wash and North Norfolk Coast marine candidate Special Area of Conservation.

9.42 The Solway Firth Partnership[26] has also produced a draft shellfish management plan to ensure sustainable management of the Solway's shellfisheries. The draft management plan details a number of measures including a licensing scheme, the establishment of a maximum sustainable yield, a minimum landing size, a code of conduct for fishermen and a closed season to allow the stocks to rebuild each year. The needs of wildlife, in particular birds such as the oystercatcher, have been incorporated into the plan and areas that are considered highly sensitive will be closed to fishing. As a result of the management plan an application to Defra and the Scottish Executive is underway for a Shellfish Regulatory Order for the Solway Firth. In the case of cockle fishing, a number of techniques have been tried. First, suction dredging was banned, followed by the use of vessels, and tractors; finally collection was limited to hand gathering. To some extent this was designed to reduce the overall level of harvest, but it was also motivated by the need to minimise collateral damage to the environment.

9.43 Many of the environmental impacts of fisheries can be addressed with technical solutions but there is a need to consider the extent to which they might result in competitive disadvantage for fishers who implement such measures appropriately. More effective enforcement and inspection regimes will be needed to ensure compliance with gear restrictions/regulations as existing enforcement schemes are costly, and are prone to high levels of non-compliance.[27] Many technical regulations are routinely flouted, for example if mesh size is increased on trawl net cod ends to reduce by-catch of undersized fish a tractor tyre placed in the net will cause the cod end to elongate and close the mesh or it can be blocked with plastic bags.[28] There is a strong argument for strong but simple gear regulations that are easily enforced, but equally there could be an important opportunity to improve compliance by developing better cooperation with the fishing industry in the design and implementation of environmental measures if appropriate funding is in place.

Table 9.1

Fisheries management in the Sound of Arisaig Special Area of Conservation[29]

Activity	Assessment	Action
Creel fishing for *Nephrops* (velvet crab)	A well established fishery, maerl is sensitive to damage from intensive creeling but not thought to pose any risk at current level of effort.	No management action required from statutory authorities of voluntary agreement.
Shellfish harvesting by diving (scallops; razorfish)	Well established with SAC with most harvesting occurring at less than 35 m depth; features of interest not thought at risk.	Encourage commercial divers to help build a database to assist monitoring.
Suction/hydraulic dredging (scallops)	Has occured in past though only to a limited extent; causes significant disturbance of seated sediments and likely to damage or destroy features of interest (maerl beds) in sea lochs.	No activity permitted in waters less than 20m, together with a buffer zone from 20-35m to avoid smothering features of interest by suspended sediments
Benthic dredging (scallops)	Little if any dredging occurs in less than 20 m or in highly sensitive sea loch locations; maerl is easily damaged by scallop dredges and is slow to recover.	
Benthic trawling (*Nephrops*)	No trawling is thought to occur in highly sensitive sea loch areas; well established in waters deeper than 30 m; maerl beds susceptible to damage from *Nephrops* trawls.	
Finfish farming	Salmon farming well established in parts of SAC; maerl may be damaged though effects of finfish farming, though tend to be localised and mainly associated with water quality and effects of cage mooring.	Farms to be assessed when consents to discharge or site lease renewals are applied for; base line surveys to be developed by SNH with SEPA.
Shellfish farming	Several leases granted for shellfish farming within designated area (some remain undeveloped) intensive activity can affect features of interest through smothering of seabead.	Continued use of existing leases is acceptable and some new activity may be possible. Authorities to review leases when applications are made for renewal.

Box 9B	RAZOR SHELL FISHERY IN THE WASH AND NORTH NORFOLK COAST CANDIDATE SAC[30]

Although there is no European requirement to undertake an environmental impact assessment before opening up a new fishery, such assessments have been carried out in the UK. An example was the proposed fishery for razor shells (*Ensis* species), trough shells (*Spisula* species) and carpet shells (*Tapes* species) in the Wash and North Norfolk coast candidate Special Area of Conservation.

When stocks of cockles and mussels are low the razor shell fishery is a profitable alternative, since a valuable market for these shellfish exists in Spain. While trough shells and carpet shells can be harvested by modifying the suction dredges used for cockle fishing, razor shells require specialised dredges due to their deep burrowing habit. These dredges vary in design, but can have significant impacts on non-target shellfish and other invertebrates.

The direct and indirect impacts of suction dredging were a significant concern for English Nature and the Royal Society for Protection of Birds in the Wash candidate Special Area of Conservation, and there was no clear evidence to suggest that this fishing technique would not cause unacceptable environmental impacts. As a result, fishing for razor shells, trough shells and carpet shells was prohibited within the Wash under Statutory Instrument 1998, No. 1276, which came into force on the 20 May 1998 pending an assessment of the ecology of the razor shells and the potential environmental effects of suction dredging.

DEEP-SEA FISHERIES

9.44 Deep-sea fisheries target bottom-dwelling species below 400 m. The expansion of deep-sea fisheries has been driven by the depletion of fisheries on the continental shelf, the tightening of regulations in those fisheries and the consequent creation of excess fishing capacity for which a financial return is sought,[31] together with technological developments that have made deep-sea fishing increasingly viable economically.

9.45 The majority of bottom trawling in the deep sea takes place within countries' exclusive economic zones (i.e. within 200 nm). But bottom trawling also takes place on the high seas, mostly along the edge of the continental shelf, over the seamounts, oceanic ridges and plateaux of the deep-sea floor. In 2001, 13 countries (mostly members of the Organisation for Economic Co-operation and Development) were responsible for over 95% of the reported bottom trawl catch on the high seas. Only a very few vessels are involved in this sector (not more than 250 to 300 full-time equivalent boats per year), constituting about 0.2% of global marine fisheries capture production. While the industry is small and concentrated in a few hands, its environmental impact is significant and widely dispersed.

9.46 There are few regulations concerning bottom trawling's impact on the biodiversity of the high seas. In the north-east Atlantic, the relevant fisheries body, the North-East Atlantic Fisheries Committee (NEAFC), has only just begun to attempt to regulate these fisheries.[32] ICES has also recommended that TACs should not be used for deep-water species, although the European Council of Fisheries Ministers has set TACs for several such species. Eight deep-sea species in the NEAFC area are subject to EC TACs, while there is no regulation at all for the large number of other deep-sea species.

9.47 Deep-sea species are very long lived (in some species individuals can be more than 100 years old), late to mature, slow growing, of low fecundity and prone to congregating in dense aggregations for spawning and/or feeding which makes them easier to catch. Their low reproductive rates make them highly vulnerable to overfishing and slower to recover than the more resilient inshore species. As a result, high seas bottom trawl fishing has often led to the serial depletion of targeted deep-sea fish stocks.[33] There are high rates of by-catch in deep water fisheries, a problem exacerbated by the fact that discarding deep-water fish always results in their death due to the difference in pressure at the surface. Those fish that do escape from nets at depth are also more susceptible to damage, as they lack the protective slippery mucus covering that is present on many shallow-water species.

9.48 Seamounts and other habitats in the deep-sea support rich assemblages of organisms like corals, sea-fans, hydroids and sponges. Like deep-water fish, many of these organisms reach great ages – some corals and sea-fans are estimated to be more than 1,000 years old – and the communities themselves may have taken several millennia to develop. The three major gear types used in bottom fishing – gillnets, longlines and bottom trawls – all have an impact on corals and bottom-dwelling organisms. Trawl fisheries, which involve dragging heavy chains, nets and steel plates across the ocean bottom, have caused great damage to the vulnerable and unique animals living on seamounts, which may be irreparable on human time-scales.

9.49 Almost all of the fish caught by bottom trawling on the high seas are straddling stocks (i.e. migratory species that cross international jurisdictions), and are therefore subject to the 1995 UN Straddling Stocks Agreement (4.20). This agreement obliges signatory states to assess the impact of fishing on non-target species belonging to the same ecosystem, minimise the impact of fishing on non-target species, protect habitats of special concern, and protect biodiversity in the marine environment. In other words, fully to implement the precautionary approach and ecosystem-based fisheries management.

9.50 Quota schemes and effort controls are unlikely to control deep-sea fisheries adequately, because fishing pressure can deplete fish populations faster than control measures can take effect. There are also significant shortcomings in our knowledge of deep-sea species and habitats, which would make the setting of quotas difficult. Monitoring and enforcement pose significant difficulties on the high seas. Nor will quotas be able to prevent damage to seabed habitats. Indeed, it has been suggested that "There is probably no such thing as an economically viable deep-water fishery that is also sustainable."[34]

9.51 **We therefore recommend that the UK government should immediately halt any deep-sea trawling taking place in UK waters or being carried out by UK vessels. We also recommend that the UK government should press the European Commission to ban bottom trawling, gillnetting and long-lining for deep-sea species in EU waters.**

9.52 Chapter 8 made proposals for the introduction of marine protected areas and reserves. This protection should be extended to the deep seas. In 2004, the 7th Conference of parties of the Convention on Biological Diversity called on the UN General Assembly and other relevant organisations to urgently take the necessary short, medium and long-term measures to eliminate destructive deep sea fishing practices. Possible measures were identified, such as the interim prohibition of destructive practices that adversely impact the marine biological diversity associated with vulnerable deep-sea ecosystems, such as seamounts and cold-water coral reefs. Protection of the deep seas, on the basis of sound scientific analysis including the application of precaution, needs to be urgently incorporated into international law.[35]

9.53 **We recommend that the UK government should promote measures to prohibit destructive deep-sea fishing practices and promote the establishment of a system of marine protected areas on the high seas. In addition, it should press for international controls on high seas bottom trawling, and for their proper implementation and enforcement under, for example, the UN Straddling Stocks Agreement and the UN Convention on the Law of the Sea.**

BY-CATCH MONITORING AND MANAGEMENT PLANS

9.54 This section covers two main issues; by-catch of marine mammals, sharks and rays, seabirds, seabed-dwelling organisms and non-target fish; and the discarding or high grading of over-quota/under-sized fish. We believe that the UK government should introduce measures to encourage the uptake of technical measures designed to increase the size and species selectivity of fishing gears. It should also promote the adoption, at a European level, of successful by-catch reduction technical measures as developed and trialled by the SeaFish Industry Authority.[36]

BY-CATCH

9.55 By-catch is generally accepted to be one of the most serious threats facing populations of small cetaceans. ASCOBANS (the Agreement on the Conservation of Small Cetaceans of the Baltic and North Seas) has recommended that by-catch should be limited to 1.7% of the estimated population, assuming no uncertainty in any parameter.[37] The European Commission has proposed a new regulation limiting cetacean by-catch.[38] In the UK, Defra has produced a *'Small cetacean by-catch response strategy'*. The House of Commons Environment, Food and Rural Affairs Select Committee has also examined the issue.[39]

9.56 The species thought to be most at risk are harbour porpoises and bottle-nosed and common dolphins. Concern has focused on the use of gillnets (particularly bottom-set nets) and pelagic pair trawling. In 2004, the UK government requested that the European Commission impose emergency measures to close a seabass pair-trawling fishery off the south-west of England that had been associated with high levels of cetacean by-catch. Previous trials of an exclusion grid in the area had been encouraging, but in 2003/4 nearly 170 dolphins were caught in the area. We welcome the government's attempts to stop pair-trawling for sea bass in the English Channel and support the negotiations to ban this practice at the European level.

9.57 A number of technical and area-based measures have been developed to reduce by-catch of cetaceans, seabirds and other non-target marine species. These include escape panels, pingers on bottom-set gillnets, modified codends and separator and cut-away trawls (appendix H). However, they have not been fully implemented, and wider by-catch reduction measures are needed.

9.58   The European Commission has proposed 5-10% observer coverage of pelagic fisheries as part of their strategy to reduce cetacean by-catch. We support the widespread use of observers to reduce all forms of by-catch, as undertaken in other jurisdictions such as the US and Iceland. However, we are aware that the cost of fisheries enforcement is already high (9.71). As 100% observer coverage may thus be unlikely in the short-term, it will be necessary to use additional means of monitoring and enforcing by-catch reduction measures.

9.59   **We recommend a staged approach to reducing by-catch. Modified gears (9.57) should be introduced for the entire fleet along with a more comprehensive monitoring regime to ensure compliance and to determine the effectiveness of these measures. If target levels of by-catch reduction are not met in a particular fishery, then this fishery should be closed.**

9.60   The Prime Minister's Strategy Unit[40] recommended that environmental impact assessments be carried out before deploying new techniques or gears within a fishery, subject to guidance provided by the fishery manager. It also sought to encourage innovation in the development of less damaging gear types. We support these recommendations.

DISCARDING

9.61   The European Commission has estimated that discards may account for nearly 70% of fish mortality in some species and locations,[41] and the volume of discards poses a serious threat to the conservation of fish (chapter 5). There are several possible ways of reducing discards, including outright bans. The following section reviews international experiences with discard reduction measures and makes recommendations designed to reduce discards.

9.62   The Norwegian government introduced a ban on the discarding of some commercial species of fish in the mid-1990s. Closed areas as a means to reduce capture of under-sized fish were only introduced after technical gear measures had taken effect, increasing the momentum towards selective gear, and further reducing the numbers of discards. By the time the discard ban was introduced, the most important problem, eliminating the capture of fish that would be discarded, had already been largely tackled.[42] Box 9C describes international experiences with discard bans.

9.63   Implementing a discard ban in European waters would be complicated because of the mixed nature of many fisheries (e.g. the North Sea demersal whitefish catch) and the geographical separation of landing ports from fish processing facilities, meaning that potential discards would have to be transported over long distances. Higher inspection or monitoring rates than those at present in place would also be needed to reduce unobserved discarding.

9.64 It has been suggested to us that, from an ecosystem point of view, it may be better to cycle any dead fish within the marine ecosystem (where it will be consumed by scavengers) than to bring it to land, thus removing yet more food from the sea.[43] However, the point of discard bans is to reduce the removal of fish from the ecosystem in the first place. Measures to reduce (or eliminate) by-catch may also have effects on seabird populations.[44] Indeed, research has found that due to decreased fishing effort and discarding, skuas are now eating other seabirds rather than relying on fishery discards as a food source.[45] Such findings are not, however, a justification for continuing to discard.

9.65 In 2002, as part of the reform of the Common Fisheries Policy, the European Commission issued a Communication on a Community Action Plan to reduce discards of fish.[46] The proposals did not include any completely new measures, but suggested ways in which the Common Fisheries Policy might be modified. These included actions to improve the selectivity of gear, adjustments to mesh and landing sizes, area and real-time closures and voluntary action by fishers to leave areas with high levels of juveniles. Pilot projects will be used to test different aspects of the action plan and a proposal may be developed in 2005 for implementation in 2006.

9.66 The proposals do not call for mandatory reporting on all retained catch, and consequently discards of non-commercial species may go unreported. The availability of additional information through the reporting of the total catch could contribute to improved population assessment modelling for non-target species and provide useful information with regard to ecosystem indicators.

9.67 At present, a vessel could conceivably catch many more fish than it could hold on a single trip, discarding the rest. Furthermore, discard bans provide powerful incentives for developing and using more selective fishing methods. Our view is that a discard ban should be introduced. Under such a scheme, all fish caught should be landed. This would enable any fishing areas with high percentages of juveniles present in the catch to be immediately closed to avoid jeopardising reproductive success. **We recommend that the UK government should negotiate at EU level for a mandatory full catch reporting scheme and that the data should be published annually by Defra and the relevant devolved authorities.**

9.68 Once our other proposals on the reversal of justification (7.59), spatial planning (10.22) and effort control (9.29) have been implemented, **we recommend that the UK government should negotiate for the introduction of an EU-wide discard ban.** The net result of a ban is likely to be that less biomass will be removed from the sea.

BOX 9C	INTERNATIONAL EXAMPLES OF DISCARD BANS[47]

Canada

Canada instituted a ban on discarding at sea in its Atlantic groundfish fishery that makes it illegal to return to the water any demersal fish except those specifically authorised or those caught in cod traps. Authorised release is only considered for species that are known to have high survival rates on release or where there is no practical or nutritional use for a particular species. In addition to the banning of discards, larger vessels are required to carry observers which would imply that there are now no illegal discards on these vessels. The discards ban in Canada has been backed up by regulations that allow temporary closure of areas with high by-catches and include small percentages of by-catch in quota allocations. Fishermen may market small fish or by-catches and these quantities are counted against their quotas.

Iceland

The introduction by Iceland of an individual transferable quota (ITQ) system of fisheries management across virtually all its major fisheries has now been followed by the introduction of a ban on at-sea discarding of catch. The Icelandic regulations require the retention of most fish specimens for which there are total allowable catches or species for which a market exists. Nevertheless, species for which there is no quota system and that have no commercial value may be discarded. It is compulsory to land smaller fish, but as the government does not wish to encourage their capture, there are upper limits on the percentage weight of fish that can be landed below minimum landing size. Fish kept on board under these no-discard rules may be marketed.

The Icelandic ban on discarding has been coupled with the establishment and running of a 'by-catch bank' since 1989. The primary aim of the bank was to demonstrate to fishermen and the fish trade that there were markets for unusual species of fish caught as by-catch and, where necessary, introduce and promote those new species to consumers. This has resulted in specific fisheries opening up for species previously discarded (e.g. megrim and grenadiers).

New Zealand

The quota management system instituted in New Zealand makes discarding or dumping of most species illegal, but it is still known to occur. In the multi-species inshore trawl fishery particularly, the capture of non-quota fish or fish for which the quota has been exhausted is often encountered. The ITQ system in New Zealand is a complex system where quota to cover over-run fish (fish caught over quota) can be bought from another quota holder after it has been landed or the value of the over-run catch be surrendered to the state. However, it seems that in many cases the fishermen find it easier to discard the fish at sea than go through the complex system of landing the fish and then making it legal. In addition the New Zealand system allows a quota to be overshot by 10% in one season, although this over-quota landing can be deducted from the next season's quota. Discarding still occurs, however, and this is illustrated by the fact that vessels carrying observers have reported larger catches of non-target species than vessels fishing in the same area but not carrying observers.

It has been suggested that with the introduction of the ITQ system in the New Zealand fishery there was an increase in discarding at sea even though the fishermen could receive 10% of the market price for fish landed outside quota. In an attempt to discourage discarding the percentage of market price was increased to 50%.

MORE EFFECTIVE ENFORCEMENT

9.69 Enforcement is one of the major practical issues associated with any framework for fisheries management. A real obstacle to effective enforcement is the level of non-compliance within the fishing industry. The situation is not helped by the complexity of fisheries regulations and their frequent changes. Technical advances such as the increased use of remote sensing and vessel monitoring systems are helping regulators to know whether vessels are fishing or not, but not necessarily which types of gear are being used. In addition to this technology, it is likely that further on-board monitoring, such as on-board observers or video cameras to survey the catch as it is brought aboard, will be needed for enforcement to be effective in UK waters.

9.70 The Sea Fisheries Inspectorate carries out the enforcement of EU and national sea fisheries legislation under the framework of the Common Fisheries Policy within British Fishery limits adjacent to England and Wales out to 200 nm or the median line with neighbouring countries. This body co-ordinates and directs inspections and surveillance at sea, using aerial and satellite surveillance and carries out land based inspections of fish and vessel documentation of catches.

9.71 The main role of Sea Fisheries Committees in England and Wales, whose present jurisdiction extends out to 6 nm, is the conservation and management of shellfish and some finfish stocks through local byelaws (4.139). They also enforce some national legislation and some measures under the Common Fisheries Policy such as minimum landing size. In 2003/04, the total cost of fishing enforcement in England and Wales was around £18 million. This included £12 million for the Sea Fisheries Inspectorate, £5 million for Sea Fisheries Committees and around £1 million for the Environment Agency (for activities relating to salmon, sea trout, eel and freshwater fisheries in all inland waters and out to the 6 nm limit). These enforcement costs in England and Wales amount to 10% of the value of the fish landed.

9.72 Administering and enforcing the Common Fisheries Policy cost the United Kingdom as a whole, some £87 million in 2002/03.[48] These sums exclude other substantial public costs, including any expenditure by the Royal Navy on its fisheries protection squadron, not met by payments from Defra, and the cost running fisheries advisory services such as CEFAS, which requires around £30 million pounds annually. They also exclude the costs incurred by the industry in complying with regulation. Expenditure on fisheries conservation and enforcement represents more than 17% of the value of fish landed in the UK.[49] Hence, in their evidence to the Commission, SeaFish suggested that some degree of simplification of fisheries regulations was desirable.[50]

9.73 In March 2004, Defra wrote to vessel owners and other interested organisations advising them of action being taken to improve monitoring, control and surveillance in the United Kingdom.[51] This was in response to the National Audit Office report on fisheries enforcement and concerns raised by the European Commission (see chapter 5). Defra noted that steps were being taken to:

- move away from criminal proceedings, in favour of administrative sanctions for breaches of fisheries regulations;

- designate markets and register sellers and buyers of first sale fish, and obligation to account for the provenance of fish by passing markets to help improve the traceability of fish landings;

- improve the weighing and boxing of fish on landing;

- strengthen the arrangements for designated landing ports and prior notification of catch;

- the wider use of single areas and single species licensing;

- fit tamper proof satellite position reporting terminals to all vessels over 15 m; and

- better utilise resources, including greater emphasis on land based inspections and wider application of satellite surveillance. This will included targeted inspections of distribution chains and processing of fish species with threatened stocks. In the longer term there will be wider use of electronic communication systems to reduce the burden of paper based systems.

9.74   In some areas the UK fisheries departments have already taken action; in others there will be further consultation with the industry and other interests before measures are taken. There is also scope for consultation and collaboration with other countries to make fisheries enforcement more effective throughout EU waters. To this end the European Commission is proposing to set up a centralised Community Fisheries Control Agency. The aim of the new agency would be to ensure a more effective use of EU and national means of fisheries inspection and surveillance by pooling available resources within a joint inspection structure. This more centralised control may result in a more level playing-field in this area.

9.75   Some countries have successfully used on-board observers to monitor compliance with regulations (9.58). Additionally, satellite transponders can be fitted to the larger vessels so that fishers' movements can be monitored from the comfort of the port, with independent accuracy. According to the European Commission timetable, all vessels greater than 15 m should have been fitted with a satellite transponder by 2003 and all vessels greater 10 m by 2004. In addition, a remote sensing vessel detection system must be fitted to both classes of vessel.[52]

9.76   **We recommend that the UK government should pursue a policy of installing tamper-proof vessel position monitoring devices on licensed fishing vessels over 8 metres in length. The aim should be to complete this installation within three to five years. We also recommend that UK fisheries departments should commission work to trial video recording of catch on board vessels.**

9.77   A recent review of fisheries enforcement measures in England and Wales has proposed that the roles and responsibilities of the Sea Fisheries Inspectorate and the Sea Fisheries Committees should be merged, and that the new body should be restructured as national agency, or possibly as part of a national marine agency. Although it could be structured as regional organisation, the review recommended that on the basis of cost and efficiency, the new enforcement body should be run on national basis despite the loss of local input into enforcement measures in the inshore areas. At the time of writing, no final decision has been made on the future shape of the enforcement authorities.

9.78 The recent Review of Marine Nature Conservation also concluded that role of enforcing environmental measures in the marine environment should fall to the Sea Fisheries Inspectorate and the Sea Fisheries Committees, including conservation and use of marine resources and enhanced spatial planning. Such responsibilities would be easier to fulfil within a nationally-managed agency. It is notable, however, that Nordic countries have successfully developed a more co-operative regional style of management to introduce measures to protect sensitive marine habitats from the effects of dragged gear. The prospects for more regionalised management are discussed in chapter 10.

SUBSIDIES

9.79 Chapter 4 described the role of subsidies within the Common Fisheries Policy. These are administered through the Financial Instrument for Fisheries Guidance (FIFG) and are available for a variety of purposes. Subsidies are subject to clearance by the European Commission for compatibility with EC rules on provision of state aid.

9.80 Historically, subsidies have been used for tie-up schemes that have provided a strong incentive for fishers to remain in the industry when they might otherwise have left it. Decommissioning schemes have therefore increasingly been targeted at buying out quotas, etc. As well as direct compensation for fishers leaving the industry, indirect compensation (through regional development or similar funding) may be needed in order to encourage diversification.

9.81 Under the Common Fisheries Policy subsidies will continue to be made available for building new vessels until the end of 2004 (but not in the UK). Public money will also continue to be made available for modernisation of vessels more than five years old. There is a provision that the modernisation grant must not lead to an increase in the catching ability of the vessel, but this will have to be monitored and controlled rigorously.[53]

9.82 Even modernisation for safety reasons could lead to an increase in catching ability. For example, if a vessel was deemed unsafe enough or unseaworthy enough to warrant modernisation, then the subsequent improvement in seaworthiness could allow the fishers to fish under a wider range of weather conditions, which could be construed as an increase in catching power. More precise satellite positioning and depth sounding technologies could arguably improve safety, but they would inevitably improve fish-finding ability too. Thus these subsidies will lead to enhanced fishing power, regardless of the wording of the conditions, or the rigour of monitoring.

9.83 **We recommend that the UK government press at EU level for an end to all subsidies that can result in increased fishing pressure, including vessel modernisation, and improving port and fish processing facilities.**

9.84 Subsidies can and should play an important role in the reduction of the environmental damage due to fisheries, particularly by reducing fishing capacity, adopting selective and environmentally friendly fishing gears, and providing the information necessary for sound fisheries management. Under the second pillar of the Common Agricultural Policy, subsidies can be used to encourage diversification and environmental schemes. The Common Fisheries Policy should be reformed in a similar manner to use subsidies to encourage fishers to use environmentally friendly equipment and practices in fishing.

9.85 Subsidies are rarely used in this way at present. But if fishers could be shown to be complying with minimum legal standards, then it would seem reasonable to pay for extra environmental services above and beyond this. These could include gathering monitoring and research data, or extra environmental safeguards in terms of gear restrictions or 'good fishing practice'. Subsidies should not, however, be paid for meeting existing legal requirements.

9.86 The Common Fisheries Policy has already been reformed in this direction to some extent. Article 4 includes incentives for the adoption of less damaging gears and pilot projects using alternative management regimes.

9.87 In its evidence to the Commission,[54] the Joint Nature Conservation Committee observed that there were various areas of fisheries management where financial incentives could be used to the benefit of both fisheries and biodiversity, including:

- payments through individual agreements relating to the use of specific fishing gears, restrictions on boat size, fishing periods, etc;

- payments to promote and facilitate collective management, such as the funding of management plans or studies, or support to attend meetings/group facilities/ coordination of services;

- charges for licences for certain fishing practices which have an adverse environmental footprint;

- payments for carrying out new duties such as monitoring non-target species or policing closed or restricted areas;

- investment aids to assist small-scale operations to increase viability and reduce their dependence on commodity output, for example, via more sustainable production systems, environment-related marketing, green labelling schemes, etc.

9.88 We have already recommended (7.35) that the UK government put in place an incentive scheme to reward financially fishers for the accurate recording of all catch data and relevant environmental data where this data will be used for monitoring purposes and to inform the development of ecosystem models and indicators.

EMISSIONS FROM FISHING VESSELS

9.89 By the turn of the nineteenth century, steam powered vessels were beginning to dominate the more traditional sail powered fishing vessels. By the 1930s, heavy-oil powered fishing vessels were at sea and by the 1940s diesel power was the preferred choice of propulsion; this situation still pertains. The change from sail to fuel-powered propulsion had a huge impact on fishing vessels, expanding fishing grounds and enlarging fisheries throughout the world and allowing vessels to tow much larger and heavier fishing gears than previously possible. The transition from sail to diesel propulsion was therefore fundamental in the global exploitation of fish stocks.

9.90   Diesel fuel is now the second biggest expense for the great majority of UK fishing vessels, typically amounting to around 10-20% of vessel earnings, while crew expenses account for about 30%. However, fuel costs vary considerably between fisheries depending on the type and size of gear deployed, vessel size and efficiency. Fuel expenditure is highest for beam trawling (at about 30% of earnings) and lowest for static gears at less than 10%.[55]

9.91   Many of the recent advances in fishing vessel and propulsion design technology, have focused on increasing efficiency by reducing fuel consumption, particularly in vessels using mobile fishing gears. Advances have often been made through the uptake of existing under-utilised technologies, such as bulbous bows and propeller nozzles, rather than radical new developments. In contrast, the fuel consumption and efficiency of small marine diesel engines has not significantly improved over the past few decades.

9.92   Although fuel is an important element of running costs, fishing benefits from lower fuel costs than other sectors because fishing vessels are exempt from fuel duty. Low energy costs have helped drive the expansion of highly mobile and long-range fisheries. Fuel efficiency values in the fisheries studied have decreased over time, and have become increasingly variable due to depletion of fish stocks and subsequent recruitment variability, such that it requires an average of 1 tonne of fuel to catch 1.5-1.8 tonnes of fish.

9.93   The decreasing fuel efficiency of many fisheries and the large contribution that fuel makes to overall running costs, has led to suggestions that this may be the Achilles' heel of the industry.[56] The rising costs of fuel are likely to impact heavily on certain fishing methods that have low fuel:fish efficiency ratios, such as trawling, in fisheries with diminished target populations, to the point that these fisheries will no longer be economic.

9.94   Our Twenty-second Report noted the contribution of shipping to rising greenhouse emissions. Marine emissions from fuels used for international journeys (which are known as bunker fuels) remain outside international agreements to control greenhouse gases. A significant proportion of UK marine greenhouse gas emissions (17%) are related to fishing, but the sector contributes a relatively small amount to overall national emissions (0.01%).[57]

9.95   Marine fuels are also high in sulphur, a problem that still causes concern despite international moves to reduce sulphur levels in marine fuels.[58] Marine fuel has an average sulphur content of 27,900 parts per million, compared with petrol for cars that contains around 50 parts per million. Ship emissions also contribute to acid rain, ground-level ozone (smog), other forms of air pollution and marine eutrophication. A recent study for the European Commission examined all shipping journeys starting or finishing in Europe, and estimated the emissions of various pollutants. The results were startling. For example, by 2010, emissions of sulphur dioxide in EU sea areas are likely to equal 75% of total land-based emissions, including those from all cars, trucks and industrial plants. The picture for nitrogen oxides is not much better, with ship emissions likely to equal two thirds of land emissions by 2010.[59]

9.96   **We recommend that the UK should promote efforts at the European and international levels to bring marine emissions of greenhouse gases within international agreements and to control other atmospheric emissions from ships.**

CONCLUSIONS

9.97 Within the EU, a complex system of fisheries management has developed over many decades, but it has failed to meet many of the social, economic and environmental objectives that the Common Fisheries Policy was intended to provide. Improvements in the system are now essential.

9.98 Taken as a whole, fisheries in the north-east Atlantic are characterized by the over-fishing of target species, high levels of by-catch, food web perturbations and the destruction of marine habitat, with detrimental impacts not only on marine ecosystems and biodiversity but the future sustainability of fisheries as well. It is now necessary to move to a model of marine management that takes greater account of the environmental effects of marine capture fisheries.

9.99 Taking measures related to catch and gear controls has so far failed to stop severe reductions in fish populations and damage to marine ecosystems in the OSPAR area. Overcapacity in fishing fleets and poor enforcement of controls also remain to be tackled. Both the ecosystem approach and the precautionary approach require a more holistic, pro-active approach to management, opposed to a reactive approach based on management plans for limited areas or species, which effectively gives priority to species and habitats in the worst state.

9.100 We require a robust new system of regulation that involves not simply a new raft of regulations but a shift in the basic nature of the system from maximising the output from fish stocks to reducing to acceptable levels the harm fisheries do to the marine environment, thereby maintaining fish catches at a sustainable level. We consider that the most important management measures among those proposed in this chapter should be a programme of effort reduction, in order to reduce the overcapacity that has hampered previous environmental and fisheries management efforts. Progress in this area will also improve the effectiveness of other measures.

9.101 Moving to an effort based management system, particularly in demersal fisheries, could address many of the difficulties identified in the existing quota allocation and catch control system. Changing to a system of effort management will have implications for all the parties involved and decision-making processes at the EU level. This should not, however, prevent such measures being introduced in as short a time as possible, within the next three to five years.

9.102 Significant environmental benefits from these measures will only be gained if they are carried out in conjunction with the closure of areas to fishing. Furthermore, the benefits of such protected areas will increase over time, rather than be eroded by improvements in fishing technology, as all effort reduction schemes are currently. We are also concerned that certain gear types are particularly destructive and strongly recommend that these be phased out. This transition should be aided by subsidies designed to improve the environmental performance of fisheries and to reward fishers for the provision of necessary management data. The outcomes of management action are often critically dependent on the reaction of fishers to the measures. Chapter 10 considers institutional changes that seek to foster a co-operative approach to the management of fisheries. Subsidies and complex management regimes are endangering the fishing industry's survival. A simpler system that protects the environment is needed to help make the industry more sustainable.

Chapter 10

BRINGING ABOUT RADICAL CHANGE

Major reform is needed to integrate fisheries management into the wider framework of policies to protect the marine environment. What measures could help achieve this? How can a wider section of society be involved in the process?

INTRODUCTION

10.1 The seas around us play a major role in sustaining life, yet our approach has been largely driven by the desire to maximise short-term gains. Policies need to evolve so that they protect the long-term future of the marine environment and promote a sustainable fishing industry. As with the challenge of climate change, a major shift in thinking will be required to respond to the scale and urgency of the problem.

10.2 In this chapter we explore two key issues. First, we examine the management changes needed to improve the protection of the marine environment. In particular, we consider the case for introducing a marine planning system. Such a system would provide the basis for a truly precautionary approach to management since no activity would be allowed unless it was part of an agreed management plan. The planning system could also help to integrate a range of other measures designed to protect the environment and make fishing more sustainable. These would include a network of marine protected areas, changes to the size of the fishing fleet, effort control and technical conservation measures.

10.3 In the second part of this chapter, we consider the role of institutions in bringing about change and the scope for more co-operative forms of regional management. We also examine the need to broaden the debate to involve the public and a wide range of stakeholders in the development of policies and measures to manage human impacts on the marine environment. Education and consumer information will be critical in enabling people to participate in the process.

10.4 As a society we have given much less priority to protecting the seas compared with the land. This needs to change. Engagement with the public and stakeholders should feature strongly in the development of new measures to protect the seas, and we begin with the case for a system of marine spatial planning.

A SYSTEM OF MARINE SPATIAL PLANNING FOR THE UK

10.5 The seas are subject to many different and often conflicting pressures as a result of human activities. Our report has focused on fishing because there is good evidence that this activity subjects the marine environment to some of the greatest stresses. But other activities cause problems too, such as pollution from land, aggregates extraction, pollution from shipping and impacts from oil and gas extraction. In future, there are likely to be increasing efforts to produce energy from offshore wind farms and wave and tidal stream devices.

THE SITUATION TODAY

10.6 At present, activities affecting the marine environment are regulated through different agencies and in different ways. At the international level, fishing policies are largely determined by the European Community's Common Fisheries Policy, while, in the northeast Atlantic, OSPAR deals with pollution, oil and gas extraction as well as marine conservation, where it is taking a lead in establishing networks of marine protected areas.

10.7 In the UK, the management of the marine environment cuts across the responsibilities of many government departments, devolved administrations and advisory bodies. Fisheries and conservation sit with Defra and its devolved equivalents, with contributions from advisory organisations such as the Joint Nature Conservation Committee. Meanwhile, the Department for Trade and Industry is responsible for a range of activities including implementing OSPAR agreements on oil and gas extraction and regulating offshore energy generation from the wind, waves and tides. The Department for Transport and its agencies oversee ports, harbours and shipping. Other key departments include the Office of the Deputy Prime Minister (land use planning and aggregates extraction) and the Ministry of Defence. There are also over a hundred Acts of Parliament[1] governing the marine environment with often confusing, and sometimes overlapping, spatial jurisdictions (figure 10-I and figure 4-I).

Figure 10-I
Geographical extent of principal marine controls in England and Wales[2]

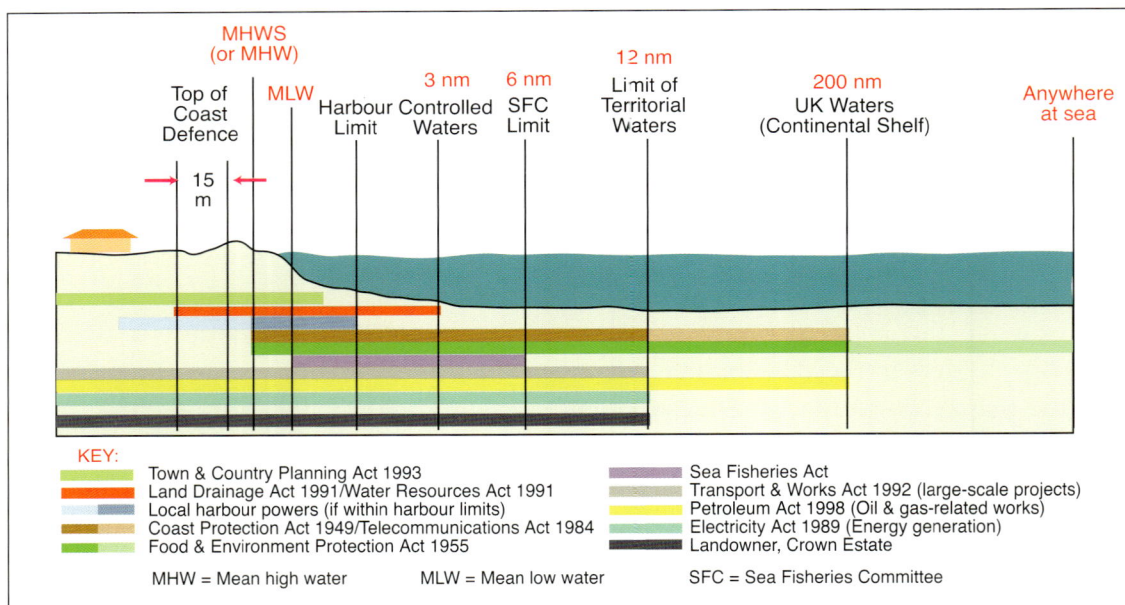

KEY:

Town & Country Planning Act 1993	Sea Fisheries Act
Land Drainage Act 1991/Water Resources Act 1991	Transport & Works Act 1992 (large-scale projects)
Local harbour powers (if within harbour limits)	Petroleum Act 1998 (Oil & gas-related works)
Coast Protection Act 1949/Telecommunications Act 1984	Electricity Act 1989 (Energy generation)
Food & Environment Protection Act 1955	Landowner, Crown Estate

MHW = Mean high water   MLW = Mean low water   SFC = Sea Fisheries Committee

10.8 Despite this intricate web of legislation, there is no spatial vision for UK waters that sets out the top-level objectives for the management of the seas, the principles that should guide its use, or how the demands of different sectors should be integrated. This leads to difficulties in trying to determine a rational basis for the day-to-day management of competing uses. At a practical level, the lack of a clear policy framework makes it hard for regulators to check compliance.[3] It also makes life more difficult for marine developers and users by adding to the uncertainty about which proposals are likely to succeed. The situation continues to grow more complex as novel enterprises such as offshore wind farms place new demands on the environment.

10.9 The danger in such a situation is that the environment will not receive adequate priority or protection. It also represents a missed opportunity to set out long-term goals for the protection of the marine environment and measures to deliver them.

10.10 The 1947 Town and Country Planning Act was established in response to a similar situation on land. It has subsequently been developed and refined to include controls on areas of special environmental value where there is a presumption against development. As yet no comparable system has been established for the seas. Some of the principles of spatial planning are beginning to be reflected in various European regulations and commitments. For example, Integrated Coastal Zone Management is being introduced in the UK in response to an EU Recommendation.[4] This initiative focuses on protecting the functioning of natural ecosystems, but it also aims to help coastal zones develop their social and economic potential. Since Integrated Coastal Zone Management crosses the land-sea divide, it has been put forward as an important link between land-use planning and any future system of marine planning.[5]

10.11 Further out to sea, the OSPAR Bergen Declaration[6] commits its signatories to investigate spatial planning in the North Sea. The EU Marine Thematic Strategy[7] takes the issue further by investigating whether a form of marine spatial planning, embodied in eco-region management plans, could be applied to all EU waters.[8] However, none of these initiatives represents a firm commitment by the UK to a fully-fledged system of marine planning. In contrast, countries such as Australia (box 10A) and the US have already begun applying planning or zoning principles to the ocean.[9,10]

BOX 10A   **MARINE PLANNING IN AUSTRALIA**

Australia's system of marine planning is based on three key principles – ecologically sustainable development, ecosystem-based management and integrated oceans management.[11] It lays out clear steps for identifying and managing risks (figure 10-II) and contains strong provisions for involving stakeholders and providing good governance.

The first draft marine plan was published in 2003 for the South-east region around Victoria, Tasmania, parts of South Australia and New South Wales. South-east Australia provides a good demonstration of the complex factors affecting the environment of a single

Figure 10-II
Developing objectives for marine planning in the South-east marine region of Australia[12]

Objectives for the South-east Marine Region	Is each objective for the Region currently being met? If not, why not, and what is the cause of the issue?
1. Relevance Test	Is the issue relevant to regional marine planning?
2. Issue Scoping and Risk Analysis	What is the extent of the issue?
3. Action (Strategy) Development	What can we do about the issue? Does the issue require a further management response?

region, consisting as it does, of highly populated coastal regions around Melbourne, as well as areas of high biodiversity, commercial fishing activity, tourism and oil exploration.

10.12 A number of recent reports have recommended that the UK follows suit by extending planning principles to the seas so that human pressures can be handled in a strategic fashion.[13,14,15,16] We concur with this view and strongly support the extension of spatial planning to UK waters.

What should a future system of marine spatial planning include?

10.13 To be truly effective a marine planning system should take an integrated approach that encompasses all major uses of the seas. By definition, this would include the fishing industry since it has an important impact on the environment and is affected in turn by other developments such as wind farms.

10.14 In chapter 7, we pointed out that at present it is assumed that fishing can take place unless it is shown to be damaging to the environment. We made the case that the pressures are now so great that the presumption in favour of fishing should not continue. Instead, the burden of justification should be reversed for fisheries so that it would be up to those seeking to carry out the activity to demonstrate that it would not cause harm. This could be done by developing plans that would be consistent with the principles of the proposed planning system and subject to strategic environmental assessment (10.25).

10.15 It is therefore essential that any future planning system describes the key principles and long-term goals that should guide management and that it sets standards against which plans would be judged.

Who should be involved in marine spatial planning?

10.16 We have not considered in detail the question of which body should manage the planning system, particularly as we are aware of ongoing discussions about the creation of a Maritime Agency.[17] A single body may well be the most appropriate way forward, but it is clear that **whichever organisation manages the system it needs to have access to expertise on environmental matters (for example, through liaison with the Joint Nature Conservation Committee), clear objectives to protect the marine environment and a strong co-ordinating role so that it can work with the many bodies that have an interest in the seas**.

10.17 It is also vital that the planning system allows the public and stakeholders to have a voice in the design of spatial management plans and in the wider decision-making process. However, there is a danger of stakeholder fatigue where the link between consultation body and ultimate policy outcomes is unclear, and this has been identified as a problem for parts of the stakeholder consultation for Integrated Coastal Zone Management.[18] **We would therefore wish to see the role of the stakeholders in the planning system clearly defined, specific opportunities for public involvement identified and the opportunity for the public to suggest candidate sites for protection**.

10.18 Greater involvement of the public and stakeholders in the design and implementation of marine spatial planning is likely to create new demands for access to information. There are many cases where government has set up organisations to share experience and promote networking on environmental topics. Such outreach may need to be part of the remit of the planning body mentioned in paragraph 10.16.

10.19 We would expect the planning system to operate on a scale that would be meaningful from an ecological perspective, capturing broadly similar regions and environmental processes, but not so large as to make local participation difficult. As recommended by the Joint Nature Conservation Committee,[19] the concept of regional seas (2.55) may provide a useful building block for the planning system that would match the likely approach of the EU Marine Thematic Strategy.[20]

THE POLICY RESPONSE

10.20 The UK government is considering aspects of marine spatial planning as part of the Review of Development in Coastal Waters and we urge it to publish this report as soon as possible. In 2004, the Scottish Executive issued a consultation document that included consideration of marine spatial planning.[21] In the same year, the Review of Marine Nature Conservation concluded that strategic and spatial planning was essential for nature conservation and for ensuring compliance with other policy goals.[22] It recommended a trial of marine spatial planning on the regional seas scale. Defra has since announced plans for a pilot study on a voluntary approach to marine spatial planning.[23]

10.21 A voluntary planning system is unlikely to be sufficient in a context where many of the other policies affecting the marine environment have a legal basis and are enforced by strict sanctions and penalties. Additionally, the commercial drive for development and extractive uses of the environment is so strong that it seems doubtful that a planning system would be effective in moderating these without a statutory basis. A statutory system would also provide greater clarity and regulatory certainty, which are important for industry. A marine planning system should also be able to deliver long-term goals for the marine environment, particularly as it may take some years for the full benefits of measures such as marine protected areas to develop (8.96). A voluntary system that could easily be dismantled, and might have patchy coverage, would not be the best vehicle for delivering such important strategic targets.

10.22 We are therefore strongly of the view that a comprehensive, statutory planning system is needed, rather than a voluntary one. The new system needs to be put in place quickly to provide a framework for other measures to protect the marine environment. In line with our Twenty-third Report that favoured extending planning principles to the seabed,[24] **we recommend that the UK government should develop a comprehensive system of marine spatial planning that:**

- **sets out the principles and long-term goals for protecting the marine environment and promoting the sustainable use of the seas;**

- **develops integrated regional management plans to guide all major uses of the seas, including fishing. These should ensure high standards of marine protection, and be subject to strategic environmental assessment; and**

- **has a statutory basis and a clear framework for public participation.**

We also recommend that the UK government should promote the principle of marine spatial planning at the European and international levels.

A NEW FRAMEWORK FOR PROTECTING THE MARINE ENVIRONMENT

10.23 One of the main advantages of a marine planning system would be the introduction of a spatial element that would allow activities to be managed in relation to their local impact. This could be a key element in helping to implement our proposals for protecting the marine environment and improving fisheries management, as described in chapters 7, 8 and 9.

10.24 Currently, the spatial management of fisheries is rather broad-brush. Existing controls operate largely through the quota and licensing systems that limit the catch of commercial species within the management areas established by the International Council for the Exploration of the Seas. In addition, some types of fishing are excluded from certain areas by the regulations under the Common Fisheries Policy or by domestic measures (8.47). We have argued that these give insufficient protection to the environment and so will not ensure the fishing industry has a sustainable future. Bringing fishing under a spatial planning system would help address this by providing the framework for new measures to tackle specific problems, and by allowing the sector to be managed in a way that takes account of other activities in the same region.

10.25 We therefore recommend that regional management plans should be introduced for UK waters to guide environmental protection and fishing, as well as the development of other sectors. The plans would need to take account of assessments of fish populations and character of the local marine environment. They would also be informed by decisions on where to site marine protected areas. Hence the plans would map out areas where (i) fishing would be allowed and (ii) there would be no fishing, or there would be restrictions related to the use of particular types of gear. Fishing licences would only be granted if they were in compliance with the management plan. The overall plan would be subject to a strategic environmental assessment (SEA) (figure 10-III).

10.26 Evidence from our visit to the US suggests that environmental assessments are less effective at delivering improvements if they feed into a decision-making process that is ultimately dominated by fishing interests.[25] Therefore it would be vital for the management plans and the SEAs to be reviewed by an organisation that had marine conservation, not commercial fisheries, as its primary responsibility.

Figure 10-III

A new framework for protecting the marine environment. SEA = strategic environmental assessment; EIA = environmental impact assessment; MPA = marine protected area (see paragraphs 10.25 to 10.27)

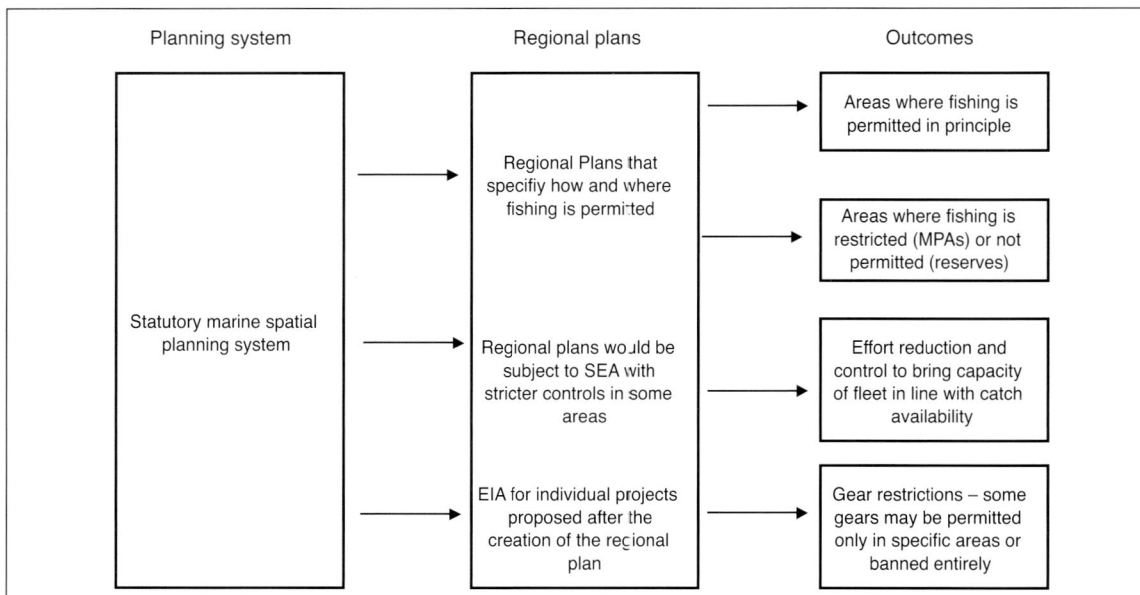

10.27 Environmental impact assessment (EIA) applies to individual projects rather than programmes. Its use in fisheries would therefore be limited to discrete, area-based fisheries proposals, but it would provide a mechanism for assessing projects that were not envisaged when the SEA of the plan was first undertaken. We would however expect the whole plan and the SEA to be revised on a regular basis, for example, to take account of scientific assessment of the effectiveness of the measures in each region to conserve the marine environment.

10.28 As noted in chapter 4, the SEA and EIA Directives do not currently apply to capture fisheries. Recent reports from the Prime Minister's Strategy Unit, the Royal Society of Edinburgh and the Joint Nature Conservation Committee have recommended extending these principles to this sector.[26,27,28] We concur with these views since the assessment processes would be an important way of improving environmental protection and function as a central plank of a new management framework. Fishing would thus be subject to some of the same regulations that govern other extractive industries. **We recommend that the UK government and devolved administrations should apply strategic environmental assessment and environmental impact assessment to fishing, amending the legislation as necessary.**

LINKS BETWEEN MARINE SPATIAL PLANNING AND MARINE PROTECTED AREAS

10.29 Applying the principles of spatial planning could help identify particularly vulnerable areas of the sea and others that are suited to some form of sustainable use. This information could be a key tool in helping to select sites as part of a network of marine protected areas (MPAs). These areas could then be reflected in regional management plans.

10.30 While we have recommended setting goals for a large-scale MPA network at a national level, the size and form of individual sites would need to be tailored to their location. The planning system could provide a forum for stakeholders and the public to participate in the design process.

POSSIBLE STEPS TOWARDS A NETWORK OF MARINE PROTECTED AREAS

10.31 The overall goal of the MPA network outlined in chapter 8 should be to protect the marine environment for the long-term, as part of an overall package of measures to improve the management of fishing. We recognise that it will take time to establish the full network, but equally there is a real danger that an incremental approach to implementing protected areas will fail to address both the scale and the urgency of the problem. Instead, we recommend that there is a commitment to a comprehensive solution, which could draw on the elements below.

10.32 **Strategic assessment of the scope for a large-scale MPA network:** We describe below a number of elements that could plausibly form such a network, and bodies such as the Joint Nature Conservation Committee have already identified some sites that could form part of its core. However, rather than be driven by an ad-hoc selection of sites, some of which may be small or only weakly protected, we suggest that the starting point should be a full strategic assessment of the optimum design of a network of MPAs within the UK's exclusive economic zone (EEZ). As we have argued in chapter 8, a sizeable proportion of the sea needs to be protected in order to allow ecosystems and fish populations to recover, and so we have recommended that 30% of the UK's EEZ should be protected by no-take reserves (8.96).

10.33 **Existing protected sites:** A pilot MPA network could be created from existing sites within UK waters (such as Natura 2000 sites, marine nature reserves, voluntary no-take zones and less obvious candidates such as exclusion zones around offshore energy installations, etc). This would not require new primary legislation. However, the protection could be improved by making greater use of strategic management plans to control fishing within these areas, and by amending the EC Habitats Directive to expand the representation of species, habitats and linked habitat features, as we recommend in paragraph 8.44.

10.34 **Inshore MPA network:** A more extensive inshore network could be created under the bye-law making powers of the Sea Fisheries Committees, in consultation with bodies such as the Joint Nature Conservation Committee. This has been helped by the fact that there is now an opportunity to extend the UK's territorial limit from 6 to 12 nm under the reformed Common Fisheries Policy. The network would be enhanced if it were developed within an overarching system of marine spatial planning, so that sites could be identified and managed within the context of other uses of the seas. In our view, a Marine Act will be necessary to provide the statutory basis for both a marine planning system and a larger MPA network protected from the effects of fishing (10.42).

10.35 **Offshore MPA network:** Because the European Community has competence in this area, developing an offshore MPA network would require action at the European level. In theory, there is scope for measures to be introduced under the Common Fisheries Policy to protect the environment from the impacts of fishing. However these are often temporary, and the burden of scientific proof needed to demonstrate that a particular area is at risk is high. Rather than depend on such reactive measures, our preference would be for new European legislation, specifically designed to allow countries to establish a protected network of MPAs and no-take reserves within their EEZs. This would also help deliver existing commitments under the Bergen Declaration.

10.36 **Management of fishing effort:** The MPA network will not fully achieve its objectives if fishing activity is simply displaced. Overall pressure must be reduced. We do not believe that the system of quota-setting based on total allowable catches is a reliable method for managing demersal fisheries and so in chapter 9 we have proposed replacing it with a system of effort control. We also recommend that, for the first time, the EC Directives on SEA and EIA should be applied to fishing so that an assessment can be made before new fisheries are established, areas are opened up to fishing for the first time, or fishing licences are granted. Even with reduced effort, some fishing methods may cause unacceptable effects on non-target species and damage to the seabed and we recommend that the more damaging gears are restricted in particular areas or banned completely, particularly in deep seas areas.

SYNERGIES BETWEEN MARINE PROTECTED AREAS AND INDUSTRIAL USES OF THE SEAS

10.37 The marine planning system could identify potentially beneficial synergies between different sectors as well as minimise conflicts. For example, areas around navigational hazards or energy installations could be candidate sites for MPAs.

10.38 Our recommendation to establish large marine protected areas with major no-take reserves (8.96) should be considered in the wider context of the many competing pressures on the seas. For example, as offshore wind farms are established, they will be surrounded by fishing exclusion zones for reasons of safety, etc. It is possible that these could form part of no-take zones established for conservation purposes, and the Joint Nature Conservation Committee has suggested that these areas be considered as trial areas for nature conservation.[29] Similarly, there are exclusion zones around North Sea oil and gas installations. There is some evidence that these have a 'reef effect' that helps harbour fish and other organisms.[30]

10.39 Many of the UK's oil and gas installations are due to be decommissioned over the next ten years. It is not yet clear what form this process will take, but it is possible that the Best Practicable Environmental Option will be to leave some part of the largest structures, and the piles of drill cuttings on the seabed, in place.[31] If this is so, the relevant areas will need to remain off-limits to fishing, and it may make sense to include these areas in larger no-take zones. The areas will also need monitoring; an activity for which the oil and gas industry would be expected to pay. This monitoring could be combined with more general monitoring of marine protected areas, to gather much needed data about how the environment, and fish populations, are responding to the protection, and so reduce overall monitoring costs.

10.40 We have not explored the links between industrial and conservation uses in detail, but as a step towards strategic planning, **we recommend that an investigation should be made of the possible synergies between the various regulatory and marine protection regimes.** This should be carried out in the context of the UK government and the Scottish Executive developing their strategies for the marine environment[32,33] and as part of any future development of a system of marine spatial planning. This should lead to constructive proposals being brought forward as part of current strategy development, and protection of the marine environment being considered as a more integrated whole.

CONCLUSIONS ON A FRAMEWORK FOR PROTECTING THE MARINE ENVIRONMENT

10.41 Taken together, our proposals for a system of marine spatial planning (figure 10-III), an extensive network of marine protected areas and no-take reserves (8.96) and measures to reduce the level and impact of fishing (chapter 9) lay out elements of an ambitious new framework aimed at radically improving the quality and protection of the marine environment so that it can recover and flourish into the future. Their effect would be to change the burden of justification so that instead of the regulator having to demonstrate damage after the fact, fishing would only be carried out in accordance with an integrated management plan. Some of the components that would help establish the framework already exist, while other important pieces are missing. It will take time to complete the whole jigsaw but there is an urgent need for a commitment to a comprehensive solution.

THE NEED FOR A UK MARINE ACT

10.42 One way of implementing the new framework and its components would be through a Marine Act that would enshrine an integrated approach to the marine environment in law and set out a long-term strategic framework for the marine environment.

10.43 Defra's *Safeguarding our Seas*,[34] sets out the government's vision for protecting the marine environment, but it has no a basis in law. During consultation on the document, Defra argued that this was not necessary, although it recognised that some elements of the programme might require new legislation.

10.44 Other countries have taken a rather different view. In the US, the Pew Oceans Commission recommended a wide-ranging act to bring together the management of the oceans.[35] Such legislation is already in place in Canada in the form of the Oceans Act (1996). The latter provides a national strategy to manage estuarine, coastal and marine ecosystems based on the principles of sustainable development, integrated management and the precautionary approach. These principles will be put into effect through a series of plans and programmes developed with various national and regional stakeholders. The Oceans Act also establishes criteria for creating marine protected areas and the powers to protect them.

10.45 Our view is that there would be substantial benefits associated with a statutory framework that sets strategic objectives and provides continuity, clarity and regulatory certainty. Such an act could also provide the necessary basis for establishing a large-scale network of marine protected areas and no-take zones which is currently missing.[36] Given the urgency and importance of protecting the marine environment, we support calls for a Marine Act, and indeed recognise that this has been a recent topic for consultation in Scotland.[37] However, the development of new legislation should not hinder action in the meantime, either under current legislation or measures, or as part of voluntary approaches.

10.46 **We recommend that the UK government and the devolved administrations should introduce Marine Acts in their areas that:**

- **set out the principles for managing human impacts on the marine environment, with the primary objective of the enhancement and long-term protection of the environment; and**

- **establish a statutory basis for marine spatial planning and targets for marine protected areas and no-take reserves.**

EFFECTIVE INSTITUTIONS

10.47 The fishing industry's long-term survival depends on a healthy environment. In the short term, however, there is often intense conflict between fishing's economic needs and the protection of the environment, with commercial interests frequently triumphing. There needs to be a transition to a new model where environmental concerns are able to influence fisheries policy to a far greater extent. This means giving the environment a much higher priority than it has had previously.

10.48 The sharp divide between the environment and fisheries is reinforced by the separation of policy responsibility for these areas within key institutions such as the European Commission, European Councils of Ministers and UK fisheries departments (for example, Defra's policy on fisheries and nature conservation is separate from its policy on the protection of the high seas). Such compartmentalisation is a barrier to a more coherent vision emerging in the future.

10.49 To-date, the European Commission can only point to a few examples where environmental concerns have been integrated into fisheries policy. These include the closure of an area of sandeel fisheries to protect the food source for sea bird colonies, the ban of drift nets to reduce by-catches of sea mammals, and the protection of the Darwin Mounds from trawling. Recent reform of the Common Fisheries Policy has created the right climate for the introduction of further environmental measures. It will, however, rely on the Council of Fisheries Ministers being willing to implement them.

10.50 The transition to a healthier marine environment requires a coherent approach that delivers both environmental and fisheries objectives. A first step to ending the division between the two, and signalling a change in priorities in favour of greater sustainability, **would be to bring together ministerial responsibilities for fishing and the marine environment within the UK, and to change the way such portfolios are described**.

10.51 In line with the principles expressed in recent Pew Report in the US,[38] **we recommend that the objective of UK policy in this area should be to protect the marine environment.** The protection of ecosystems is ultimately the only route to overall sustainability. Fisheries management should therefore focus on fostering the long-term sustainability of fish populations and thereby the future of the fishing industry. **We recommend that the allocation of resources should reflect these new policy priorities.**

10.52 To encourage greater coherence between policy objectives at the European level **we recommend that the UK government should encourage the European Commission and European Councils of Ministers to co-operate in the development of joint fishery and environment objectives within the Common Fisheries Policy, EU Marine Thematic Strategy and other relevant policies.**

10.53 We recognise that as environmental standards are raised in Europe, on the way to an improved marine environment and more sustainable fish populations, there may be a temptation to transfer unsustainable practices and pressures elsewhere, for example, into the waters around developing countries through poorly regulated access agreements (chapter 4), or onto high seas fisheries.

10.54 **We recommend that the UK government should strongly promote action at the European level to ensure that, outside its home waters, the European fleet does not fish to standards that would be unacceptable within the EU. In particular, access agreements should restrict fishing to sustainable levels that respect the environment and the livelihoods of people in developing countries, and we recommend that effective measures are rapidly developed to monitor and police such fishing.**

PROMOTING CO-OPERATION IN FISHERIES MANAGEMENT

10.55 Promoting greater co-operation between those with a direct interest in the running of fisheries is often seen as a promising route to improving sustainability. On the one hand, bringing fishers into a system of regional co-management could be a positive step – offering a route to more sympathetic management by people who feel a degree of ownership for a particular fishery and co-operate to improve its management and the enforcement of regulations. On the other hand, there is a danger that fishing industry interests could race each other to capture the management process, opening the door to over-exploitation and the sidelining of environmental objectives.

CO-MANAGEMENT IN THE US

10.56 An early form of co-management was established in the US by the 1976 Magnuson-Stevens Act. This established eight Regional Fishery Management Councils to develop regional fishery management plans, set stock limits and hold public hearings on their plans.[39] Representation on the Councils is heavily weighted to the fishing industry and, consequently, many have come to be seen as lobby groups managing the seas for maximum exploitation.[40]

10.57 While some Councils have been more successful than others in promoting sustainable management, an adversarial atmosphere between fishers and scientists is widespread, with Regional Fishery Management Councils often downplaying scientific advice and increasing catch limits.[41] The result has been a decline in many fish stocks and a sharp increase in litigation over disputed stock assessments and catch limits. In 2002, there were over 110 outstanding lawsuits involving the National Marine Fisheries Service. A subsequent review by the US National Academy of Public Administration concluded that fisheries management was in disarray.[42] As a result, in 2004, the US Commission on Ocean Policy[43] recommended that the Regional Fishery Management Councils should be 'refined' to take more account of advice from their Scientific and Statistical Committees, and that new members should be chosen from a field that included the public and recreational fishers, as well as the commercial fleet.

10.58 Although the US Commission on Ocean Policy suggested that the regional fisheries management system merely needed fine-tuning, it recommended a more radical shake-up of wider ocean policy. This would include a new management framework, overseen by a new National Ocean Council and a number of Regional Ocean Councils. The latter would be non-regulatory bodies drawing on a wider stakeholder base of local government and NGO expertise to produce advice, raise awareness, mediate disputes and even design marine protected areas. However, if these are established, it is not clear how effective they will be given that they would be relying on the co-operation of the existing Regional Fishery Management Councils, without having any control over them.

10.59 In summary, while the US Regional Fishery Management Councils were a step towards greater regional flexibility and control, it is clear that a regional structure and a stakeholder process are not enough by themselves to ensure that fishing is well managed. In our view, systems of regional co-management need to involve a broad stakeholder base, have a system that takes proper account of scientific advice and have a clear mandate to manage fisheries sustainably, with controls in place to prevent harm to the environment and over-exploitation.

CO-MANAGEMENT IN THE EU

10.60 We have examined the experience from the US regional management system in some depth because a similar system has been proposed for the EU as part of the recent reform of the Common Fisheries Policy. Under the new system, seven Regional Advisory Councils (RACs) will be established. Five will be area-based, the sixth will cover pelagic populations (blue whiting, mackerel, horse mackerel and atlanto-scandic herring) and the seventh will be concerned with the EU's involvement in distant water fisheries. Each area-based RAC will involve stakeholders from at least two Member States and the intention is to cover all of the EU's fishing activity (figure 10-IV).[44]

10.61 The RACs will comprise a general assembly and an executive committee. Two-thirds of the membership of both bodies will be reserved for the fisheries sector (including the catching and processing sectors). The remaining one-third of seats will be open to other groups including environmental NGOs.

Figure 10-IV
Proposed Regional Advisory Councils. The Regional Advisory Council for distant water fisheries is not shown.

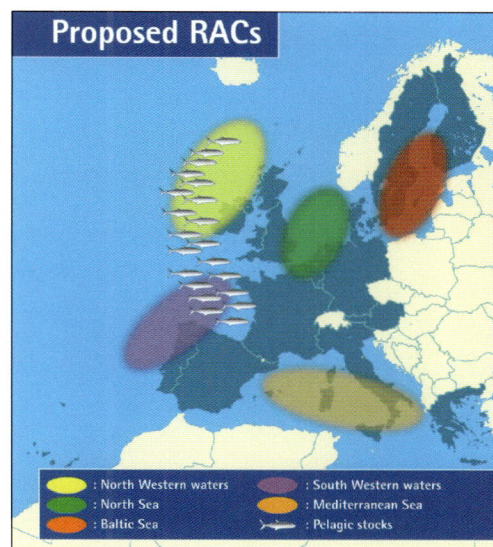

10.62 The RACs will cover large, heterogeneous areas that encompass different ecosystems. They thus bear little relation to the concept of 'regional seas' described earlier (2.55) and appear to be poorly co-ordinated with the regional eco-management plans proposed under the EU Marine Thematic Strategy. Local representatives may also find it costly in time and money to participate at this level.

10.63 There are provisions[45] to allow smaller areas (such as the Irish Sea) to be considered as subdivisions of RACs, and to be represented by working groups reporting to the RAC. However, at the present time, the European Commission believes that a large number of RACs would be too complex and expensive to be acceptable to Member States, although the door remains open to more RACs in the future. It is also worth noting that while the European Commission will provide each RAC with a contribution of up to €250,000 for the first year, the proportion of central funding will drop from 90% to 50% over the first five years, and in the long term the RACs are meant to be self-supporting. This may affect the number and long-term sustainability of the groups that are set up.

10.64 As well as concerns about their size, there are questions about the ability of RACs to deliver change. They are essentially consultative bodies with a duty to advise and inform. It is difficult to see how a limited number of stakeholders at RAC level will be able to reflect concerns of local fishers and exert influence over the European Commission and the Council of Ministers. However, the European Commission anticipates that if RACs are successful, they could evolve into more influential Regional Management Councils.

10.65 In the UK, the Prime Minister's Strategy Unit[46] has proposed an additional strand of regional management mirroring the structure of the RACs. These more powerful individuals, known as Regional Fisheries Managers, would have the ability to develop management strategies, monitor compliance and agree controls.

10.66 Experience, such as that from the US, shows that fisheries need to be managed sustainably, not simply in the short-term interests of the fishing industry. **If a regional management system (such as RACs or Regional Fisheries Managers) is established with decision-**

243

making powers we recommend that safeguards should be put in place at every level to ensure that it:

- takes proper account of scientific advice;

- involves a balanced range of stakeholders and is not dominated by fisheries' interests;

- has a clear mandate to protect the environment and to manage fisheries for long-term sustainability; and

- is established on a regional seas scale (2.55). This would make them more locally representative and effective as well as better integrated with the EU Marine Thematic Strategy; and

- has independent oversight and is subject to an early review of its scope, success and environmental impact.

CO-MANAGEMENT OF INSHORE FISHERIES IN THE UK

10.67 While it remains to be seen whether RACs or Regional Fisheries Managers can help improve fisheries management, the proposals do not appear to offer any clear benefits for other aspects of the environment. We believe other measures will have to be taken (for example, through a new planning system) which we have outlined elsewhere (10.22 to 10.46). We do, however, see a more positive role for other systems of local management within inshore waters.

10.68 The inshore zone is under heavy pressure from a variety of sources, but it is here that many of the most highly valued parts of the marine ecosystem are found, such as Special Areas of Conservation, wildlife reserves and inshore fishing grounds. It is also the area where the UK retains the most discretion in management, and so provides an opportunity to demonstrate how environmental and fishery objectives can be better managed in future. Chapter 8 describes several examples where inshore fishers have agreed to voluntary management measures to reduce their impact on the marine environment.

10.69 To allow the UK the maximum opportunity to improve the management of inshore waters, **we recommend that the UK government should seek to extend its powers to regulate UK and foreign vessels out to 12 nm** (from the current 6 nm), as is now possible for environmental purposes under the reformed Common Fisheries Policy. The UK's territorial waters extend to 12 nm, although these limits are reviewed periodically. We also agree with the Royal Society of Edinburgh[47] in **recommending that the UK government should press for the 12 nm territorial limit to be made permanent.**

10.70 Parts of the UK already benefit from a formal system of localised fisheries management through the Sea Fisheries Committees established in England and Wales at the end of the 19th Century (4.138). These Committees draw in local authorities, fishers and fish processors. As well as the ability to manage aspects of fishing in inshore waters, they have the power to make environmental bye-laws. Evidence submitted to us[48] noted that these are seldom invoked to regulate fishing on purely environmental grounds, except where fish stocks are directly concerned. It was also suggested that the situation is unlikely to alter without additional legislative changes, and the Committees having a greater say in the management of fisheries in important habitat areas.

10.71 We suggest that Sea Fisheries Committees should be able to reach local agreements to refine enforcement regimes, and effort and gear controls within their jurisdiction, if they can show that the changes would be in the interest of the environment and/or conservation of fish populations. We also support calls made by the Royal Society of Edinburgh[49] and the Prime Minister's Strategy Unit[50] to extend the principle of inshore management committees to Scotland. **We therefore recommend that the Scottish Executive should establish inshore management committees in Scotland. We also recommend that the powers of Sea Fisheries Committees to protect the environment are examined and upgraded as described above as part of the current review of their role.**

PUBLIC PARTICIPATION

10.72 One of the most important policies affecting Europe's marine environment, the Common Fisheries Policy, has been developed over the last 20 years by a small circle of politicians, experts and industry stakeholders. Progress towards a more sustainable policy has often been hampered by narrow national interests, and heavily influenced the fishing industry which has a direct financial interest in the outcome. There are moves to decentralise the process by offering the fishing industry the opportunity to become involved in a limited form of regional co-management. However, the general public and groups representing wider interests have had little chance to influence the way the seas are managed.

10.73 Does this matter? In the UK, no-one lives more than 70 miles from the sea. Many of us take our holidays there, or enjoy the seafood it provides. The seas are also an astonishing source of biodiversity – accounting for over 90% of the biosphere. Wherever we live, we all benefit from the role they play in wider processes, such as moderating climate and buffering changes in atmospheric gases, that are vital for a healthy planet. Future generations as well as our own have a stake in preserving a properly functioning ecosystem.

10.74 In 2004, a report from the US Commission on Ocean Policy[51] set out thirteen principles to guide US oceans policy. Prominent among these were sustainability, stewardship and participatory governance. All three underline the role of the general public, not as bystanders or passive consumers, but as key players in the ownership and management of ocean resources. To quote the report "people must understand the role the oceans have on their lives and livelihoods and the impacts they themselves have on the oceans". We see this inclusive approach as critical to developing better policies for the marine environment in future.

10.75 The promotion of a wider debate on the management of the seas is long overdue, but there is a valuable window of opportunity as the new Common Fisheries Policy and the EU Marine Thematic Strategy begin to take shape, and, at the UK level, as responses are developed to Defra's *Safeguarding our Seas*,[52] the Prime Minister's Strategy Unit report and the Review of Marine Nature Conservation.[53] These initiatives have all involved some degree of stakeholder consultation. However, as our Twenty-first Report made clear, while traditional forms of interaction such as paper-based consultation exercises are useful, they are not the best way of seeking public values,[54] and the level of public participation has been very low in some cases.[55]

10.76 We advocate the use of more participatory forms of dialogue that focus on helping to articulate public values at an early stage in the policy process. Although these methods have yet to be widely adopted, there is evidence that where public values have been sought in this way, they have changed the terms of the debate and inspired organisations to revise their most basic assumptions.[56,57]

10.77 In this report, we recommend putting much greater emphasis on protecting the marine environment. The aim is to help the ecosystems recover and preserve them for the future, so that they can sustain the wider environment and the activities that depend upon it. The wider ramifications of this approach make it vital that the Government engages directly with the public, the fishing industry and the broad base of interested stakeholders in the debate over how the seas are managed in future. For example, such a strategy will require adjustments in the size and structure of the fishing fleet, and this will have implications for the industry and the communities in which it is based. However, the 'do nothing' approach is not an option. The Prime Minister's Strategy Unit report has already highlighted that the industry has a poor future if fish populations cannot be made more sustainable.[58]

10.78 Creating a new approach to the marine environment represents an important opportunity to energise public interest and debate with a clear purpose in mind. Major policy developments in Australia (such as the rezoning of the Great Barrier Reef and marine planning system) and high profile reviews of marine policy in the US, show how this can be done successfully to deliver a practical outcome (box 10B).

10.79 As part of the development of a new approach to the marine environment, **we recommend that the UK government and the devolved administrations should:**

- **establish a process that will provide an opportunity for a broad cross-section of the public and civil society to engage in informed debate about the management of human activity in the marine environment;**

- **from an early stage, use the process to inform the development new policies on the marine environment and fisheries, including the development of marine spatial planning and a UK network of marine protected areas; and**

- **encourage greater use of these methods in the formulation of policy at the European level.**

BOX 10B	PUBLIC PARTICIPATION IN REZONING THE GREAT BARRIER REEF MARINE PARK[59]

The rezoning of the park was a major opportunity to raise public interest in marine issues as well as to take practical measures. While the reef was widely recognised as being of immense value ecologically, culturally and economically, there was less awareness of the more 'invisible' components of the ecosystem. For example, when protection was proposed for areas lying between the coast and the coral reefs the question was raised "why...when there isn't anything there?" The relative invisibility of parts of the ocean and their role in supporting the wider ecosystem can thus be a barrier to communicating the very real nature of the threats to the marine environment.

One of the most important lessons learned from the exercise was the need to engage all the people in an area who have an interest – a plan without a degree of public acceptance is a plan that cannot be enforced.[60] The Park Authority went much further than the statutory consultation requirements and conducted over 600 meetings in at least 90 locations with thousands of people. The level of public interest exceeded all expectations and was the largest for any environmental planning process in Australia. After the zoning exercise was complete, literature was made widely available within the park and on the Authority's website to publicise and explain the new plan.

THE ROLE OF THE MEDIA, EDUCATION AND RESEARCH

10.80 It seems unlikely that more people will be able to take full part in the policy process, or be motivated to do so, unless the value of the seas is better understood. Recent debates about genetic modification and public health indicate that the public has an appetite for better access to reliable scientific and environmental information.[61]

10.81 The media have been highly influential in raising the profile of environmental issues, but could do more to expose the harm that is being done to the seas, as well as highlight their innate beauty. A broad cross-section of civil society – non-governmental organisations, academic institutions and the private sector – could also help stimulate the debate about the future management of the seas.

10.82 Schools, universities, museums and aquaria have an important role to play in increasing the profile of the marine environment and the understanding of its natural wonders. We should avoid raising a generation of schoolchildren who worry about the fate of the tropical rainforest but have no appreciation of the value of the local shoreline, the life under the surface of the seas or the source of the fish we eat. Developing a strategy in the UK to promote formal and informal education on marine issues, as has been recently recommended in the US,[62] could be an important milestone on the road to increasing public engagement and a route to improving learning and training. **We therefore recommend that the UK government should develop and implement strategies for:**

- **improving education on marine matters by including these issues into key stages 2 and 3 of the national curriculum; and**

- **communicating marine issues to stakeholders and the public.**

BETTER INFORMATION FOR THE CONSUMER

10.83 One obvious role in which the public is already active is as consumers. Currently, the effects of overfishing can be more or less invisible to the consumer. As one source of fish becomes heavily depleted, for example, North Sea cod, producers and retailers switch to another source, another species or even a completely different type of product.

10.84 Present legislation recognises that consumers should have access to food that is fit to eat, but consumers are also increasingly interested in how and where food has been produced. There have been many instances where the method of food production has led to the highly visible 'branding' of certain products, such as organic foods, that have become popular with some consumers. Examples of such schemes in the fisheries sector are given in box 10C.

BOX 10C   LABELLING, STANDARDS AND CONSUMER INFORMATION

'Dolphin friendly' tuna

This is an area where misleading schemes have been allowed to proliferate. The term 'dolphin friendly' embraces a wide range of standards administered by different international bodies. Some of these are so weak that they allow fishing methods that are in fact 'dolphin deadly' such as the encircling of fish and dolphins by nets.[63]

Others have also criticised the schemes for their unintentional side-effects. For example, the switch to different types of fishing gear to reduce impacts on dolphins has been blamed for increased by-catch of other species.[64] None of this is obvious to the consumer and underlines the need for a strong system of international standards.

Product labelling schemes

Producers and retailers can have an important influence on public attitudes and buying habits. One of the best-known labelling schemes for fish products is certification by the Marine Stewardship Council (MSC). This was jointly founded in 1997 by WWF and Unilever, the world's largest buyer of frozen food, with the aim of creating economic incentives for sustainable fishing. Producers are assessed against the MSC's environmental standard for sustainable fishing, which is based on the Food and Agriculture Organization's Code of Conduct for Responsible Fishing and reviewed by scientists on the MSC's Technical Advisory Board.[65] If a company passes an independent assessment it wins the right to use the MSC logo on its products. The MSC now has 100 member organisations in more than 20 countries. It has been criticised by some in the industry for being too restrictive.[66]

Information for consumers

Information for consumers does not need to be tied to particular products or retailers. The Good Fish Guide[67] produced by the Marine Conservation Society provides a wide range of detailed information to help people choose fish that come from healthy stocks, are sustainably managed, are caught using methods which minimise damage to the environment and support UK and local fishing communities.

Response of retailers

Representatives from one UK supermarket chain[68] told us that they make considerable efforts to ensure that their fish products were ethically sourced and took account of environmental concerns. Nonetheless, the company was aware of concerns over the reliability of categorising certain wild fisheries as sustainable and reported difficulties in sourcing such fish.

They considered the Marine Stewardship Council certification to be imperfect. Nevertheless they do not sell the 20 species of fish on the Marine Conservation Society campaign list of unsustainable fisheries. They also pay a premium to skippers for larger fish and for changing gear and fish grounds to avoid by-catch. Issues surrounding farmed fish were complicated and there was, in their view, a requirement for external endorsement of a set of standards. The supermarket's aim is to have a better quality product that supports the brand, the cost of which was considered to be indistinguishable from the cost of doing good business.

Environmental management systems

Defra supports the EC Eco-Management and Audit Scheme (EMAS) that awards a logo for environmental management systems. Some UK fishery organisations are investigating whether such systems, linked to a British Standard, can be adopted to improve environmental performance by, for example, requiring the use of certain types of gear such as dolphin 'pingers' on nets. The International Standards Organization (ISO) also plays an important role in developing standards for environmental management systems and for environmental labelling and declarations used by manufacturers.

10.85 Some of the approaches described in box 10C are potentially powerful, but their impact will depend on whether they are able to affect a significant share of the market (at present, the Marine Stewardship Council only certifies 4% of the world's wild fish), and the degree to which they contribute to a real increase in environmental protection.

10.86 There may also be opportunities for new and more imaginative schemes that highlight a number of positive images associated with a particular product. This could include emphasising both the health benefits of eating fish and the sustainability of production. Nevertheless, it is clear that retailers, caterers and consumers have concerns over environmental claims that are sometimes vague and may have only a flimsy basis.[69] If such schemes are to have a positive effect, our Twenty-first Report[70] highlighted the fact that they need to be based on sound standards and reliably enforced and audited.

10.87 We recognise that there can be problems with labelling and awareness raising schemes, but they can be a force for good in encouraging the uptake of sustainable production methods, which is urgently needed, and providing the consumer with a more informed choice. We note that the European Commission is supportive of eco-labelling and is intending to launch a debate on the merits of voluntary versus statutory systems. **We therefore recommend that within the next two years the UK government and devolved administrations should work with producers, retailers and caterers to produce a strategy that would:**

- **increase the proportion of seafood and aquaculture products produced under environmentally-accredited schemes;**

- **improve the quantity and quality of environmental information available to consumers through labelling and awareness schemes in the retail and catering sectors; and**

- **ensure the reliability of such schemes by developing appropriate standards and auditing procedures.**

CONCLUSIONS

10.88 Society has become familiar with the arguments for sustainable development on land, and is beginning to take steps to deliver that vision. But in the case of the marine environment we are lagging far behind. It has taken many years to acknowledge the problem and there is still disagreement about causes and solutions.

10.89 Yet urgent action is undoubtedly needed to reverse the impoverishment of marine ecosystems and address the problem of falling fish populations and the precariousness of the fishing industry's future. The European Commission and Member States have accepted the need for widespread reform of the Common Fisheries Policy and are making welcome efforts in this direction. But these will not be enough on their own, since a fisheries policy can never hope to address all the complex demands of protecting the marine environment.

10.90 **What is needed, above and beyond a reformed Common Fisheries Policy, is a set of policies and measures that recognise the dependence of one part of the marine environment upon another. This means setting out a strategic vision that places most emphasis on protecting the ecosystem, both for its own sake and to sustain the goods and services that flow from it.**

10.91 As part of establishing this more integrated vision, we have proposed a statutory system of marine spatial planning to provide a rational basis for managing and co-ordinating human activities in the seas. This would, for the first time, introduce area-based management for fishing and subject it to some of the same environmental standards and assessment procedures that are applied to other extractive industries.

10.92 A marine planning system, backed up by a new Marine Act, would set the context for a system of marine protected areas and no-take reserves. It would also set out the principles that would underlie a much more precautionary approach to the marine environment, whereby activities are not carried out except on the basis of an agreed regional plan. Taken together, these measures would set-aside areas of the sea for recovery and preservation, while fishing continues elsewhere.

10.93 There will continue to be an important place for traditional fishery management measures to moderate the impact of gears, and reduce by-catch and discards. Further measures are needed to remove over-capacity in the fishing fleet, without which it will be impossible to achieve a turn around in the condition of the marine environment, or indeed of the fishing industry. We call upon the UK government to examine urgently the opportunities for further decommissioning and diversification in the industry.

10.94 The seas are the province of the many, not the few. The way that they are protected should be of concern to all of us, since we have a direct stake in the outcome. Our report stresses the need for policy on the marine environment to take more account of the wider public interest, and find new ways to reflect public values in the development and management of the system. Consumers also need better information about the health benefits and risks of the seafood they eat, and the overall environmental impacts associated with its production.

10.95 Policies for the marine environment and fisheries management have traditionally been developed largely in isolation from one another. This lack of coherence and underplaying of the role of the natural environment in sustaining human activities has helped contribute to the current crisis in our seas. We have suggested a new approach that relies on bringing these different, but equally important, strands of policy together. Such steps are both necessary and possible, to move from a situation where our seas are at risk to one where society can be confident that their future is secure.

Chapter 11

RECOMMENDATIONS

We bring together the recommendations that appear in bold type elsewhere in this report.

BRINGING ABOUT RADICAL CHANGE

11.1 As a society we have assigned much lower priority to protecting the seas compared with the land. This needs to change urgently.

11.2 Over-fishing is a global problem that has damaged the marine environment and led to the collapse of fisheries in many areas. Our report is one of several worldwide to bring the problem to the attention of governments and call for urgent action. OSPAR has identified fishing as the cause of three of its top six risks to the environment of the North Sea. We therefore focus on the effects of fishing, and we call for radical change. Present policies have failed, and incremental improvements will not deliver what is needed. We face further collapses in fisheries and harm to the marine environment unless there is significant and urgent action.

11.3 We recognise that, in the short-term, this will be painful to those in the fishing industry, but government has to look at the wider picture, including society's stewardship of the environment. Technology has brought enormous changes in the ability to exploit the sea and society needs to accept the consequent burden of responsibility. The industry will need support to adjust, but in the longer term we believe that change will be in the best interests of the industry itself. A continued regime of too little, too late will ultimately leave many sectors of the fishing industry without a future.

11.4 Many of our recommendations will require action at the European level as well as in the UK. However, the UK government has an important opportunity to demonstrate international leadership by improving the protection of the seas for the benefit of this country and the nations with which we share them. A better marine environment is also an essential part of securing a brighter future for the fishing industry. The first step will be to place fisheries management firmly within the context of wider management of human activities in the marine environment.

BETTER MANAGEMENT OF HUMAN ACTIVITIES IN THE MARINE ENVIRONMENT

11.5 The principles of an ecosystem approach are well established. They change the focus from the management of single fish stocks to the conservation of ecosystems within the wider marine environment. While there is broad consensus that this holistic approach is a better way forward, there has been little progress in implementing the concept. We advocate a pragmatic response that would put the emphasis on robust, practical steps to halt the degradation of the marine environment, rather than a slow, incremental approach to implementation.

11.6   Fisheries managers have so far failed to fully apply the precautionary approach and, at present, fisheries management does not take sufficient account of the large uncertainties that exist in scientific advice, whether on fish population assessments, or the effects of fishing on the wider marine environment. **We therefore recommend that:**

- **human impacts on the marine environment should be managed in a fully precautionary manner. Fishing should only be permitted where it can be shown to be compatible with the framework of protection set out in this report** (7.54); and

- **the above principle would reverse the current presumption in favour of fishing. In future, applicants for fishing rights (or aquaculture operations in the marine environment) should have to demonstrate that the effects of their activity would not harm the sea's long-term environmental sustainability** (7.59).

DEVELOPING A NETWORK OF MARINE PROTECTED AREAS

11.7   Firm evidence exists that marine protected areas and reserves can provide habitat protection and form part of an effective response to the effects of over-fishing.[1] We already have sufficient information to identify some sites that could form the basis of future networks. There is a strong case for establishing large-scale protected areas and **we recommend that the UK government should:**

- **develop selection criteria for establishing a network of marine protected areas so that, within the next five years, a large-scale, ecologically coherent network of marine protected areas is implemented within the UK. This should lead to 30% of the UK's exclusive economic zone being established as no-take reserves closed to commercial fishing; and**

- **develop these proposals in consultation with the public and stakeholders** (8.96).

11.8   For maximum effectiveness, marine protected areas should be part of a suite of measures, which should include reduction in fishing effort to sustainable levels, gear modification and other forms of marine protection. **We recommend that the marine protected area network referred to above should be implemented as part of a balanced package of measures to improve the management of human impacts on the marine environment and to reduce fishing effort** (8.98). They should also be developed in the context of a new marine spatial planning system described below.

INTRODUCING MARINE SPATIAL PLANNING

11.9   The development of marine spatial planning in UK waters would be a major step forward. It would provide a plan-based system that would allow current and future pressures to be handled in a strategic fashion. We are strongly of the view that a comprehensive, statutory marine planning system is needed.

11.10 It is clear that whichever organisation manages the marine planning system, it needs to have access to expertise on environmental matters, clear objectives to protect marine ecosystems and a strong co-ordinating role so that it can work with the many bodies that have an interest in the seas. We would also expect the system to be based on meaningful ecological areas. The concept of regional seas may provide a useful administrative unit.

11.11 **We recommend that the UK government should develop a comprehensive system of marine spatial planning that:**

- **sets out the principles and long-term goals for protecting the marine environment and promoting the sustainable use of the sea;**

- **develops integrated regional management plans to guide all major uses of the sea, including fishing. These should ensure high standards of marine protection, and be subject to strategic environmental assessment; and**

- **has a statutory basis as well as a clear framework for public participation.**

We also recommend that the UK government should promote the principle of marine spatial planning at European and international levels (10.22).

11.12 A marine planning system would help manage conflicts between sectors, but it could also identify potentially important synergies, for example, between candidate sites for marine protected areas and closed areas around navigational hazards, wind farms, and oil and gas installations. **We recommend that an investigation should be made of the possible synergies between various regulatory and marine protection regimes** (10.40).

11.13 In addition, **we recommend that the UK government and devolved administrations should apply strategic environmental assessment and environmental impact assessment to fishing, amending the legislation as necessary** (10.28). Incorporating these principles into the marine planning system would strengthen environmental protection and bring fishing into line with the regulations governing other extractive processes.

PROVIDING A STATUTORY FRAMEWORK

11.14 In our view a new statutory framework is needed to set strategic objectives for the marine environment and provide continuity, clarity and regulatory certainty. This could take the form of a Marine Act that would also provide the basis for establishing a large-scale system of marine protected areas and no-take reserves, which is currently missing. **We therefore recommend that the UK government and the devolved administrations should introduce Marine Acts in their areas that:**

- **set out the principles for managing human impacts on the marine environment, with the primary objective of the enhancement and long-term protection of the environment; and**

- **establish a statutory basis for marine spatial planning and targets for marine protected areas and no-take reserves** (10.46).

EFFECTIVE INSTITUTIONS

11.15 The transition to a healthier marine environment requires coherent policies that deliver both environmental and fisheries objectives. A first step to ending the division between the two, and signalling a change in priorities in favour of greater sustainability, would be to bring together ministerial responsibilities for fishing and the entire marine environment within the UK, and to change the way such portfolios are described.

11.16 **We recommend that the principal objective of UK policy in this area should be to protect the marine environment** (10.51).[2] The protection of ecosystems is ultimately the only route to overall sustainability.[3] Fisheries management should therefore focus on fostering the long-term sustainability of fish populations and thereby the future of the fishing industry. **We recommend that the allocation of resources should reflect these new policy priorities** (10.51).

11.17 To encourage greater coherence between policy objectives at the European level **we recommend that the UK government should encourage the European Commission and European Councils of Ministers to co-operate in the development of joint environment and fishery objectives within the Common Fisheries Policy, the EU Marine Thematic Strategy and other relevant policies** (10.52).

11.18 We recognise that as environmental standards are raised in Europe, there may be a temptation to transfer unsustainable practices and pressures elsewhere, for example, into the waters around developing countries through poorly regulated access agreements, or onto high seas fisheries.

11.19 **We recommend that the UK government should strongly promote action at the European level to ensure that, outside its home waters, the European fleet does not fish to standards that would be unacceptable within the EU. In particular, access agreements should restrict fishing to sustainable levels that respect the environment and the livelihoods of people in developing countries, and we recommend that effective measures are rapidly developed to monitor and police such fishing** (10.54).

SHORT-TERM MEASURES TO PROTECT THE MARINE ENVIRONMENT

11.20 The package of measures we recommend represents radical change and will take time to deliver. But the need for change is urgent. The EC Habitats and Birds Directives afford some protection to parts of the marine environment, but the conservation outcome could be considerably enhanced if the Directives reflected the interconnected nature of the marine environment and included more species and habitats. This would also increase the chance that the Natura 2000 sites could help in establishing the core of a UK network of marine protected areas. **We therefore recommend that the UK government should:**

- **amend legislation to allow UK Marine Nature Reserves to be designated even where there are objections;**

- **introduce measures to protect all designated sites (such as Natura 2000 sites) from the adverse effects of fishing. If such measures cannot be agreed under the Common Fisheries Policy, the UK should introduce unilateral measures to protect these sites;**

- **review the operation of the EC Habitats Directive and consider how the Directive's ability to protect the marine environment could be improved and extended to the wider environment as well as vulnerable areas;**

- **with the Joint Nature Conservation Committee, develop proposals to extend the Annexes of the EC Habitats Directive to provide adequate coverage of important marine species and habitats (see paragraphs 8.40-8.42); and**

- **use the findings of the above review to negotiate with the EU to amend the EC Habitats Directive** (8.44).

11.21 This action is necessary now, but is not of itself sufficient to deliver the change that is needed.

FISHERIES MANAGEMENT

DECOMMISSIONING PART OF THE UK FLEET

11.22 The overcapacity in the European fishing fleet has been well documented and leads to unsustainable pressures on fish populations, making them more difficult to manage. It also reduces the profitability and stability of the industry, by leaving it highly exposed to short-term fluctuations in price. Decommissioning fishing vessels can be an effective way of reducing capacity if schemes are properly designed. The required scale of decommissioning remains a matter of debate and will vary between the various fisheries sectors and areas, but is likely to be largest in the case of the demersal fleet.[1]

11.23 Introducing large-scale marine protected area networks, including the 30% no-take reserves recommended in paragraph 11.7, is likely to have a socio-economic impact because some fishers will have to adjust their fishing. Better catches in future could improve the financial security of individual fishers and the industry in the longer-term, but in the short-term some may struggle. It may be appropriate to consider ways of assisting the fishing industry through the transition to protected areas in the UK. This could be done in tandem with the measures described above to reduce overall capacity. Such efforts will also help avoid large-scale displacement of activity from these areas (5.82).

11.24 **We recommend that the UK government and fisheries departments should initiate a decommissioning scheme to reduce the capacity of the UK fishing fleet to an environmentally sustainable level and ensure similar reductions are made in EU fleets that fish in UK waters** (9.15).

11.25 **We recommend that funds should be made available to help the transition of the industry during the establishment of the UK network of marine protected areas and no-take reserves** (9.16).

11.26 To reduce impacts on communities, **we recommend that the UK government should review arrangements for EU Structural Funds and other funds to promote economic diversification in fisheries dependent areas** (9.16).

11.27 **We also recommend that the UK government should press at EU level for an end to all subsidies that can result in increased fishing pressure, including vessel modification and improving port and fish processing facilities** (9.83).

INTRODUCING EFFORT CONTROL

11.28 The current system of European fisheries management relies on controlling the amount of fish landed. The system is based on limits for particular species (known as total allowable catches) and has many problems. In contrast, effort control is a direct measure of fishing pressure, and so may be a more effective form of control, particularly in mixed demersal fisheries. **We therefore recommend that the UK government should move towards managing fisheries on the basis of effort controls (in terms of kilowatt-hours at sea) within the next three to five years** (9.29)

11.29 **We recommend that the UK government should take steps to ensure that appropriate effort controls are introduced throughout EU waters in the shortest possible time frame** (9.29).

TECHNICAL CONSERVATION MEASURES

11.30 Gear regulations, in tandem with a system of spatial planning (10.22), could provide the key to protecting sensitive habitat types. **We therefore recommend that the UK government should make greater use of renewable fishing licences to regulate UK fisheries, by linking licensing to marine spatial plans, reductions in fishing effort, gear restrictions and improvements in vessel monitoring technology** (9.34).

11.31 **We recommend that the UK government should rank the impacts of gear specific to UK fisheries in relation to their impact on habitats and press at EU level for the introduction of appropriate policy responses.** Such a review should take account of the marine landscape classification and seabed mapping exercises being carried out by the Joint Nature Conservation Committee (9.37).

11.32 Some habitats in the sea – e.g. maerl beds and reefs – are so sensitive that they should receive total protection from trawling and other destructive techniques. **We recommend that the Joint Nature Conservation Committee should develop a list of potentially damaging operations, which should be avoided in all areas of marine conservation importance** (9.39).

11.33 **We recommend that the UK government should introduce plans to give complete protection to sensitive marine habitats from destructive fishing techniques in specific areas through a new process of marine planning and strategic environmental assessment (10.22-10.28) that would approve the use of gears only in those areas where they will not cause significant environmental harm** (9.40).

11.34 Bycatch involves the accidental catching of marine mammals, sharks, rays, sea birds, organisms from the seabed and non-target fish. Bycatch is one of the most serious threats facing populations of small cetaceans, yet a number of measures are available that can help reduce it. **We recommend a staged approach to reducing by-catch. Modified gears (9.57) should be introduced for the entire fleet along with a more comprehensive monitoring regime to ensure compliance and to determine the effectiveness of these measures. If target levels of by-catch reduction are not met in a particular fishery, then this fishery should be closed** (9.59).

11.35 Discarding involves throwing away over-quota or undersized fish. The European Commission has estimated that discards may account for nearly 70% of fish mortality for some species and locations.⁵ The volume of discards poses a serious threat to the conservation of fish. **We recommend that the UK government should negotiate at EU level for a mandatory full catch reporting scheme and that the data should be published annually by Defra and the relevant devolved authorities** (9.67).

11.36 Once our other proposals have been implemented, **we recommend that the UK government should press for the introduction of an EU-wide discard ban** (9.68).

11.37 A real obstacle to effective enforcement is the current level of non-compliance within the fishing industry. The situation is not helped by the complexity of fisheries regulations and their frequent changes. We recognise that there are practical reasons why fishing regulations have cut-off points that exempt smaller fishing vessels from regulation. We are also aware, however, that smaller vessels with sizeable capacity have been built to benefit from cut-off points. **We therefore recommend that the UK government should review the activities and environmental impact of smaller vessels that do not fall under the full set of fishing controls to ensure that the benefits of our recommendations are not reduced** (9.8).

11.38 In addition, **we recommend that the UK government should pursue a policy of installing tamper-proof vessel position monitoring devices on licensed fishing vessels over 8 metres in length. The aim should be to complete this installation within three to five years** (9.76).

11.39 **We also recommend that UK fisheries departments should commission work to trial video recording of catch on board vessels** (9.76).

11.40 Bottom trawling has a major environmental impact across wide areas of the sea. High seas bottom trawling has led to the serial depletion of deep-sea fish stocks.⁶ There are also high rates of bycatch in deep-water fisheries and trawling can cause irreparable damage to important seabed features such as seamounts. **We therefore recommend that the UK government should immediately halt any deep-sea trawling taking place in UK waters or being carried out by UK vessels** (9.51).

11.41 **We also recommend that the UK Government should press the European Commission to ban bottom trawling, gillnetting and long-lining for deep-sea species in EU waters** (9.51).

11.42 **We recommend that the UK government should promote measures to prohibit destructive deep-sea fishing practices and promote the establishment of a system of marine protected areas on the high seas. In addition, it should press for international controls on high seas bottom trawling, and for their proper implementation and enforcement under, for example, the UN Straddling Stocks Agreement and the UN Convention on the Law of the Sea** (9.53).

11.43 Marine emissions from fuels used for international journeys (known as bunker fuels) remain outside international agreements to control greenhouse gases. We note that marine fuel used in fishing vessels is exempt from UK fuel duty. **We recommend that the UK should promote efforts at the European and international levels to bring marine**

emissions of greenhouse gases within international agreements and to control other atmospheric emissions from ships (9.96).

CO-MANAGEMENT OF FISHERIES

11.44 Recent reform of the Common Fisheries Policy has resulted in the fishing industry being given a greater role in the regional management of fisheries. While a more co-operative management model may have benefits, experience has shown that a regional structure and a stakeholder process are not enough by themselves to ensure that fisheries are sustainable. If a regional management system (such as Regional Advisory Councils or UK Regional Fisheries Managers) is established with decision-making powers **we recommend that safeguards are in place at every level to ensure that it:**

- **takes proper account of scientific advice;**

- **involves a balanced range of stakeholders and is not dominated by fisheries' interests;**

- **has a clear mandate to protect the environment and to manage fisheries for long-term sustainability;**

- **is established on a regional seas scale (2.55). This would make it more locally representative and effective as well as better integrated with EU Marine Thematic Strategy; and**

- **has independent oversight and is subject to an early review of its scope, success and environmental impact** (10.66).

11.45 To allow the UK the maximum opportunity to improve the management of inshore waters **we recommend that the UK government should seek to extend its powers to regulate UK and foreign vessels out to 12 nm** (from the current 6 nm), as is now possible for environmental purposes under the reformed Common Fisheries Policy (10.69).

11.46 Currently, the UK's territorial waters extend to 12 nm, although these limits are reviewed periodically under the Common Fisheries Policy. **We recommend that the UK government should press for the 12 nm territorial limit to be made permanent** (10.69).

11.47 **We recommend that the Scottish Executive should establish inshore management committees in Scotland. We further recommend that the powers of Sea Fisheries Committees to protect the environment are examined and upgraded as part of the current review of their role** (10.71).

PUBLIC PARTICIPATION

11.48 Wherever we live we all have an interest in a thriving marine environment. Future generations as well as our own have a stake in preserving a properly functioning marine environment. Yet past policies have been determined largely by policymakers and the fishing industry. The general public and wider interests have had little chance to influence the way the seas are managed. **We recommend that the UK government and the devolved administrations should:**

- **establish a process that will provide an opportunity for a broad cross-section of the public and civil society to engage in informed debate about the management of the marine environment;**

- **use the above process to inform from an early stage the development of policy on the marine environment and fisheries, including the development of marine spatial planning and marine protected areas; and**

- **encourage greater use of these methods in the formulation of policy at the European level** (10.79).

11.49 Developing a UK strategy to promote formal and informal education on marine issues, as has been recently recommend in the US,[7] could be an important step on the road to increasing public engagement and a route to improving learning and training. **We therefore recommend that the UK government should develop and implement strategies for:**

- **improving education on marine matters by including these issues into key stages 2 and 3 of the national curriculum; and**

- **communicating marine issues to stakeholders and the public** (10.82).

11.50 Present legislation recognises that consumers should have access to food that is fit to eat, but consumers are also increasingly interested in how and where food has been produced. Labelling and awareness raising schemes can have their problems but they can also be a force for good in encouraging sustainable production methods and providing the consumer with a more informed choice. **We recommend that within the next two years the UK government and devolved administrations should work with producers, retailers and caterers to produce a strategy that would:**

- **increase the proportion of seafood and aquaculture products produced under environmentally-accredited schemes;**

- **improve the quantity and quality of environmental information available to consumers through labelling and awareness schemes in the retail and catering sectors; and**

- **ensure the reliability of such schemes by developing appropriate standards and auditing procedures** (10.87).

11.51 Recognising the health benefits of the long chain n-3 polyunsaturated fatty acids found in fish, the Food Standards Agency has recommended that people should eat two portions of fish a week, one of which should be oily. Higher levels of fish consumption will increase pressure on already depleted fish populations. There are also concerns over the contamination of fish and fish oil with toxic products such as dioxins and polychlorinated biphenyl compounds, and over consuming excessive levels of fat-soluble vitamins. **We therefore recommend that:**

- **studies are undertaken to examine the full environmental implications of the Food Standards Agency's advice on eating fish;**

- **every effort is made to introduce alternative sources of long chain polyunsaturated fatty acids (n-3 PUFAs) from biological sources other than fish;**

- **an urgent effort is made to discover efficient chemical synthetic pathways to generate the fatty acids, EPA and DHA;**

- **further consideration is given to providing advice to the public about adding long-chain n-3 PUFAs as dietary supplements rather than relying solely upon an increase in oily fish consumption; and**

- **further research is undertaken to discover the mechanisms by which long chain n-3 PUFAs benefit human development and health** (3.52).

RESEARCH TO UNDERSTAND THE MARINE ENVIRONMENT

11.52 **We recommend that Government should encourage universities, research councils and others to fund research on the marine environment and consider ways of improving the dissemination and use of marine data. The issue should also be considered by the Natural Environment Research Council (NERC) as part of its review of marine science in 2004/5** (7.32).

11.53 NERC is currently working with other bodies to establish a joint venture to further develop understanding of the marine environment. **We recommend that the Natural Environment Research Council makes it a priority of the above initiative to fund research on the environmental impacts of marine capture fisheries and to ensure that this knowledge is transferred to policymakers, regulators, fisheries managers and others** (7.31).

11.54 **We recommend that the UK government should adopt a suite of indicators that reflect the state of marine ecosystems in UK territorial waters and measure progress in conserving the marine environment** (7.70).

11.55 **We recommend that fisheries subsidies should support research and monitoring schemes that use information provided by fishers in order to supply data for modelling and management** (7.35).

REDUCING THE ENVIRONMENTAL IMPACT OF AQUACULTURE

11.56 In Europe, marine aquaculture relies on compound diets based in part on fishmeal and fish oil derived from wild caught fish. This boosts production and economic performance, while improving flesh quality and reducing local environmental impacts. However, supplies of fishmeal and fish oil are under pressure, with significant consequences for the wider ecosystem. A number of routes exist to reduce aquaculture's dependency on wild fish populations.

11.57 **We recommend that the UK government and the Scottish Executive should promote a strategy to improve the sustainability of fishmeal and fish oils supplies. This should include steps to:**

- **increase the efficiency with which fish meal and oil are used within the aquaculture industry;**

- **encourage the trend away from the use of fishmeal and oil in the livestock industry, so that the aquaculture industry is given preference of supply;**

- **accelerate the development and use of viable alternatives within aquaculture. This should include research into the feasibility of substituting fishmeal and fish oil with alternatives, the farming of non-carnivorous fish and consideration of a tax or other economic instrument on the use of fishmeal and fish oil** (6.49).

11.58 Freshwater fish require less fish protein in their diet than carnivorous species like salmon. However, the farming of kinds of fish could produce problems similar to those experienced with Atlantic salmon. Causes for concern include the impacts of farmed fish on wild populations through interbreeding, and the transfer of diseases and parasites. **We recommend that appropriate controls should be put in place at the start of farming of new species** (6.45).

11.59 It is important to protect genetic diversity in wild fish populations because they harbour gene complexes capable of responding to changing evolutionary forces in natural environments. **We recommend that:**

- **the UK government and the Scottish Executive should publish an action plan to describe how they will meet their obligations under the North Atlantic Salmon Conservation Organization's Williamsburg Resolution;**

- **the Scottish Executive and Scottish Environment Protection Agency should fund research into the design of protection zones to separate cage farms from salmon rivers, including cage location based away from migratory routes of wild salmon, and apply the findings;**

- **the Scottish Executive should continue to work with the fish farming industry to strengthen its Code of Containment and to make the Code mandatory. In addition, the Guidelines on the Containment of Farm Salmon developed by the North Atlantic Salmon Conservation Organization should be reflected in the minimum standard for the construction and operation of fish farms;**

- **Scottish Environment Protection Agency and the fish farming industry should collaborate to carry out further research to improve technical and operational standards on fish farms so as to reduce escapes. The findings of this research should be reflected in the Code of Containment; and**

- **the Scottish Executive should introduce regulations to prevent the outflow from smolt rearing units flowing into salmon rivers** (as is already the case in Norway) (6.69).

11.60 UK regulatory bodies have taken a strict line on commercialisation of genetically modified fish. Given the widespread concerns **we recommend that genetically modified fish should not be released or used in commercial aquaculture in the UK for the foreseeable future** (6.81).

11.61 Fish farms often have a detectable environmental impact on their immediate vicinity, but our knowledge of the fate and effect of common pollutants is far from adequate. **We recommend that the Scottish Executive and Scottish Environment Protection**

Agency should commission further research and monitoring into the long-term environmental effects of using chemical therapeutants and copper antifoulants in aquaculture and into alternatives to such compounds (6.103).

11.62 **We recommend that the Scottish Executive and Scottish Environment Protection Agency should develop a set of indicators to describe the pollutant load from fish farms, and that the performance of fish farms against these indicators should be monitored and published.** The indicators could include aspects such as organic load on the seafloor and the capacity for nutrient processing (6.121).

11.63 **We also recommend that an environmental impact assessment should be carried out for every application for a new or significantly modified fish farm** (6.142).

ALL OF WHICH WE HUMBLY SUBMIT FOR
YOUR MAJESTY'S GRACIOUS CONSIDERATION

Tom Blundell *Chairman*

Roland Clift

Paul Ekins

Brian Follett

Ian Graham-Bryce

Stephen Holgate

Brian Hoskins

Jeffrey Jowell

Susan Owens

Jane Plant

Steve Rayner

John Speirs

Janet Sprent

Tom Eddy   *Secretary*

Diana Wilkins

Jonathan Wentworth } *Assistant Secretaries*

Philippa Powell

MEMBERS OF THE ROYAL COMMISSION

CHAIRMAN

Sir Tom Blundell FRS FMedSci

Chair of School of Biological Sciences, University of Cambridge

Sir William Dunn Professor and Head of Department of Biochemistry, University Cambridge and Professorial Fellow, Sidney Sussex College

Chief Executive, Biotechnology and Biological Sciences Research Council 1994-96

Director General, Agricultural and Food Research Council 1991-94

Member, Advisory Council on Science and Technology 1988-90

Honorary Director, Imperial Cancer Research Fund Unit in Structural Biology, Birkbeck College, University of London 1989-96

Professor of Crystallography, Birkbeck College, University of London 1976-90

MEMBERS

Professor Roland Clift OBE MA PhD FREng FIChemE Hon FCIWEM FRSA

Distinguished Professor of Environmental Technology and Director, Centre for Environmental Strategy, University of Surrey

Visiting Professor, Chalmers University of Technology, Göteborg, Sweden

Member, Research Advisory Committee, Forest Research and Forestry Commission 2004

Member, UK Ecolabelling Board 1992-99

Chairman, Clean Technology Management Committee, Science and Engineering Research Council 1990-94

Professor Paul Ekins MPhil MSc PhD

Head, Environment Group, Policy Studies Institute

Professor of Sustainable Development, University of Westminster

Associate Director, Forum for the Future

Senior Consultant, Cambridge Econometrics

Specialist Advisor, House of Commons Environmental Audit Committee

Member, Sustainable Energy Policy Advisory Board

Member, Judging Panel, Alsden Awards for Sustainable Energy

Member, Environmental Advisory Group, OFGEM

Trustee and Special Adviser, Right Livelihood Awards Foundation

Sir Brian Follett FRS

Chair, Teacher Training Agency

Chair, Arts and Humanities Research Board

Professor, Department of Zoology, University of Oxford

Fellow, Wolfson College

Chair, Inquiry into Infectious Diseases of Livestock 2001-02 (for government under auspices of the Royal Society)

Chair, Review into Research Library Provision 2002-03 (for UK universities under auspices of the Funding Councils and the British Library)

Vice-Chancellor, University of Warwick, 1993-2001

Professor of Zoology and Head of Biological Sciences, University of Bristol 1978-93

Vice-President and Biological Secretary, The Royal Society 1987-93

Dr Ian Graham-Bryce CBE DPhil FRSC FRSE

President, Scottish Association for Marine Science

Principal Emeritus, University of Dundee

Chairman, East Malling Trust for Horticultural Research

Principal and Vice-Chancellor, University of Dundee 1994-2000

Convener, Committee of Scottish Higher Education Principals 1998-2000

President, British Crop Protection Council 1998-2000

British Crop Protection Medal 2000

Member, Natural Environmental Research Council 1989-96

Head, Environmental Affairs Division, Shell International 1986-94

President, Association of Applied Biologists 1988-89

Chairman, Agrochemical Planning Group, International Organisation for Chemistry in Development (IOCD) 1985-88

Director, East Malling Research Station 1979-86

President, Society of Chemical Industry 1982-84

Member of Editorial Board of Journal 'Pesticide Science' 1978-1980

Chairman of Pesticides Group of the Society of Chemical Industry 1878-80

Head of Insecticides and Fungicide Department, Rothamstead Experimental Station 1972-79

Senior Research Officer, ICI Plant Protection Division 1970-1972

Special Lecturer in Pesticide Chemistry, Imperial College of Science and Technology 1970-72

Personal research on pesticides (efficacy and behaviour in the environment) 1964-79; author of book and various papers

Professor Stephen Holgate MD DSc FRCP FMedSci FRSA

Medical Research Council Clinical Research Professor of Immunopharmacology, University of Southampton

Member of the Polish Academy of Arts and Sciences

Honorary Consultant Physician, Southampton University Hospital and the Royal Bournemouth Hospital Trusts

Former Adviser, House of Lords Select Committee on Science and Technology

Chairman of Expert Panel on Air Quality Standards

Seat on various Department of Health Advisory Committees

Professor Brian Hoskins CBE FRS

Royal Society Research Professor 2001- and Professor of Meteorology, University of Reading (Head of Department 1990-1996)

President, Royal Meteorological Society 1998-2000

President, International Association of Meteorology and Atmospheric Sciences 1991-95

Chair, Royal Society Global Environmental Research Committee

Chair, Meteorological Office Science Advisory Committee and Member, Meteorological Office Board

Vice-Chair, Joint Scientific Committee for the World Climate Research Programme

Member, Scientific Review Committees for Hadley Centre and European Centre for Medium Range Weather Forecasts (Chair 1985-88)

Medium Range Weather Forecasts (Chair 1985-88)

Foreign Associate, US National Academy of Sciences

Foreign Member, Chinese Academy of Sciences

Professor Jeffrey Jowell QC

Professor of Public Law, University College London and Barrister

Dr Susan Owens OBE AcSS FRSA

Reader in Environment and Policy, University of Cambridge, Department of Geography, and Fellow of Newnham College

Member, Defra's Horizon Scanning Sub-Group 2003-04

Member, Countryside Commission 1996-99

Member, UK Round Table on Sustainable Development 1995-98

Member, Deputy Prime Minister's Panel during preparation of 1998 Transport White Paper 1997-98

Member, Foresight Agriculture, Natural Resources and Environment Panel, Office of Science and Technology 1994-96

Professor Jane Plant CBE FRSA FRSE FIMM CEng FGS

Professor of Geochemistry, Imperial College, London (from March 2005)

Chief Scientist, British Geological Survey (Natural Environment Research Council)

Professor, Geochemistry, Imperial College, London

Visiting Professor, University of Liverpool

Council Member, Parliamentary and Scientific Committee

Chairman, Advisory Committee on Hazardous Substances

Member, Chemicals Stakeholder Forum

Professor Steve Rayner FRAI, FRSE, FAAAS, FSfAA

James Martin Professor of Science and Civilisation, Saïd Business School, Oxford University

Professor of Environment and Public Affairs, Columbia University, 1999-2003

Chief Scientist, Pacific Northwest National Laboratory, USA 1996-99

Director, ESRC Science in Society Programme

Member, Intergovernmental Panel on Climate Change

Past President, Sociology and Social Policy Section of the British Association

John Speirs CBE LVO MA MBA FRSA

Non-executive Director of the Carbon Trust

Member of the Chemistry Leadership Council 2004 and Chairman of its Futures Group Committee 2003

Past President, National Society for Clean Air and Environmental Protection

Director, The Carbon Trust 2001-

Member, Management Committee of the Prince of Wales's Business and Environment Programme 1996-, and Chairman of its UK faculty 1994-2002

Member, Advisory Committee, Kleinwort Benson Equity Partners

Chairman, Dramgate Ltd 1991-

Managing Director, Norsk Hydro (UK) Ltd 1981-2001

Past President, the Aluminium Federation 1997-98

Member of the Chemical Industry Association Council 1993-2002 and Chairman of its Public Affairs Committee 1993-2000

Member, Science and Engineering Research Council 1993-94

Member, Advisory Committee on Business and the Environment 1991-95

Chairman, Merton, Sutton and Wandsworth Family Health Services Authority 1989-95

Divisional Director, The National Enterprise Board 1976-81

Professor Janet Sprent OBE DSc FRSE FRSA FLS

Emeritus Professor of Plant Biology, University of Dundee

Honorary Research Professor, Scottish Crop Research Institute

Board Member, Scottish Natural Heritage

Member SHEFC 1992-96

Member Natural Environment Research Council 1991-95

Governor, Macaulay Land Use Research Institute 1990-2000; Chairman 1995-2000

Honorary Member, British Ecological Society

REFERENCES

CHAPTER ONE

1   Vincent, M.A., Atkins, S.M., Lumb, C.M., Golding, N., Lieberknecht, L.M. and Webster, M. (2004). *Marine Nature Conservation and Sustainable Development – the Irish Sea Pilot.* Report to Defra by the Joint Nature Conservation Committee (JNCC), Peterborough, UK. Basking sharks *Cetorhinus maximus* grow to 10 m in length. They filter plankton, taking advantage of productive near-surface waters, and occur regularly off the coasts of the Isle of Man and Arran. © Naturepl.com.

2   Kindly supplied by Professor S. Scott, University of Toronto.

3   Supplied by apexnewspix©apexnews

4   OSPAR website www.ospar.org.

5   OSPAR Commission (2000). Quality status report, 2000. OSPAR Commission, London

6   Myers, R.A. and Worm, B. (2003). Rapid worldwide depletion of predatory fish communities. *Nature* **423**, 280-283.

7   Kindly supplied by English Nature © Paul Knapman/English Nature.

8   Degnbol, P., Carlberg, A., Ellingsen, H., Tonder, M., Varjopura, R. and Wilson, D. (2003). Integrating Fisheries and Environmental Policies – Nordic Experiences. *Tema Nord* 2003:521, Nordic Council of Ministers, Copenhagen.

9   Institute of Marine Research website http://www.imr.no/coral/fishery_impact.php. Video photographs from the Norwegian continental break. The image on the right shows a barran landscape with crushed remains of *Lophelia*-skeleton spread over the area. This is a region subject to considerable bottom trawling. A path can be seen stretching from bottom-left to top-right of the photograph, indicating the path of a trawl. © Institute of Marine Research, Bergen.

10   Department for Food and Rural Affairs (Defra), Scottish Executive and Welsh Assembly Government (2002). *Safeguarding our seas: a strategy for the conservation and sustainable development of our marine environment.* Defra, London.

11   Kindly supplied by Dr Brian Bett, Deepseas Group, Southampton Oceanography Centre. (©SOC).

12   Kindly supplied by Greenpeace. Large cod © Greenpeace/Germain; Small cod © Greenpeace/Cobb

13   Prime Minister's Strategy Unit. (2004). Net Benefits. *A sustainable and profitable future for UK fishing.* Cabinet Office, London.

14   Kindly supplied by Greenpeace (2004) © Greenpeace/Morgan.

15   First Report, 1971.

16   Pauly, D. and Maclean, J. (2003). *In a perfect ocean – the state of fisheries and ecosystems in the North Atlantic Ocean.* Island Press, London.

17   Clover, C. (2004). *The end of the line – how overfishing is changing the world and what we eat.* Ebury Press, London.

18   Culliney, J.L. Wilderness Conservation, Sept-Oct 1990.

CHAPTER TWO

1   Adapted from OSPAR Commission (2000).

2   International Council for the Exploration of the Seas (ICES) (2003). *Environmental Status of the European Seas*. Available on ICES website:
     http://www.ices.dk/reports/germanqsr/23222_ICES_Report_Samme.pdf

3   Adapted from O'Dor, R.K. (2003). *The Unknown Ocean: The Baseline Report of the Census of Marine Life Research Program*. Consortium for Oceanographic Research and Education: Washington DC, 28pp.

4   OSPAR Commission (2000).
     ICES (2003).

5   OSPAR Commission (2000);
     ICES (2003).

6   Personal communication from Professor Shimmield, August 2004.

7   ICES (2003).

8   Adapted from material supplied by Professor Ed Hill, NERC, Proudman Oceanography Laboratory.

9   United Nations Convention on Biological Diversity (1992). Article 2. Use of Terms. Available on CBD website at http://www.biodiv.org/convention/articles.asp?lg=O&a=cbd-02

10   Pauly and Maclean (2003).

11   DeYoung, B., Peterman, R.M., Dobell, A.R., Pinkerton, E., Breton, Y., Charles, A.T., Fogarty, M.J., Munro, G.R. and Taggart, C. (1999). Canadian marine fisheries in a changing and uncertain world. *Canadian Special Publications of Fisheries and Aquatic Science*, No.129.

12   Adapted from Pauly, D. and Maclean, J. (2003). *In a perfect Ocean: The state of fisheries and ecosystems in the North Atlantic Ocean*. Island Press, London.

13   Supplied by Wim Van Egmond ©. Thalassa website,
     http://thalassa.gso.uri.edu/plankton/diatoms/general/eucamp19/genus/eucomtext.htm;
     SAFHOS (2004) Ecological status report. Technical Report ISSN174L-0750

14   Department of Trade and Industry (DTI) (2001). *Overview of Plankton Ecology in the North Sea*. Technical Report TR_005. Technical Report produced for Strategic Environmental Assessment – SEA2AN. Produced by SAHFOS, August 2001.

15   *Ibid.*

16   Beaugrand, G. (2004). The North Sea regime shift: Evidence, causes, mechanisms and consequences. *Progress in Oceanography,* **60**, 245–262;
     Edwards, M. and Richardson, A. (2004). *Ecological Status 2001/2002 North Atlantic (including the North Sea)*. Available at:
     http://192.171.163.165/PDF_files/ecological%20status%20north%20atlantic%20(low).pdf

17   Supplied by Wim Van Egmond ©; DTI (2001).

18   DTI (2001).

19   Grassle, J.F. and Maciolek, N.J. (1992). Deep-sea species richness: Regional and local diversity estimates from quantitative bottom samples. *American Naturalist*, **139**, 313–341;
     Nybakken, J.W. (2001). *Marine Biology: An ecological approach*. Benjamin Cummings, San Francisco.

20   Supplied by © Wim Van Egmond.

21   Supplied by © Wim Van Egmond; SAC website:
     www.ukmarinesac.org.uk/communities/subtidal/brittlestar/bs31.htm

22   Dacey, W.H. and Wakeham, S.G. (1986). Oceanic dimethylsulfide: Production during zooplankton grazing on phytoplankton. *Science*, **233**, 1314–1315; Oceans online.com website: http://www.oceansonline.com/oceanicfoodwebs.htm

23   Beaugrand, G., Reid, P.C., Ibañez, F., Lindley, J.A. and Edwards, M. (2002). Reorganization of North Atlantic marine copepod biodiversity and climate. *Science*, **296,** 1692–1694;

Beaugrand, G. (2003). Long-term changes in copepod abundance and diversity in the north-east Atlantic in relation to fluctuations in the hydroclimatic environment. *Fisheries Oceanography*, **12**(4/5), 270–283.

24   Beaugrand (2004).

25   Cook, R.M., Sinclair, A. and Stefansson, G. (1997). Potential collapse of North Sea cod populations. *Nature*, **385**, 521–522;

Beaugrand (2003).

26   Beaugrand *et al.* (2002).

27   Brander, K. (2004). *Consequences of changing climate for North Atlantic cod populations and implications for fisheries management*. ICES Symposium on the Influence of Climate Change on North Atlantic Fish Populations. Bergen, Norway.

28   Clarke, R.A., Fox, C.J., Viner, D. and Livermore, M. (2003). North Sea cod and climate change – Modelling the effects of temperature on population dynamics. *Global Change Biology*, **9**, 1669–1680.

29   Figure adapted from Beaugrand, G., Brander, K.M., Lindley, J.A., Souissi, S. and Reid, P.C. (2003). Plankton effect on cod recruitment in the North Sea. *Nature*, **426**, 661-664.

30   Edwards, M. and Richardson, A.J. (2004). The impact of climate change on marine pelagic phenology and trophic mismatch. *Nature*, **430**, 881–884.

31   Kindly supplied by JNCC, © Rohan Holt/JNCC.

32   OSPAR Commission (2000).

33   Greenstreet, S.P.R. and Hall, S.J. (1996). Fishing and the ground-fish assemblage structure in the north-western North Sea: An analysis of long-term and spatial trends. *Journal of Animal Ecology*, **65**, 577–98.

34   Kindly supplied by Keith Hiscock, © *MarLIN*.

35   Ocean Blue website at: http://www.oblue.utvinternet.com/ob_irelands1.htm

36   Walday, M. and Kroglund, T. (2002). The North Sea – Bottom trawling and oil and gas exploration. In: *Europe's Biodiversity – Biogeograpical regions and seas, seas around Europe*. European Environment Agency.

37   OSPAR Commission (2000).

38   Kindly supplied by Keith Hiscock, *MarLIN*.

39   *Ibid.*

40   Defra (2004). *Review of Marine Nature Conservation. Summary of Working Group Report to Government*. Defra, London.

41   *Ibid.*

42   Heip, C., Basford, D., Craeymeersch, J.A., Dewarumez, J.M., Dorjes, J., Dewilde, P., Duineveld, G., Eleftheriou, A., Herman, P.M.J., Niermann, U., Kingston, P., Kunitzer, A., Rachor, E., Rumohr, H., Soetaert, K. and Soltwedel, T. (1992). Trends in biomass, density and diversity of North Sea macrofauna. *ICES Journal of Marine Science*, **49,** 13–22.

43   Pearson, T. and Manvik, H.P. (1998). Long-term changes in the diversity and faunal structure of benthic communities in the North Sea: Natural variability or induced instability? *Hydrobiologia*, **375/376**, 317–329.

44   Kindly supplied by Keith Hiscock, © *MarLIN*.

45   Kindly supplied by Keith Hiscock, © *MarLIN*; Marshall, C.E. and Wilson, E. 2004. *Pecten Maximus*. Great Scallop. Marine Life Information Network: Biology and Sensitivity Key information sub-programme (online). Plymouth Marine Biological Association of the United Kingdom. Available from http://www.marlin.ac.uk/

46   Ocean Blue website at: http://www.oblue.utvinternet.com/ob_irelands1.html

47   SAC website at: http://www.ukmarinesac.org.uk/marine-communities.htm

48   Kindly supplied by Keith Hiscock, © *MarLIN*; Ager, O.E.D. (2004). *Parazoanthus axinellae*. Yellow cluster anemone. Marine Life Information Network: Biology and Sensitivity Key Information Sub-programme (on-line). Plymouth: Marine Biological Association of the United Kingdom. Available frcm: http://www.marlin.ac.uk/species/parazoanthuisainallae.htm

49   SAC website at: http://www.ukmarinesac.org.uk/marine-communities.htm

50   Marine Life Network for Britain and Ireland Network (Marlin) website at: http://www.marlin.ac.uk/

51   Supplied by Paul Kay © Marine Wildlife Photo Agency.

52   Kindly supplied by Keith Hiscock, © *MarLIN*; Tyler-Walters, H (2001). *Modiolus modiolus*. Horse mussel. Marine Life Informatior Network: Biology and Sensitivity Key information sub-programme (online). Plymouth Marine Biological Association of the United Kingdom. Available from http://www.marlin.ac.uk/species/Modiolusmodiolus.htm

53   Gage, J. and Gordon, J. (1995) Sound bites, Sciences and the Brent Spar: Environmental considerations relevant to the deep sea disposal option. *Marine Pollution Bulletin* **30**(12), 772–779.

54   Ocean Blue website at: http://www.oblue.utvinternet.com/ob_irelands1.html

55   Kindly supplied by Professor Steve Scott, University of Toronto.

56   Kindly supplied by Ivor Rees © University of Wales, Bangor.

57   Bett, B.J. (2000). Signs and symptoms of deep-water trawling on the Atlantic margin. In: *Man-Made Objects on the Seafloor 2000*. The Society for Underwater Technology, London. Pages 107–118;
Fosså, J.H., Mortensen, P.B. and Furevik, D.M. (2002). The deep-water coral *Lophelia pertusa* in Norwegian waters: Distribution and fishery impacts. *Hydrobiologia*, **471**; 1–12.
Gubbay, S., Baker, M., Bett, B. and Konnecker, G. (2002). *The offshore directory, review of a selection of habitats, communities and species of the north-east Atlantic*. World Wide Fund for Nature. See website at: http://www.ngo.grida.no/wwfneap/Projects/Reports/Offshore.pdf;
Hall-Spencer, J., Allain, V. and Fosså, J. H. (2002). Trawling damage to North-east Atlantic ancient coral reefs. *Proceedings of the Royal Society, London*, **269,** 507–511;
Grehan, A. J., Unnithan, V., Olu-Le Roy, K. and Opderbecke, J. (in press). Fishing impacts on Irish deep-water coral reefs: Making the case for coral conservation. In: Thomas, J. and Barnes, P. (Eds) (in press). *Proceedings from the Symposium on the Effects of Fishing Activities on Benthic Habitats: Linking Geology, Biology, Socioeconomics and Management*. American Fisheries Society, Bethesda, Maryland, USA.

58   Kindly supplied by Professor Steve Scott, University of Toronto.

59   Konnecker, G. (2002). Sponge Fields. In: *Offshore Directory. Review of a selection of habitats, communities and species of the North-East Atlantic.* Ed. by S. Gubbay. WWF-UK. North East Atlantic Programme;
      Johnston, C.M. (2004). *Scoping Study: Protection of vulnerable high seas and deep oceans biodiversity and associated oceans governance.* Joint Nature Conservation Committee, Peterborough, UK;
      Klitgaard, A.B. and Tendel, O.S. (2001). *"Ostur" – "cheese bottoms" – sponge dominated areas in Faroese shelf and slope areas. In Marine biological investigations and assemblages of benthic invertebrates from the Faroe Islands, pp. 13–21.* Ed. by G. Guntse and O.S. Tendal. Kaldbak Marine Biological Laboratory, the Faeroe Islands.

60   UNEP/CBD/COP/5/INF/7
      Grassle, J.F. (1991). Deep-sea benthic biodiversity. *BioScience,* **41**,464–469;
      Grassle and Maciolek (1992);
      Angel, M.V. (1993). Biodiversity of the pelagic ocean, *Conservation Biology,* **7**, 760–772;
      Heywood, V.H. and Watson, R.T. (Eds.) (1995). *Global Biodiversity Assessment.* Cambridge University Press for the United Nations Environment Programme.

61   Levinton, J.S. (1995). *Marine Biology. Function, biodiversity, ecology.* Oxford University Press, Oxford.

62   OSPAR Commission (2000).

63   Kindly supplied by Dr Brian Bett, Deepseas Group, Southampton Oceanography Centre (© DTI).

64   © Crown copyright 2004. Reproduced by permission of CEFAS.

65   See the Marine Life Network for Britain and Ireland Network (Marlin) website at: http://www.marlin.ac.uk/

66   OSPAR Commission (2000).

67   See ASCOBANS website at: http://www.ascobans.org/;
      see Whale and Dolphin Conservation Society website at: http://www.wdcs.org/;
      see Sea Mammal Research Unit website at:
      http://smub.st-and.ac.uk/GeneralInterest.htm/general_interest.htm;
      JNCC (2001). *JNCC Marine Monitoring Handbook.* Eds. J. Davies (senior editor), J. Baxter, M. Bradley, D. Connor, J. Khan, E. Murray, W. Sanderson, C. Turnbull and M. Vincent;
      Reid *et al.* (2003).

68   Kindly supplied by English Nature, © Andy Rouse/English Nature.

69   See Sea Mammal Research Unit website at:
      http://smub.st-and.ac.uk/GeneralInterest.htm/generalinterest.htm;
      JNCC (2001).

70   Kindly supplied by JNCC.

71   ICES (2003).

72   See JNCC website at:
      http://www.jncc.gov.uk/Publications/seabird_pops_britain_ireland/default.htm

73   Supplied by RSPB images © Mike Lane/RSPB.

74   *The Observer.* North Sea birds dying as waters heat up. June 2004; RSPB website: http://www.rspb.org.uk/scotland/action/disaster/index.asp

75   Supplied by RSPB images © David Tipling/RSPB.

76   Kindly supplied by Graham Shimmield.

77 Platt, T. and Sathyendranath, S. (1988). Oceanic primary production: Estimation by remote sensing at local and regional scales. *Science*, **241**, 1613–1620;
Longhurst, A.R. (1995). Seasonal cycles of pelagic production and consumption. *Progress in Oceanography*, **36**,77–167;
Longhurst, A.R. (1998). *Ecological Geography of the Sea*. Academic Press, San Diego;
Platt, T. and Sathyendranath, S. (1999). Spatial structure of pelagic ecosystem processes in the global ocean. *Ecosystems*, **2**: 384–394;
Pauly, D., Christensen, V., Froese, R., Longhurst, A., Platt, T., Sathyendranath, S., Sherman, K. and Watson, R. (2000). *Mapping fisheries onto marine ecosystems: a consensus approach for regional, oceanic and global integrations*. ICES 2000 Annual Science Conference;
Pauly and Maclean (2003).

78 Platt and Sathyendranath (1988);
Longhurst (1995);
Longhurst (1998);
Platt and Sathyendrauath (1999);
Pauly *et al.* (2000).

79 *Ibid.*

80 Sherman, K. and Duda, A.M. (1999). An ecosystem approach to global assessment and management of coastal waters. *Marine Ecology Progress Series*, **190**, 271–287.

81 Sherman, K. and Duda, A.M. (2002). A new imperative for improving management of large ecosystems. *Ocean & Coastal Management*. **45,** 797–833.

82 Platt and Sathyendranath (1988);
Longhurst (1995);
Longhurst (1998);
Platt and Sathyendranath (1999);
Pauly *et al.* (2000).

83 Laffoley, D.d'A., Burt, J., Gilliland, P., Baxter, J., Connor, D.W., Davies, J., Hill, M., Breen, J., Vincent, M., and Maltby, E. (2003). *Adopting an ecosystem approach for the improved stewardship of the maritime environment: some overarching issues*. English Nature, Peterborough, UK English Nature Research Reports, No.538, 20pp.
Lieberknecht, L.M., Vincent, M. and Connor, D.W. (2003). *Criteria for the identification of nationally important marine features*. Irish Sea Pilot. Regional Sea Management Project. Interim Report for consultation. Available on line at http://www.jncc.gov.uk

84 Connor, D.W., Breen, J., Champion, A., Gilliland, P.M., Huggett, D., Johnston, C., Laffoley, D.d'A., Lieberknecht, L., Lumb, C., Ramsay, K. and Shardlow, M. (2002). *Rationale and criteria for the identification of nationally important marine conservation features and areas in the UK*. Version 02.11. Available on line at http://www.jncc.gov.uk

85 Roff, J.C. and Taylor, M.E. (2000). National frameworks for marine conservation – a hierarchical geophysical approach. *Aquatic conservation: Marine and Freshwater Ecosystems,* **10**,209–223.

86 Vincent *et al.* (2004).

87 *Ibid.*

88 *Ibid.*

89 *Ibid.*

90 Defra (2002).

References

91   CBD (2001). *The value and effects of marine and coastal protected areas (MCPAs) on marine and coastal biodiversity – a review of available information.* UNEP/CBD/AHTEG-MCPA/1/2

Gray, J. (1997). *Marine biodiversity: patterns, threats and conservation needs.* GESAMP Reports and Studies No. 62. International Maritime Organization.

92   See World Resources Institute website at:

http://pubs.wri.org/pubs_content_text.cfm?ContentID=535

93   CBD (2001).

Norse, E.A. (1993). *Global marine biological diversity: A strategy for building conservation into decision making.* Island Press, Washington DC;

Gray (1997).

94   Grassle (1991);

Grassle and Maciolek (1992);

Corliss, J.O. (1994). An interim utilitarian ('user friendly') hierarchical classification and characterization of the protists. *Acta Protozoologic,* **33**, 1–51;

National Research Council (1995). *Understanding Marine Biodiversity: A research agenda for the nation.* National Academy Press.

95   Turner, R.D. (1973). Wood-boring bivalves, opportunistic species in the deep sea. *Science,* **180**,1377–1379;

Turner, R.D. (1981). 'Wood islands' and 'thermal vents' as centers of diverse communities in the deep sea. *Biologiya Morya,* **1**, 3–10;

Grassle, J.F. (1986). The ecology of deep-sea hydrothermal vent communities. *Advances in Marine Biology,* **23**,301–362;

Williams, A.B. (1988). New marine decapod crustaceans from waters influenced by hydrothermal discharge, brine, and hydrocarbon seepage. *Fisheries Bulletin,* **86**, 263–287;

Kennicutt, M.C., Brooks, J.M., Bidigare, R.R., McDonald, J.J., Adkison, D.L. and Macko, S.H. (1989). An upper slope 'cold' seep community: Northern California, 1989. *Limnology & Oceanography,* **34**, 635–640;

Smith, C.R., Kukert, H., Wheatcroft, R.A., Jumars, P.A. and Deming, J.W. (1989). Vent fauna on whale remains. *Nature,* **341,** 27–28;

Southward, A.J. (1989). Animal communities fuelled by chemosynthesis: life at hydrothermal vents, cold seeps and in reducing sediments. *Journal of Zoology,* **217**, 705–709;

MacDonald, I.R., Reilly II, J.F., Guinasso Jr. N.L., Brooks, J.M., Carney, R.S., Bryant, W.A. and Bright, T.J. (1990). Chemosynthetic mussels at a brine-filled pockmark in the northern Gulf of Mexico. *Science,* **248**:1096–1099;

Tunnicliffe, V. (1991). The biology of hydrothermal vents: Ecology and evolution. *Oceanography and Marine Biology Annual Review,* **29**, 319–407;

National Research Council (1995).

96   Census of Marine Life Project (CoML) website at: http://www.coml.org/coml.htm

97   National Research Council (1995).

CHAPTER THREE

1   Based on Charles, A. (2001). *Sustainable Fishery Systems*. Blackwell Science, Oxford.

2   Countryside Council for Wales (2001). *A Glossary of Marine Nature Conservation and Fisheries*. Countryside Council for Wales, Bangor.

3   United Nations Food and Agriculture Organization (FAO) (2002). Part 1: Review of world fisheries and aquaculture. In: *The State of World Fisheries and Aquaculture*. FAO, Rome.

4   *Ibid.*

5   European Communities (2004). *Facts and Figures on the Common Fisheries Policy*. European Communities, Luxembourg.

6   *Ibid.*

7   FAO (2002).

8   *Ibid.*

9   *Ibid.*

10   FAO Fishery information, data and statistics unit (2003). *Overview of fish production, utilisation, consumption and trade*. Vannuccini, S. FAO, Rome.

11   International Food Policy Research Institute (IFPRI) and WorldFish Centre (2003). *Fish to 2020: Supply and demand in changing global markets*. C.L. Delgado, N. Wada, M.W. Rosegrant, S. Meijer and M Ahmend. IFPRI, Washington DC, WorldFish Centre, Penang, Malaysia.

12   *Ibid.*

13   *Ibid.*

14   Duda, A.M. and Sherman, K. (2002). A new imperative for improving management of large marine ecosystems. *Ocean and Coastal Management* **45** 797-833. Based on Watson, R. and Tyedmers, P. Projected decline in per capita seafood availability. Fisheries Centre, University of British Columbia (www.data.fisheries.ubc.ca).

15   FAO (2002).

16   European Communities (2004).

17   Website of the European Union (EU). See (EU) statistics at: http://europa.eu.int/comm/fisheries/policy_en.htm

18   *Ibid.*

19   *Ibid.*

20   Evidence from SeaFish, July 2003.

21   *Ibid.*

22   European Communities (2004).

23   Department for Environment, Food and Rural Affairs (Defra) statistics at: http://statistics.defra.gov.uk/esg/publications/fishstat/uksfs02.pdf

24   Evidence from Centre for Environment, Fisheries and Aquaculture Science (CEFAS), May 2003.

25   Figure compiled from UK sea fisheries statistics at: http://statistics.defra.gov.uk/esg/publications/fishstat/default.asp

26   *Ibid.*

27   Prime Minister's Strategy Unit (2004).

28   Defra statistics at: http://statistics.defra.gov.uk/esg/publications/fishstat/uksfs02.pdf

29   Defra (2004a). *Review of marine fisheries and environmental enforcement*. Defra, London.

30   *UK Marine Industries World Export Market Potential.* A report for the Foresight Marine Panel by Douglas-Westwood Associates, October 2000.

31   Prime Minister's Strategy Unit (2004).

32   Defra statistics at http://statistics.defra.gov.uk/esg/publications/fishstat/uksfs02.pdf

33   Figure compiled from UK sea fisheries statistics at:
      http://statistics.defra.gov.uk/esg/publications/fishstat/default.asp

34   Prime Minister's Strategy Unit. (2003). *UK Fisheries Industry – Current situation analysis.* Cabinet Office, London.

35   Adapted from Prime Minister's Strategy Unit (2004).

36   Evidence from SeaFish, July. 2003.

37   Prime Minister's Strategy Unit. (2004).

38   *Ibid.*

39   Figure from Prime Minister's Strategy Unit (2003), derived from the 1970 -2000 National Food Survey and the 2001/02 Expenditure and Food Survey (which replaced the National Food Survey from April 2001). Note:* includes value added chilled fish products.

40   Food Standards Agency (FSA) Scotland (2002). *Consumption and Purchase of Fish Survey.* April 2002.

41   Calder, P.C. (2004). n-3 Fatty acids and cardiovascular disease: evidence explained and mechanisms explored. *Clinical Science*, **107**, 1–11.

42   Adapted from Calder, P.C. (2004).

43   Burr, G.O. and Burr, M.M. (1929). A new deficiency disease produced by rigid exclusion of fat from the diet. *Journal of Biology and Chemistry*, **82**, 345–367.

44   Calder (2004).

45   *Ibid.*

46   Sanderson, P., Finnegan, T.E., Williams, C.M., Calder, P.C., Burdge G.C., Wooston, S.A. Griffin, B.A., Millward, D.A., Pegge, N.C. and Bemmelmans, W.J.E *et al* (2002). UK Food Standards Agency ALA workshop report. *British Journal for Nutrition*, **88**, 573-579.

47   Adapted from Calder (2004).

48   See www.nutrition.org.uk/medianews/pressinformation/n3fatty.htm

49   See www.iger.bbsrc.ac.uk/Publications/Innovations/In2003/Ch7.pdf

50   Dewhurst, R.J., Scollan, N.D., Lee, M.R.F., Ougham, H.J. and Humphries, M.O. (2003). Forage breeding and management to increase the beneficial fatty acid content of ruminant animals. *Proceedings of the Nutrition Society.* **62**, 329–336.

51   See www.wvu.edu/~agexten/forglvst/humanhealth.pdf

52   See www.csuchico.edu/agr/grsfdbef/health-benefits/ben-o3-o6.html

53   Wood, J. (2004). Better beef and lamb. *Research Review*, University of Bristol, at:
      http://www.bristol.ac.uk/university/publications/research/issue-7/beef.pdf

54   Article by N. Hasting in *Planet Earth*, Autumn 2003. Natural Environment Research Council (NERC) publication.

55   Science Blog (2004). Modified linseed produces healthier omega-3 and 6- fatty acids; at:
      http://scienceblog.com/community/article-print-4113.html

56   See: http://www.ars.usda.gov/is/AR/archive/dec02/oil1202.pdf

57   See: http://www.foodstandards.gov.uk/multimedia/presentations/fsabhfpresentation.ppt

58   Connor, W.E. (2000). Importance of n-3 fatty acids in health and disease. *American Journal of Clinical Nutrition*, **71**,171S–175S.

59   FSA (2002). National diet and nutrition survey volume 1: *National Diet and Nutrition Survey of British Adults aged 19-64 years 2000-01; types and quantities of food consumed.* Henderson, L., Gregory, J., Swan, G. FSA and ONS, London.

60   FSA website at: www.foodstandards.gov.uk

61   FSA issues new advice on oily fish consumption. *Food Standards Agency Press Release,* 24 June 2004. At www.foodstandards.gov.uk

62   *Quarterly Journal of Medicine,* **96**, 465–480; available at: doi:10.1093/qjmed/hcg092

63   *Ibid.*

64   Letter by: Rembold, C.M. (2004). The health benefits of eating salmon. *Science,* **305**, 475-476.

65   Letter by: Tuomisto, J., Tainio, M., Wittynen, M., Verkasalo, P., Vartianinen. T., Kiuiranta, H., Pekkanen, J. *et al.* (2004). Risk-benefit analysis of eating farmed salmon. *Science,* **305**, 476-477.

66   FSA issues new advice on oily fish consumption. *Food Standards Agency Press Release,* 24 June 2004.

67   CEFAS (2003). *A study on the consequences of technological innovation in the capture fishing industry and the likely effects upon environmental impacts.* CEFAS Contract Report C1823. Available at: http://www.rcep.org.uk/fisheries.htm; cited as CEFAS (2003).

68   Jennings S., Kaiser M.J. and Reynolds J.D. (2001). *Marine Fisheries Ecology.* Blackwell Science, Oxford. 432 pp.

69   CEFAS (2003).

CHAPTER FOUR

1   Grotius (1633). *The Freedom of the Seas, or the Right Which Belongs to the Dutch to Take Part in the East Indian Trade*, A Dissertation – Translated with a version of the Latin text of 1633 by Dr R. Magoffin. Oxford University Press, New York, 1916.

2   United Nations (2002). *Oceans: The Source of Life. United Nations Convention on the Law of the Sea 20th Anniversary* (1982 – 2002). UNCLOS II was held in 1960, but did not produce a treaty text. Addressed to the vexed question of the maximum breadth of the territorial sea – a compromise Canadian formula of 6+6 territorial sea and contiguous zone failed narrowly. Hence the 1982 LOSC also established for the first time a maximum breadth for the territorial sea, at 12 nautical miles.

3   *Ibid.*

4   *Ibid.*

5   Kimball, L.A. (2001) *International Ocean Governance Using International Law and Organizations to Manage Marine Resources Sustainably.* IUCN, Gland, Switzerland.

6   United Nations Convention on the Law of the Sea (UNCLOS), Article 194, paragraph 5.

7   Scovazzi, T. (2003) *Marine Protected Areas on the High Seas: Some Legal and Policy Considerations.* Paper Presented at the World Parks Congress, Durban, South Africa, 11 September 2003.

8   UNCLOS Art. 211(6).

9   UNCLOS Art 118.

10   UNCLOS Articles 61 and 119.

11   Pew Oceans Commission (2002). Managing Marine Fisheries in the United States. In: *Proceedings of the Pew Oceans Commission workshop on fisheries management, Seattle, Washington 18-19 July 2001.* Pew Oceans Commission, Arlington, Virginia.

12   *The 1995 UN Agreement For The Implementation Of The Provisions Of The United Nations Convention On The Law Of The Sea Of 10 December 1982 Relating To The Conservation And Management Of Straddling Fish Stocks And Highly Migratory Fish Stocks (UN Fish Stocks Agreement).* Text available at: http://www.un.org/Depts/los/convention_agreements/convention_overview_fish_stocks.htm. Article 297(3)(a)

13   *Ibid.* Article 298(1)(b).

14   *Ibid.* Article 290.

15   Sands, P. (2003). *Principles of International Environmental Law.* 2nd Edition. Cambridge University Press, Cambridge, UK.

16   *Ibid.*

17   UN Oceans and Law of the Sea website; http://www.un.org/Depts/los/convention_agreements/convention_overview_fish_stocks.htm.

18   *The UN FAO International Plan of Action for the Management of Fishing Capacity, adopted by the 23rd Session of the UN FAO Committee on Fisheries, February 1999.* http://www.fao.org/fi/ipa/ipae.asp

19   Institute of European Environmental Policy (IEEP) (2000). *Subsidies to the European Union Fisheries Sector.* Paper Commissioned by WWF Europe Fisheries Campaign.

20   Website of the Food and Agriculture Organisation of the United Nations (FAO); http://www.fao.org/docrep/003/x9066e/x9066e01.htm#a.

21   FAO (2001). *International Plan of Action to Prevent, Deter and Eliminate Illegal, Unreported and Unregulated Fishing.* Twenty-fourth Session of the Committee on Fisheries (COFI), Rome.

22   FAO (1999). *IPOA for reducing incidental catch of seabirds in longline fisheries*

23   *Ibid.*

24   *Ibid.*

25   United Nations Convention on Biological Diversity (1995). COP Decision II/10 on conservation and sustainable use of marine and coastal biodiversity.

26   United Nations (2002a). *Report of the World Summit on Sustainable Development Johannesburg, South Africa, 26 August-4 September 2002.* A/CONF.199/20.

27   See IWC website at: http://www.iwcoffice.org/iwc.htm#History.

28   Sands, P. (2003).

29   Website of BBC news, http://news.bbc.co.uk/1/hi/world/asia-pacific/3160682.stm.

30   OSPAR (1998). *1992 OSPAR Convention ANNEX V The Protection and Conservation of the Ecosystems and Biological Diversity of the Maritime Area.* Reference number 1998-15.1.

31   IUCN (1994). *Guidelines for protected area management categories.* IUCN, Cambridge, UK and Gland, Switzerland.

32   OSPAR (2003). *Joint Ministerial Meeting of the Helsinki and OSPAR Commissions (JMM), Bremen 25-26 June 2003 – Declaration of the Joint Ministerial Meeting of the Helsinki and OSPAR Commissions.*

33   Council Directive 79/409/EEC on the Conservation of Wild Birds.

34   Council Directive 92/43/EEC on the Conservation of Natural Habitats and of Wild Fauna and Flora.

35   Information on the Bern Convention from the JNCC website; http://www.jncc.gov.uk/legislation/conventions/bern.htm.

36   Johnston, C.M., Turnbull, C.G., Tasker M.L. (2003). *NATURA 2000 in UK Offshore Waters: Advice to support the implementation of the EC Habitats and Birds Directives in UK offshore waters.* Report 325, JNCC, Peterborough, UK.

37   Communication from the Commission to the Council and the European Parliament COM(1998) 42 final.

38   COM(2001) 162 final Biodiversity Action Plan for Fisheries. European Commission, Brussels.

39   European Commission (2004)
      The European Marine Strategy – policy document. SGO(2)04/4/1.European Commission, Brussels, Belgium.

40   EC Directive 2001/42/EC on the assessment of the effects of certain plans and programmes on the environment. OJ 21.7.2001.

41   Postnote July 2004, No. 223. Parliamentary Office of Science and Technology (2004) Strategic Environmental Assessment (SEA). Postnote No. 223. Post, London.

42   Coffey, C. and Richartz, S. In press. *UK Offshore Marine Protected Areas: Unblocking the Process.* IEEP Report For Greenpeace.

43   Adapted from Coffey, C. and Richartz, S. (In Press).

44   Website of Joint Nature Conservation Council. Information on the Convention on Biological Diversity at: http://www.jncc.gov.uk/legislation/conventions/biodiversity.htm.

45   Green paper on the future of the Common Fisheries Policy. COM (2001) 135 Final. European Commission Brussels.

46   *Ibid.*

47   Defra evidence, July 2004.

48   COM (2001) 135 final.

49   Council Regulation (EC) No.2371/2002 on the conservation and sustainable exploitation of fisheries resources under the Common Fisheries Policy.

50   Hatcher, A.C. (1997). Producers' Organisations and Devolved Fisheries Management in the United Kingdom: Collective and Individual Quota Systems. *Marine Policy*, **21** (6), 519-533.

51   ICES (2003).

52   COM (2002) 181 final Communication from the Commission on the reform of the Common Fisheries Policy ("Roadmap"). Brussels, 28.5.2002.

53   Fischler, F. (2000). *Management of fleet capacity and fishing effort.* Hearing on "Reducing fleet overcapacity and fishing effort in the EU", Brussels, 21 September 2000. [available online from: europa.eu.int/comm/fisheries/news-corner/discours/speech4_en.html].

54   Pascoe, S., Tingley, D. and Mardle, S. (2002). *Appraisal of Alternative Policy Instruments to Regulate Fishing Capacity.* CEMARE Report ER0102/6.

55   European Commission DG Fisheries (2003). First application of the reformed Common Fisheries Policy: Commission proposes long-term recovery plan for cod. Press release available at: http://europa.eu.int/comm/fisheries/news_corner/press/inf03_14_en.htm.

56   *Ibid.*

57   Council Regulation COM 2287/2003. Fixing for 2004 the fishing opportunities and associated conditions for certain fish stocks and groups of fish stocks. *Official Journal* **L344/1**. 31.12.2003.

58   Evidence to the Royal Commission from CEFAS, 2002.

59   IEEP (2000).

60   *Ibid.*

61   National Audit Office (2003). Fisheries Enforcement in England. HC 563, Session 2002-2003. The Stationery Office, London.

62   European Commission (2004). Commissioner Fischler: "EU Fisheries Control Agency in Vigo will boost enforcement". Press release 26.01.04.

63   SFPA (2004) Annual Report and accounts. SFPA, TSO, Edinburgh

64   European Commission DG Fisheries (2001). European Distant Water Fishing Fleet.

65   IEEP (2000).

66   *Ibid.*

67   IEEP (2002). *Fisheries Agreements with third countries. Is the EU moving towards Sustainable Development?* Report to WWF.

68   Evidence from Defra, 2004.

69   Regulation No 2847/93.

70   IFREMER (1999). Evaluation of the Fisheries Agreements Conclude by the European Community – Summary Report. Community Contract 97/S 240-152919.

71   Hatcher, A.C. (1997).

72   Sea Fisheries (Conservation) Act 1995.

73   Website of Eastern Sea Fisheries, http://www.esfjc.co.uk/Environ.htm.

74   COM (2002). 511 final *A Strategy for the Sustainable Development of European Aquaculture.*

75   Directive 85/337/EC on Environmental Assessment as amended by Directive 97/11/EC

76   Scottish Executive Environment and Rural Affairs Department (SEERAD) (2002). *Locational Guidelines for the Authorisation of Marine Fish Farms in Scottish Waters.* SEERAD, Edinburgh.

77   Under the terms of the Salmon Act 1986.

78   Diseases of Fish Act 1937, as amended.

79   Under the Zetland County Council Act 1974.

80   Nautilus Consultants Ltd and EKOS Economic Consultants Ltd (2000). *Study into Inland and Sea Fisheries in Wales.* National Assembly for Wales. Administrative Department, Cardiff.

81   Magnuson-Stevens Fishery Conservation and Management Act Public Law 94-265 As amended through October 11, 1996.

82   *Ibid.* Article 104-297.

83   Pascoe *et al* (2002).

84   Institute of Marine Research website; http://www.imr.no/coral/news.php.

85   Canada's Oceans Act (1996). Available at; http://laws.justice.gc.ca/en/O-2.4/index.html.

86   Great Barrier Reef Marine Park Authority (2003). *Great Barrier Reef Marine Park Zoning Plan 2003. Great Barrier Reef Marine Park Act 1975.*

CHAPTER FIVE

1   Pitcher, T. (2001). Fisheries managed to rebuild ecosystems? Reconstructing the past to salvage the future. *Ecological Applications* **11**(2), 601-617.

2   Cushing, D.H. (1988). *The Provident Sea.* Cambridge University Press, Cambridge UK.

3   Pauly, D., Christensen, V., Dalsgaaard, J., Froese, R., and Torres, F. (1998). Fishing down marine food webs. *Science* **279**, 860-863.

4   FAO (2002).

5   House of Lords Select Committee on Science and Technology, (1995). *Fish Stock Conservation and Management.* Evidence received.
     Prime Minister's Strategy Unit (2004).

6   *Ibid.*

7   Defra website:
     http://www.defra.gov.uk/environment/statistics/coastwaters/cwfishstock.htm

8   Defra (2004a).

9   Adapted from ICES (2003).

10   Kaiser, M.J., Collie, J.S., Hall, S.J., Jennings and S., Poiner, I.R. (2002). Impacts of fishing gear on marine benthic habitats. In: *Responsible Fisheries in the Marine Ecosystem.* Eds. M. Sinclair and G. Valdimarsson. FAO, Rome, p. 197-218.

11   Gislason, H. (1994). Ecosystem effects of fishing activities in the North Sea. *Marine Pollution Bulletin* **29**, 520–527.
     Gislason, H., Sinclair, M., Sainsbury, K. and O'Boyle, R. (2000). Symposium overview: incorporating ecosystem objectives within fisheries management. *ICES Journal of Marine Science* **57**, 468–475.
     Dayton, P.K., Thrush, S.F., Agardy, M.T. and Hofman, R.J. (1995). Environmental effects of marine fishing. *Aquatic Conservation: Marine and Freshwater Ecosystems* **5**, 205–232.
     Hall, SJ. (1999). *The Effects of Fishing on Marine Ecosystems and Communities.* Blackwell Science, Oxford.
     Robinson, L.A. and Frid, C. (2003). Dynamic ecosystem models and the evaluation of ecosystem effects of fishing, can we make meaningful predictions? *Aquatic Conservation: Marine and Freshwater Ecosystems* **13**, 5-20.

12   Pope, J.G, and Macer, C.T. (1996). An evaluation of the stock structure of North Sea cod, haddock, and whiting since 1920, together with a consideration of the impacts of fisheries and predation effects on their biomass and recruitment. *ICES Journal of Marine Science* **53**, 1157–1169.
     Rogers, S.I. and Millner, R.S. (1996). Factors affecting the annual abundance and regional distribution of English inshore demersal fish populations: 1973 to (1995). *ICES Journal of Marine Science* **53**, 1094–1112.
     Serchuk, F.M., Kirkegaard, E. and Daan, N. (1996). Status and trends of the major roundfish, flatfish, and pelagic fish populations in the North Sea: thirty-year overview. ICES Journal of Marine Science **53**, 1130–1145.

13   Millner, R.S. and Whiting, C.L. (1996). Long-term changes in growth and population abundance of sole in the North Sea from 1940 to the present. *ICES Journal of Marine Science* **53**, 1185–1195.

Rijnsdorp, A.D. and van Leeuwen, P.I. (1996). Changes in growth of North Sea plaice since 1950 in relation to density, eutrophication, beam-trawl effort, and temperature. *ICES Journal of Marine Science* **53**, 1199–1213.

Gislason, H. and Rice, J.C. (1998). Modelling the response of size and diversity spectra of fish assemblages to changes in exploitation. *ICES Journal of Marine Science* **55**, 362–370.

14   Rumohr, H. and Kros, P. (1991). Experimental evidence of damage to benthos by bottom trawling, with special reference to *Arctica islandica*. *Helgoländer Meeresuntersuchungen* **33**, 340–345.

Camphuysen, C.J, Calvo, B., Durinck, J., Ensor, K., Follestad, A., Furness, R.W., Garthe, S., Leaper, G., Skov, H., Tasker, M.L. and Winter, C.J.N. (1995). *Consumption of discards in the North Sea*. Final Report EC DG XIV Research Contract BIOECO/93/10. Netherlands Institute for Sea Research, Texel, NIOZ, Rapport 5.

Thrush, S.H., Hewitt, J.E., Cummings, V.J. and Dayton, P.K. (1995). The impact of habitats disturbance by scallop dredging on marine benthic communities: what can be predicted from the results of experiments? *Marine Ecology Progress Series* **129**(1–3), 141–150.

Tuck, I.D., Hall, S.J., Robertson, M.R., Armstrong, E. and Basford, D.J. (1998). Effects of physical trawling disturbance in a previously unfished sheltered Scottish sea loch. *Marine Ecology Progress Series* **162**, 227–242.

15   Churchill J.H., (1989). The effect of commercial trawling on sediment resuspension and transport over the Middle Atlantic Bight Continental Shelf. Continental Shelf Research **9**, 841–864.

Messieh, S. (1991). Fluctuations in Atlantic herring succession in relation to organic enrichment and pollution of the marine environment. *Oceanography and Marine Biology Annual Review* **16**, 229–311.

Riemann, B. and Hoffmann, E. (1991). Ecological consequences of dredging and bottom trawling in the Limfjord, Denmark. *Marine Ecological Progress Series* **69**, 171–178.

Auster, P.J, Malatesta, R.J. and Larosa, S. (1995). Patterns of microhabitat utilisation by mobile megafauna on the southern New England (USA) continental shelf and slope. *Marine Ecology Progress Series* **12**, 77–85.

Thrush *et al.* (1995).

Schwinghamer, P., Guigne, J.Y. and Siu, W.C. (1996). Quantifying the impact of trawling on benthic habitat structure using high-resolution acoustics and chaos theory. *Canadian Journal of Fisheries and Aquatic Science* **53**, 288–296.

Collie, J.S. and Spencer, P.D. (1994). Modelling predator-prey dynamics in a fluctuating environment. Canadian *Journal of Fisheries and Aquatic Sciences* **51**, 2665–2672.

16   Prins, T.C. and Smaal, A.C. (1990). Benthic-pelagic coupling: the release of inorganic nutrients by an intertidal bed of *Mytilus edulis*. In: *Trophic Interactions in the Marine Environment*. Eds M. Barnes and R.N. Gibson. Aberdeen University Press: Aberdeen; 89–103.

      ICES. (1998). *Report of the Working Group on the Ecosystem Effects of Fishing Activities*. ICES. Copenhagen, ICES CM 1998/ACFM/ACME:01 Ref.:E, 263pp.

      Jennings, S., Dinmore, T.A., Duplisea, D.E., Warr, K.J. and Lancaster, J.E. (2001). Trawling disturbance can modify benthic production processes. *Journal of Animal Ecology* **70**(3), 459–475.

17   Carpenter, S.R., Kitchell, J.F. and Hodgson, J.R. (1985). Cascading trophic interactions and lake productivity. *BioScience* **35**, 634–639.

18   Frid, C.L.J., Clark, R.A. and Hall, J.A. (1999a). Long-term changes in the benthos on a heavily fished ground off the NE coast of England. *Marine Ecology Progress Series* **188**, 13–20.

      Frid, C.L.J., Hansson, S., Ragnarsson, S.A., Rijnsdorp, A. and Steingrimsson, S.A. (1999b). Changing levels of predation on benthos as a result of exploitation of fish populations. *Ambio*, **28**, 578–582.

19   Pitcher T. (2001).

20   Watling, L. and Norse, E.A. (1998). Disturbance of the seabed by mobile fishing gear: a comparison to forest clear-cutting. *Conservation Biology* **12**, 1180-1197.

21   Alverson, D.L., Freeberg, M.H., Pope, J.G. and Murawski, S.A. (1994). *A global assessment of fisheries bycaatch and discards*. FAO Fisheries Technical Paper, 339. Rome, Food and Agriculture Organisation.

      Moore, G. and Jennings, S. (2000). *Commercial Fishing: The Wider Ecological Importance*. British Ecological Society, Blackwells, pp66.

22   Written evidence from JNCC.

      Garthe, S., Camphyusen, K and Furness, R.W. (1996). Amounts of discards by commercial fisheries and their significance as food for seabirds in the North Sea. *Marine Ecology Progress Series*, **136**, 1-11.

23   Kindly supplied by English Nature ©.

24   Covey and Laffoley (2002); Groot, S.J. and de Lindeboom, H.J. (1994). *Environmental Impact of Bottom Gears on Benthic Fauna in Relation to Natural Resources Management and Protection of the North Sea*. Netherlands Institute for Fisheries Research (NIOZ), Ijmuiden, The Netherlands.

25   Scottish Natural Heritage website:
      http://www.snh.org.uk/trends/trends_notes/pdf/Commercial%20fisheries/Cod%20stocks.pdf

26   Scottish Natural Heritage website

27   Covey and Laffoley (2002).

28   *Ibid.*

      Genner, M.J., Hawkins, S.J., Sims, D. and Southwood, A.J. (2000). Unpublished data from the Marine Biological Association, Plymouth.

29   Fertl, D. and Leatherwood, S. (1995). *NAFO Science Council*. Research Documents 1995, no. 95/82. 34 pp.

30   Reid, J.B. and Evans, P.G.H., Northridge S.P. (2003). *Atlas of Cetacean distribution in north-west European waters*. JNCC. Peterborough, UK.

31   ICES. (2001). *Working Group on marine mammal population dynamics.* ICES C.M.
      2001/ACE:01
      Morizur, Y., Berrow, S.D., Tregenza, N.J.C, Couper.is, A.S. and Pouvreau, S. (1999).
      Incidental catches of marine mammals in pelagic trawl fisheries of the north-east
      Atlantic, *Fisheries Research* **41**, 297-307.

32   Tregenza, N.J.C and Collet, A. (1997). Common dolphin (*Delphinus delphis*) bycatch in
      pelagic trawl and other fisheries in the Northeast Atlantic. *Forty-Eighth Report of the
      International Whaling Commission* **48**, 453-462. Annual Report. International Whaling
      Commission.

33   Reid *et al.* (2003).

34   Vinther, M. (1999) Bycatches of harbour porpoises (*Phocoena phocoena* L.) in Danish
      set-net fisheries. *Journal of Cetacean Resource Management*, **1**(2), 123-135.

35   Reid *et al.* (2003).

36   Tregenza, N.J.C., Berrow, S.D., Hammond, P.S. and Leaper R. (1997). Common dolphin
      (*Dephinus delphis*) by-catch in bottom set gill nets in the Celtic sea. *Report of the
      International Whaling Commission*, **47**, 835-839.

37   Ross, A. and Isaac, S. (2004). *The Net Effect? A review of cetacean bycatch in pelagic
      trawls and other fisheries in the north-east Atlantic.* A WDSC report for Greenpeace.

38   Defra (2003). UK small cetacean bycatch response strategy. Defra, London.

39   Ross and Isaac (2004).

40   Defra (2003).

41   Ross and Isaac (2004).

42   *Ibid.*

43   *Ibid.*

44   ICES (2002). *Working Group on seabird ecology.* ICES C.M. 2002/C:04.

45   Harrison, N., and Robins, M., (1992). The threat from nets to seabirds. *RSPB
      Conservation Review*, **6**: 51-56.

46   Laffoley, D. and Tasker, M. (2003). Marine Environment. Anlaytical paper produced to
      support the report Net Benefits – a sustainable and profitable future for UK fishing.
      Available on strategy unit website at http://www.strategy.gov.uk/files/pdf/marine2.pdf

47   CEFAS (2003). *A study on the consequences of technological innovation in the capture
      fishing industry and the likely effects upon environmental impacts.* Report to the Royal
      Commission on Environmental Pollution. Available on website at
      http://www.rcep.org.uk.

48   Dunn, E. and Steel, C. (2001). *The impact of longline fishing on seabirds in the north-east
      Atlantic: recommendations for reducing mortality.* RSPB, Sandy. NOF Rapprtserie Report
      no 5-(2001).

49   *Ibid.*

50   Kindly supplied by Marine Conservation Society.

51   Myers, R.A., Hutchings, J.A. and Barrowman, N.J. (1996). Hypothesis for the decline of
      cod in the North Atlantic. *Marine Ecology Progress Series,* **138**, 293-308.

52   Greenstreet and Hall (1996).

53   Jennings, S., Greenstreet, S.P.R. and Reynolds, ¨.D. (1999). Structural change in an
      exploited fish community: a consequence of differential fishing effects on species with
      contrasting life histories. *Journal of Animal Ecology,* **68**, 617-627.

54   Dulvy, N.K., Metcalfe, J.D., Glanville, J., Pawson, M.G., Reynolds, J.D. (2000). Fishery Stability, Local Extinctions, and shifts in community structure in skates. *Conservation Biology*, **14**(1), 283-293.
      Fogarty, M.J. and Murawski, S.A.(1998) Large scale disturbance and the structure of marine system: fishery impacts on the Georges Bank. *Ecological Applications* **8**, (S)6-S22.

55   Greenstreet and Hall (1996).

56   *Ibid.*

57   Jennings S. and Blanchard J.L. (2004). Fish abundance with no fishing predictions based on macroecological theory. *Journal of Animal Ecology*, **73**, 632-642
      Horwood, J.W. (1993). The Bristol Channel sole (*Solea Solea* L.): a fisheries case study. *Advances in Marine Biology*, **29**, 215-367.

58   Worm, B. and Myers, R.A. (2003). Meta-analysis of cod-shrimp interactions reveals top-down control in oceanic food webs. *Ecology*, **84**, 162-173.
      ICES (2003).

59   Jennings S. and Blanchard J.L. (2004). Fish abundance with no fishing predictions based on macroecological theory. *Journal of Animal Ecology*, **73**, 632-642.
      Steele, J.H. and Schumacher, M. (2000). Ecosystem structure before fishing. *Fisheries Research*, **44**, 201-205.

60   Jennings and Blanchard (2004).
      Walther, G-R., Post, E., Convey, P., Menzel, A., Parmesan, C., Beebee, T.J.C., Fromentin, J.-C., Hoegh-Guldberg, O. and Barlein F. (2002). Ecological responses to recent climate change. *Nature*, 416, 389-395;
      Daan, N. and Richardson, K. (1996). Changes in the North Sea ecosystem and their causes: Arhus 1975 revisited. *ICES Journal of Marine Science*, **53**, 879-1226;
      Pinnegar, J.K., Polunin, N.V.C., Francour, P., Badalamenti, F., Chemello, R., Harmelin-Vivien, M.L., Hereu, B., Milazzo, M., Zabala, M., D'Anna, G. and Pipitone, C. (2000); Trophic cascades in benthic marine ecosystems: lessons for fisheries and protected-area management. *Environmental Conservation, 27*, 179-200.

61   Jennings and Blanchard (2004).

62   *Ibid.*

63   Jennings, S., Greenstreet, S.P.R., Hill, L., Piet, G.J., Pinnegar, J.K. and Warr, K.J. (2002). Long-term trends in the trophic structure of the North Sea fish community: evidence from stable-isotope analysis, size-spectra and community metrics. *Marine Biology* **141**(6), 1085-1097.

64   Jennings, S. and Warr, K.J. (2003). Smaller predator-prey body size ratios in longer food chains. *Proceedings of the Royal Society London* **270**, 1413 – 1417.

65   Pauly, D. (2002). The crisis in fisheries and marine Biodiversity. In: *Sustaining seascapes; the science and policy of marine resource management*, March 7 & 8, (2002). American Museum of Natural History, New York City;
      Estes, J.A. and Duggins, D.O. (1995). Sea Otters and kelp forests in Alaska: generality and variation in a community ecology paradigm. *Ecological Monographs* **65**, 75-100.

66   Kindly supplied by Keith Hiscock, © MarLIN.
      Jackson A. (2004). *Psammechinus miliaris.* Green sea urchin. Marine Life Information Network: Biology and Sensitivly Key Information sub-programme (on-line). Plymouth: Marine Biological Association Available from:
      http//www.marlin.ac.uk/species/psammechinusmiliaris.htm

67   Myers and Worm (2003).
      Fogarty and Murawski (1998).

68   Köster, F.W. and Möllmann, C. (2000). Trophodynamic control by clupeid predators on recruitment success in Baltic cod? *ICES Journal of Marine Science*, **57**, 310–323.

69   Linley J.A., Gamble J.C. and Hunt, H.G. (1995). A change in the zooplankton of the central North Sea (55° to 58°N): a possible consequence of changes in the benthos. *Marine Ecology Progress Series* **119**, 299-303.

70   Kindly supplied by JNCC, © David Connar/JNCC.
      Hill, J.M. and Wilson, E. (2001). *Virginia mirabilis* and *Ophiuria* spp. on circa littoral sand or shelly mud. Marine Life Information Network: Biology and Sensitivity Key Information sub-programme (on-line). Plymouth: Marine Biological Association. Available from: http//www.marlin.ac.uk/biotypes/Bio_Eco_CMSVirOph.htm

71   Votier, S.C., Furness, R.W., Bearhop, S., Crane, J.W., Caldow, R.W.G, Catry, P., Ensor, K. and Hamer, K.C., Hudson, A.V., Kalmbach, E., Klomp, N.I., Pfeiffer, S., Phillips, R.A, Prieto, I. and Thompson, D.R. (2004). Changes in fisheries discard rates and seabird communities. *Nature* **427**, 727-730.

72   Pauly (2002).

73   Daskalov. G.M. and Mackinson S. (2004). Trophic modelling of the North Sea. *Paper presented at the ICES Annual Science Conference* FF, 40.

74   Jennings and Blanchard (2004).

75   Hutchings, J.A. (2000). Numerical assessment in the front seat, ecology and evolution in the back seat: time to change drivers in fisheries and aquatic sciences? *Marine Ecology Progress Series* **208**, 299-302.
      Smith, T.D. (1994). *Scaling fisheries: the science of measuring the effects of fishing, 1855-1955*. Cambridge University Press, Cambridge.

76   Roff, D.A. (1992). *Evolution of life histories, theory and analysis*. Chapman and Hall. New York.

77   Powles, H., Bradford, M.J., Bradford, R.G., Doubleday, W.G., Innes, S. and Levings, C.D. (2000). Assessing and protecting endangered marine species. *Ices Journal of Marine Science* **57**(3), 669-676.

78   Baumgartner, T.R., Souter, A. and Riedel, W. (1995). *Natural time scales of variability in coastal pelagic fish populations of the California Current over the past 1500 years: response to global climate change and biological interaction*. California Sea Grant, Biennial Report of Completed Projects, 1992-94, 31-37.

79   Sadovy, Y. (2001). The threat of fishing to highly fecund fishes. *Journal of Fish Biology*. **59** (Supplement A), 90-108.

80   Bradbury, I.R. and Snelgrove, P.V.R. (2001). Spatial and temporal distribution in benthic marine fish and invertebrates: the role of passive and active processes. *Canadian Journal of Fisheries and Aquatic Science* **58**, 811-823;
      Largier, J.L. (2003). Considerations in estimating larval dispersal distances from oceanographic data. *Ecological Applications* **13**(1) Supplement pp.S71-S89.

81   Mertz, G. and Myers, R.A. (1996). Influence of fecundity on recruitment variability of marine fish. *Canadian Journal of Fisheries and Aquatic Science*. **53**,1618-1625;
      Rickman, S.J., Dulvy, N.K., Jennings, S. and Reynolds, J.D. (2000). Recruitment variation related to fecundity in marine fishes. *Canadian Journal of Fisheries and Aquatic Science*, **57**(1), 116-124;

Hutchings J.A. (2001). Conservation of marine fishes: perceptions and caveats regarding assignment of extinction risk. *Canadian Journal of Fisheries and Aquatic Science* **58**: 108-121.

82 Myers, R.A. and Bowen, K.G. and Barrowman, N.J. (1999). Maximum reproductive rate of fish at low population sizes. *Canadian Journal of Fisheries and Aquatic Sciences*, **56**, 2404-2419.

83 Dulvy N.K, Sadovy Y. and Reynolds J.D. (2003). Extinction vulnerability in marine populations. *Fish and Fisheries*, **4**, 25-26.
Hutchings (2001).
Longhurst, A. (2002). Murphy's Law revisited: longevity as a factor in recruitment to fish populations. *Fisheries Research*, **56**, 125-131.
Sadovy (2001).

84 Hutchings J.A. (2000). Collapse and recovery of marine fisheries. *Nature*, **406**, 882-885.
Hutchings (2001).

85 Sadovy (2001).

86 Dulvy *et al.* (2003).

87 Palumbi, S. R. (2004). Why mothers matter? *Nature*, **430**, 621-622.
Berkeley, S. A., C. Chapman, and S. M. Sogard. (2004). Maternal age as a determinant of larval growth and survival in a marine fish, *Sebastes melanops. Ecology*, **85**(5),1258-1264.

88 Palumbi (2004).
Coleman, F., C. Koenig and L. A. Collins. (1996). Reproductive styles of shallow-water grouper, consequences of fishing spawning aggregations. *Environmental Biology of Fishes* **47**, 129-141.
Hannah, Robert W., and S. A. Jones. (1991). Fishery-induced changes in the population structure of pink shrimp *Pandalus jordani . Fishery Bulletin, U.S.* **89**, 41-51.

89 Myers, R.A and Worm B. (2004). Extinction, survival or recovery of large predatory fishes. *Proceedings of the Royal Society,* In press.

90 Hutchings (2000).
Hutchings (2001).

91 Mace, P. (2004). In defence of fisheries scientists, single-species models and other scapegoats: confronting the real problems. *Marine Ecology Series.* **274**, 269-303.

92 Hutchings J.A. and Myers, R.A. (1994). What can be learned from the collapse of a renewable resource? Atlantic Cod*, Gadus Morhua,* of Newfoundland and Labrador. *Canadian Journal of Fisheries and Aquatic Sciences,* **51**, 2126-2146;
Myers, R.A., Hutchings, J.A. and Barrowman, N.J. (1997). Why do fish populations collapse? The example of Cod in Atlantic Canada. *Ecological Applications* **7**(1), 91-106.

93 Finlayson, A.C. (1994). *Fishing for truth: a sociological analysis of northern cod stock assessments from 1977 to 1990.* Social and Economic Studies, Vol 52, Institute of Social and Economic Research, Memorial University of Newfoundland, St Johns.
European Environment Agency. (2001). Late Lessons from early warnings: the precautionary principle 1896-2000. *Environmental Issue Report* **22**.

94 European Environment Agency. (2001).
Hutchings, J.A. and Walters C. and Haedrich R.L. (1997). Is scientific inquiry incompatible with government information control? *Canadian Journal of Fisheries and Aquatic Science* **54**, 1198-1210.

95   Morgan, M.J. and DeBlois, E.M. and Rose, G.A. (1997). An observation on the reaction of Atlantic cod (*Gadus morhua*) in a spawning shoal to bottom trawling. *Canadian Journal of Fisheries and Aquatic Science* **54**, 477-488.
       Rowe, S. and Hutchings, J.A. (2003). Mating systems and the conservation of commercially exploited marine fish. *Trends in Ecology and Evolution* **18**(11). 567-572.

96   Rowe and Hutchings (2003).
       Hutchings, J.A. and Bishop, TD, and CR McGregor-Shaw. (1999). Spawning behaviour of Atlantic cod, *Gadus morhua*: evidence of mate competition and mate choice in a broadcast spawner. *Canadian Journal Fisheries Aquatic Science* **56**,97-104.
       Nordeide, J.T., Kjellsby, E. (1999). Sound from spawning cod at their spawning grounds. *ICES Journal of Marine Science* **56**, 326 – 332.

97   Myers and Worm (2003).
       Christensen, V., Guénette, S., Heymans, J.J., Walters, C.J., Watson, R., Zeller, D. and Pauly, D. (2003). Hundred year decline of North Atlantic predatory fishes. *Fish and Fisheries* **4** (1):1-24.

98   Adapted from Christensen, V., Guénette, S., Heymans, J.J., Walters, C.J., Watson, R., Zeller, D. and Pauly, D. 2001. Estimating fish abundance of the North Atlantic, 1950 to 1999, in Guénette, S., Christensen, V. & Pauly, D. (Eds.), Fisheries Impacts on North Atlantic ecosystems: models and analyses. *Fisheries Centre Report* **9**(4): 1-26.

99   Jennings and Blanchard (2004).

100   Kaschner K. and Pauly D. (2004). *Competition between Marine Mammals and Fisheries.* Report to the IWC.

101   Myers and Worm (2003).

102   Walters, C. (2003) Folly and fantasy in the analysis of spatial catch rate data. *Canadian Journal of Fisheries and Aquatic Science,* **60**, 1433-1436.

103   Pelagic Fisheries Research Program, Joint Institute for Marine & Atmospheric Research, University of Hawaii, website:
       http://www.soest.hawaii.edu/PFRP/pdf/Myers_comments.pdf

104   Myers and Worm (2004).

105   May, M., Lawton, J.H. and Stork, N.E. (1995). *Assessing extinction rates.* Pages 1-24 in J.H. Lawton and R.M. May editors. Extinction Rates. Oxford University Press, Oxford, UK.

106   Roberts, C.M. (2003). Our shifting perspectives on the oceans. *Oryx* **37**(2), 166-1772

107   Jackson, J.B.C., Kirby, M.X., Berger, W.H., Bjorndal, K.A., Botsford, L.W., Bourque B.J., Bradbury, R.H., Cooke, R., Erlandson, J., Estes, J A., Hughes, T.P., Kidwell, S., Lange, C.B., Lenihan, H.S., Pandolfi, J. M., Peterson, C.H., Steneck, R.S., Tegner, M.J. and Warner, R.R. (2001). Historical Overfishing and the Recent Collapse of Coastal Ecosystems. *Science,* **293** (5530), 561-748.

108   *Ibid.*

109   Pitcher T. (2001).

110   Dulvy *et al.* (2003).

111   Myers and Worm (2004).

112   Dulvy *et al.* (2003).

113   Watling and Norse (1998).

114   Roman, J. and Palumbi, S.R. (2003). Whales before whaling in the North Atlantic. *Science* **301**(5632), 508-510.

115   Jackson *et al.* (2001).

Myers and Worm (2003).

Baum, J.K., Myers, R.A., Kehler, D.G., Worm, B., Harley, S.J. and Doherty, P.A. (2003). Collapse and Conservation of Shark Populations in the Northwest Atlantic. *Science,* **299** (5605) 389-392.

116   Purvis, A., Gittleman, J.L., Cowlinshaw, G. and Mace, G.M. (2000). Predicting extinction risk in declining species. *Proc. R. Soc. London B,* **267**, 1947-1952.

117   Dulvy *et al.* (2000).

Fogarty and Murawski (1998).

118   © Crown copyright 2004, reproduced by kind permission of CEFAS; Dulvy *et al.* 2003.

119   Dulvy *et al.* (2000).

Fogarty and Murawski (1998).

120   © Crown copyright 2004, reproduced by kind permission of CEFAS. Dulvy *et al.* (2000).

121   Dulvy *et al.* (2000).

Fogarty and Murawski (1998).

122   Dulvy, N.K. and Reynolds, J.D. (2002). Predicting Extinction Vulnerability in Skates. *Conservation Biology* **16**(2), 440-450.

Hilton-Taylor, C. (2000). *IUCN red list of threatened species.* World Conservation Union, Gland, Switzerland;

Dulvy *et al.* (2000);

Carlton, J.T., Geller, J.B., Reaka-Kudla, M.L. and Norse, E.A.(1999). Historical Extinctions in the sea. *Annual Review of Ecology and Systematics,* **30**,525-538;

Casey, J. and Myers, R.A. (1998). Near extinction of a large, widely distributed fish. *Science* **281**, 690-692.

123   The Shark Trust website http://www.sharktrust.org

124   © Crown copyright 2004, reproduced by kind permission of CEFAS

125   Baum *et al.* (2003).

126   *Ibid.*

127   Baum, J. K. and Myers, R. (2004). Shifting baselines and the decline of pelagic sharks in the Gulf of Mexico. *Ecology Letters,* **7,** 135 – 145.

128   Dulvy *et al.* (2000).

Casey, J.M. and Myers, R.A. (1998). Near extinction of a large widely distributed fish. *Science* **281**, 690-691.

129   Dulvy and Reynolds (2002).

Hilton-Taylor, C. (2000);

Dulvy *et al.* (2000);

Carlton *et al.* (1999);

Casey and Myers (1998).

130   Dulvy *et al.* (2000);

Fogarty and Murawski (1998).

131   IUCN (1996). *1996 IUCN Red List of Threatened Animals.* IUCN, Gland, Switzerland.

Mace, G.M. and Hudson, E.J. (1999). Attitudes towards sustainability and extinction. *Conservation and Biology* **13**,242-246.

132   Dulvy and Reynolds (2002).

Watts, J. (2001). Loadsa tunny £570/kg price record. The Guardian (London) 1 June: p16.

133 Farrow, S. (1995). Extinction and Market forces: two case studies. *Ecological Economics* **13**, 115-123.

134 Dulvy *et al.* (2003).

135 Roberts, C.M. (2003).

136 Stokes, K. and Law, R. (2000). Fishing as an evolutionary force. *Marine Ecology Progress Series.* **208**, 307-309.

137 Rijnsdorp, A.D. (1993). Fisheries as a large-scale experiment on life-history evolution: disentangling phenotypic and genetic effects in changes on maturation and reproduction of North Sea plaice, *Plueronectes platessa* L. *Oecologica*, **96**,391-401.

138 Olsen E.M, Helno, M., Lilly G.R., Morgan, M.J., Brattey J., Ernando, B. and Dieckmann U. Maturation trends indicative of rapid evolution preceded the collapse of northern cod. *Nature*, **428**, 932-935.

139 Hutchings, J.A. (2000). Numerical assessment in the front seat, ecology and evolution in the back seat: time to change drivers in fisheries and aquatic sciences? *Marine Ecology Progress Series* **208**, 299-302.
     Smith, T.D. (1994).

140 Stokes and Law (2000).

141 Kenchington E., Henno M. and Nielsen E.E. (2003). Managing marine genetic diversity: time for action? *ICES Journal of Marine Science* **60**, 1172-1176.

142 Law R. (2000). Fishing, selection, and phenotypic evolution. *ICES Journal of Marine Science*, **57**, 659-668.

143 © Crown copyright 2004 reproduced by kind permission of CEFAS.

144 Ryman N., Utter F. and Laikre L.(1995). Protection of intraspecific biodiversity of exploited fishes. *Review of Fish Biology and Fisheries*, **5**, 417-446.

145 Hauser, L., Adcock, G.J., Smith, P.J., Bernal Ramirez, J.H. and Carvalho, G.R. (2002). Loss of microsatellite diversity and low effective population size in an overexploited population of New Zealand Snapper. *Proceedings of the National Academy of Science* **99**, 11742-11747.

146 Hedgecock, D. (1994). In: *Genetics and Evolution of Aquatic Organisms*, ed. Beaumont A.R. (Chapman and Hall, London), pp. 122-134.

147 Ryman *et al.* (1995).

148 Hutchinson W.F., van Oosterhout, C., Rogers, S.I. and Carvalho, G.R. Temporal analysis of archived samples indicates marked genetic changes in declining North Sea cod (*Gadus morhua*). *Proceedings of the Royal Society of London – Biological Sciences*, **270**, 2125-2132.

149 Dulvy *et al.* (2003).

150 Malakoff, D. (1997). Extinction on the high seas. *Science* **277**, 486-488.

151 Dulvy *et al.* (2003).

152 Smedbol, R.K. and Stephenson, R. (2001) The importance of managing within species diversity in cod and herring fisheries of the north-east Atlantic. *Journal of Fish Biology*, **59A**, 109-128.

153 Roberts, C.M. , McClean, C.J., Veron, J.E.N., Hawkins, J.P, Allen, G.R., McAllister, D.E., Mittermeier, C.G., Schlueler, F.W., Spalding, M. Wells, F., Vynne, C. and Werner, T.B. (2002). Marine biodiversity hotspots and conservation priorities for tropical reefs. *Science*, **295**, 1280-1284.

154 Dulvy *et al.* (2003).

155   Davis, G.D., Haaker, P.L. and Richards, D.V. (1996). Status and trends of white abalone at the California Channel Islands. *Transactions of the American Fisheries Society*, **125**, 42-48.

156   ICES. (2004). Working Group on the Application of Genetics in Fisheries and Mariculture. ICES CM 2004/F:04, Ref ACFM, ACMES, J.

157   Kenchington *et al.* (2003).

158   Stokes and Law (2000).

159   Law (2000).

160   Olsen E.M, Helno, M., Lilly G.R., Morgan, M.J., Brattey J., Ernando, B., Dieckmann U. Maturation trends indicative of rapid evolution preceded the collapse of northern cod. *Nature*, **428**, 932-935.

161   Kenchington *et al.* (2003).

162   Schratzberger, M., Dinmore, T.A. and Jennings, S. (2002). Marine Biology. Impacts of trawling on the diversity, biomass and structure of meiofauna assemblages. *Marine Biology*, **140**(1), 83-93, 2002.

163   Watling, L. and Norse, E.A. (1998).

164   Kaiser, M.J., Collie, J.S., Hall, S.J., Jennings, S. and Poiner, I.R. (2002). Modification of marine habitats by trawling activities: prognosis and solutions. *Fish and Fisheries*, **3**, 114-136.

165   Roberts L.M. (2003).

166   Watling and Norse (1998).

167   OSPAR Commission (2000);
Coffey and Richartz (2004).

168   Kindly supplied by Keith Hiscock, © *MarLIN*.

169   Schratzberger *et al.* (2002).

170   Schratzberger *et al.* (2002);
Groot SJ de and Lindeboom HJ (1994) *Environmental impact of bottom gears on benthic fauna in relation to natural resources management and protection of the North Sea.* Netherlands Institute for Sea Research, Texel;
Dayton *et al.* (1995);
Jennings, S. and Kaiser, M.J. (1998). The effects of fishing on marine ecosystems. *Advances in Marine Biology* **34**,201–352;
Lindeboom, H.J., Groot, S.J. de (1998). *The effects of different types of fisheries on the North Sea and Irsh Sea benthic ecosystems.* Netherlands Institute of Sea Research, Texel;
Hall, S.J. (1999). *The effects of fishing on marine ecosystems and communities.* Blackwell, Oxford;
Collie, J.S., Hall, S.J., Kaiser, M.J., Poiner, I.R. (2000). A quantitative analysis of fishing impacts on shelf sea benthos. *Journal of Animal Ecology* **69**,785–798 issues. Blackwell, Oxford;
Gislason, H., Sinclair, M., Sainsbury, K. and O'Boyle, R. (2000) Symposium overview: incorporating ecosystem objectives within fisheries management. *ICES Journal of Marine Science* **57**,468–475;
Kaiser, M.J. and Groot, S.J. de (2000) *The effects of fishing on non-target species and habitats: biological, conservation and socio-economic issues.* Blackwell, Oxford.

171   Schratzberger *et al.* (2002).

Kaiser M.J. and Spencer, B.E. (1996). The effects of beam-trawl disturbance on infaunal communities in different habitats. *Journal of Animal Ecology* **65**,348–358

Thrush, S.F., Hewitt J.E., Cummings, V.J., Dayton, P.K., Cryer, M., Turner, S.J., Funnell, G.A., Budd, R.G., Milburn, C.J. and Wilkinson, M.R. (1998) Disturbance of marine benthic habitat by commercial fishing, impacts at the scale of the fishery. *Ecological Applications* **8**,866-879;

Tuck, I., Hall, S.J., Roberston, M., Armstrong, E. and Basford, D. (1998) Effects of physical trawling disturbance in a previously unfished sheltered Scottish sea loch. *Marine Ecology Progress Series* **162**,227– 242;

Bergman, M.J.N. and Santbrink, J.W. van (2000a). Fishing mortality of populations of megafauna in sandy sediments. In: Kaiser M.J., Groot S.J. de (eds) *Effects of fishing on non-target species and habitats: biological, conservation and socio-economic issues.* Blackwell, Oxford, pp 49–68;

Bergman, M.J.N and Santbrink, J.W. van (2000b). Mortality in megafaunal benthic populations caused by trawl fisheries on the Dutch continental shelf in the North Sea in (1994). *ICES Journal of Marine Science* **57**,1321–1331;

Gislason, H. and Sinclair, M.M. (2000) Ecosystem effects of fishing. *ICES Journal of Marine Science* **57**,465–791;

Hall-Spencer, J.M., Moore, P.G. (2000). Scallop dredging has profound, long-term impacts on maerl habitats. *ICES Journal of Marine Science* **57**,1407–1415;

Kaiser, M.J., Ramsay, K., Richardson, C.A., Spence, F.E. and Brand, A.R. (2000). Chronic fishing disturbance has changed shelf sea benthic community structure. *Journal of Animal Ecology* **69**,494–503;

Collie, J.S., Escanero, G.A., and Valentine, P.C. (1997). Effects of bottom fishing on the benthic megafauna of Georges Bank. *Marine Ecology Progress Series* **155**,159-172, (1997).

172   Gubbay, S. and Knapman, P.A. (1999) *A review of the effects of fishing within UK European marine sites*. UK Marine SAC's Project, English Nature, ISBN 1857164822; Kaiser *et al.* (2000);

Jennings, S. Dinmore, T.A., Duplesea, D.E., Warr, K.J. and Lancaster, J.E. (2001) Trawling disturbance can modify benthic production processes. *Journal of Animal Ecology* **70**,459–475.

173   Jennings and Kaiser (1998).

174   Kaiser, M.J., Collie, J.S., Hall, S.J., Jennings, S. and Poiner, I.R. (2002). Modification of marine habitats by trawling activities: prognosis and solutions. *Fish and Fisheries* **3**, 114-136.

175   Kindly supplied by Keith Hiscock, © *MarLIN*.

Hiscock, K. (2001). Eunicella verrucosa. Pink Sea fan. Marine Life Information Network. Biology and Sensitivity Key Information sub-programme (on-line). Plymouth: Marine Biological Association of Great Britain. Available from http://www.marlin.ac.uk/species/Eunicellaverrucosa.htm

176 OSPAR Commission (2000).

Krost, P., M. Bernhard, F. Werner, and W. Hukriede. 1990. Otter trawl tracks in Kiel Bay (Western Baltic) mapped by side-scan sonar. *Meeresforsch* **32**,344-353;

Jennings and Kaiser (1998);

Bergman, M.J.N. and Hup, M., (1992). Direct effects of beam trawling on macro-fauna in a sandy sediment in the southern North Sea. *ICES Journal of Marine Science*, **49**, 5-11.

177 Kaiser *et al.* (2002).

178 Kindly supplied by Keith Hiscock, © *MarLIN.*

179 Thiel, H. and Schriever, G. (1990). Deep-seamining, environmental impact and the DISCOL project. *Ambio,* **19**, 245-250;

Kaiser, M.J. and Spencer, B.E. (1996). The effects of beamtrawl disturbance on infaunal communities in different habitats. *Journal of Animal Ecology* **65**, 348-358.

Auster, P.J. and Langton, R.W. (1999). The effects of fishing on fish habitat. In: *Fish Habitat: Essential Fish Habitat and Restoration, Vol. 22* Ed. L. Benaka. American Fisheries Society Symposium, Bethesda, Maryland, pp. 150-187.

180 Kaiser *et al.* (2002)

181 Riesen, W. and Riese, K. 1982. Macrobenthos of the subtidal Wadden Sea: revisited after 55 years. *Helgolander Meeeresuntersuchungen,* **35**: 409-423.

182 Professor Chris Frid, personal communication, August 2004.

183 Kaiser *et al.* (2002).

184 Kaiser, M.J., Rogers, S.I. and Ellis, J.R. (1999). Importance of benthic habitat complexity for demersal fish assemblages. *In: Fish Habitat: Essential Fish Habitat and Restoration , Vol. 22.* Ed. L. Benaka. American Fisheries Society, Bethesda, Maryland, pp.212 – 223.

185 Kaiser *et al.* (2002).

186 Kaiser *et al.* (2002).

187 Hall-Spencer and Moore (2000)

Collie *et al* (2000).

188 Figure from Freiwald, A., Fossa, J.H., Grehau, A., Koslow, T. and Roberts, J.M. (2004). *Cold-water coral reefs.* UNEP-WCMC, Cambridge, UK.

189 Coleman, F.C., Williams, S.L. (2002). Overexploiting marine ecosystem engineers: potential consequences for biodiversity. *Trends in Ecology and Evolution,* **17**, 40-44.

190 Kaiser *et al.* (2002).

191 Meeuwig, J.J. (1999) Predicting coastal eutrophication from land-use: an empirical approach to small non-stratified estuaries. *Marine Ecology Progress Series,* **176**, 231-241.

192 Kaiser *et al.* (2002).

193 Jennings and Kaiser (1998);

Collie *et al.* (2000);

Kaiser *et al.* (2002);

Kaiser, M.J. and Groot, S.J. de (2000). *The effects of fishing on non-target species and habitats: biological, conservation and socio-economic issues.* Blackwell, Oxford

Schratzberger *et al.* (2002).

194 Rumohr, H. and Kujawski, T. (2000) The impact of trawl fishery on the epifauna of the southern North Sea. *ICES Journal of Marine Science,* **57**,1389-1394.

195 Kaiser *et al.* (2002).

196 Lindeboom and de Groot (1998).

197 Duplisea, D.E., Jennings, S., Warr, K.J. and Dinmore, T. (2002). A size based model of the impacts of bottom trawling on benthic community structure. *Canadian Journal of Fisheries and Aquatic Science*, **59**,1785-1795.

198 Duplisea *et al.* (2002).
Gilkinson K., Paulin, M., Hurley S. and Schwinghamer, P. (1998). Impacts of trawl door scouring on infaunal bivalves: results of a physical trawl door model/dense sand interaction. *Journal of Experimental Marine Biology and Ecology*, **224**, 291-312.

199 Duplisea *et al.* (2002).

200 Brey T. (1999). Growth performance and mortality in aquatic macro-benthic invertebrates. *Advances in Marine Biology* **35**, 153-223.

201 Roberts, C.M. (2003);
Auster, P.J. and Langton, R.W. (1999). The effects of fishing on fish habitat. *American Fisheries Society Symposium* **22**, 150-187;
Lindholm, J.P., Auster, P.J. and Kaufman, L.S. (1999). Habitat-mediated survivorship of juvenile (0-year) Atlantic cod *Gadhus morhua. Marine Ecology Progress Series*, **180**, 247-255.

202 Kaiser *et al.* (1999)

203 Sainsbury, K.J., Campbell, R.A., Lindholm, R. and Whitelaw, A.W. (1997) Experimental management of an Australian multispecies fishery: examining the possibility of trawl induced habitat modification. In: *Global Trends: Fisheries Management* (eds K. Pikitch, D.D.Huppert and M.P. Sissenwine). American Fisheries Society, Bethesda, Maryland.

204 Kaiser *et al.* (2002).

205 Kaiser *et al.* (1999).

206 Kaiser *et al.* (2002).

207 Duplisea *et al.* (2002).

208 Kaiser *et al.* (2002).

209 Dinmore, T.A., Duplisea, D.E., Rackham, B.D., Maxwell, D.L. and Jennings, S. (2003). Impact of a large scale area closure on patterns of fishing disturbance and the consequences for benthic communities. *ICES Journal of Marine Science*, **60**, 371-380.

210 McAllister, D.E., Baquero, J. Spiller, G. and Campbell, R. (1999). *A global Trawling Ground Survey*. Marine Conservation Biology Institute, World Resources Institute and Ocean Voice International, Ottawa, Canada.

211 Figure from FSBI (2001). Marine protected areas in the North Sea. Briefing paper 1, Fisheries Society of the British Isles, Granta Information Systems, Cambridge, UK. Reproduced by Kind permission of Simon Jennings CEFAS.

212 Kaiser *et al.* (2002).

213 Dinmore *et al.* (2003).

214 Kaiser *et al.* (2002).

215 Duplisea, D.E., Jennings, S., Malcolm, S.J., Parker, R. and Sivyer, D. (2002a) Modelling the potential impacts of bottom trawl fisheries on soft sediment biochemistry in the North Sea. *Geochemical Transactions*, **14**,1-6.

216 Kaiser *et al.* (2002).

217 Duplisea *et al.* (2002a).

218 Roberts, C.M. (2003).

219 Kindly supplied by Dr Brian Bett, Deepseas Group, Southampton Oceanography Centre (© DTI).

220 Koslow, J.A., Boehlert, G.W., Gordon, J.D.M., Haedrich, R.L., Lorance, P. and Parin, N. (2000). Continental slope and deep-sea fisheries implications for a fragile ecosystem. *ICES Journal of Marine Science*, **57**, 548-557.
Koslow, J.A., Gowlett-Holmes, K., Lowry J.K., O'Hara, T., Poore, G.C.B. and Williams, A., (2001). Seamount benthic macrofauna off southern Tasmania: Community structure and impacts of trawling. *Marine Ecology Progress Series*, **213**, 111-125.

221 Large, P.A., Hammer, C., Bergstad, O.A., Gordon, J.D.M. and Lorance, P. (2003). Deep-water Fisheries of the Northeast Atlantic: II Assessment and management Approaches. *Journal of North West Atlantic Fisheries Science* **31**, 151-163.

222 Roberts, C.M. (2002). Deep impact: the rising toll of fishing in the deep-sea. *Trends in Ecology and Evolution* (Available at http://tree.trends.com).

223 Gubbay, S. (2003). *Seamounts of the North-East Atlantic.* OASIS, WWF, Germany, 37pp.

224 Rogers, A.D. (1994) The biology of seamounts. *Advances in Marine Biology* **30**,305-350.

225 Gubbay (2003).

226 Stone, G., L. Madin, K., Populations, G., Hovermale, P., Hoagland, M., Schumacher, C., Tausig, S. and Tausig, H. (2003). *Seamount Biodiversity, Exploration and Conservation.* Case Study Paper for Defying Ocean's End Conference, Cabo San Lucas, Mexico.

227 Lack, M., Short, K. and Willock, A. (2003). Report title: *Managing risk and uncertainty in deep-sea fisheries: lessons from Orange Roughy.* TRAFFIC Oceania and WWF Australia.

228 Rogers (1994);
Johnston, C.M. (2004). Scoping Study: *Protection of vulnerable high seas and deep oceans biodiversity and associated oceans governance.* Joint Nature Conservation Committee, Peterborough.

229 Bett, B.J. (1999). *RRS Charles Darwin Cruise 112C Ley2, 19 May – 24 June 1998. Atlantic Margin Environmental Survey: seabed survey of deep-water areas (17th round tranches) to the north and west of Scotland.* Southampton Oceanography Centre, Cruise report No25. 171pp
Bett, B.J. (2001). UK Atlantic Margin Environmental Survey: introduction and overview of bathyal benthic ecology. Continental Shelf Research, 21,917-956.

230 Figure from Peckett, F. (2003). *Lophelia Pertusa.* A cold water coral. Marine Life Information Network: Biology and Sensitivity Key Information Sub-programme (on-line). Plymouth: Marine Biological Association of the United Kingdom. Available from http://www.marlin.ac.uk/species/Lopheliapertusa.htm

231 Masson, D.G., Bett, B.J., Billett, D.S.M., Jacobs, C.L., Wheeler, A.J., Wynia, R.B. (2003). The origin of deep-water, coral topped mounds in the northern Rockall Trough, Northeast Atlantic. Marine Geology, 194, 159-180.

232 Fossa, J.H., Mortensen, P.B. and Furevik, D.M. (2002). The deep-water coral *Lophelia pertusa* in Norwegian Waters: distribution and fisheries impacts. *Hydrobiologia*, **471**, 1-12.
Bett, B.J., Billett, D.S.M., Masson, D.G., Tyler, P.A. (2001). RRS "Discovery" Cruise 248, 07 Jul – 10 Aug 2000. A multidisciplinary study of the environment and ecology of deep-water coral ecosystems and associated seabed facie sand features (The Darwin Mounds, Porcupine Bank and Porcupine Seabight). Southampton: Southampton Oceanography Centre. 108pp. Southampton Oceanography Centre Cruise Report; No 36.

233 Gubbay (2003).

234 Lack *et al.* (2003).

235   Southampton Oceanography Centre, (2003). *The Environmental Impact of Deep-Sea Demersal Fishing*. (Available at http://www.soc.scton.ac.uk/GDD/DEEPSEAS).

236   Hall-Spencer, J., Allain, V.& Fossa, J.H. (2002). Trawling damage to Northeast Atlantic ancient coral reefs. *Proceedings of the Royal Society of London B*, **269**, 507-511.

237   Figure from Gubbay (2003).

238   WWF/IUCN. (2001). *The status of Natural Resources on the high seas*. WWF and IUCN, Gland, Switzerland;
Lack *et al*. (2003).

239   Large, P.A., Hammer, C., Bergstad, O.A., Gordon, J.D.M. and Lorance, P. (2003).
ICES. (2001). *Report of the study group on the biology and assessment of deep-sea fisheries resources*. ICES C.M. Doc., ACFM, 23, 38 p.

240   Large *et al*. (2003).

241   Figure from Gubbay (2003).

242   Gubbay (2003).

243   Koslow, J.A., Boehlert, G.W., Gordon, J.D.M., Haedrich, R.L., Lorance, P. and Parin, N. (2000). Continental slope and deep-sea fisheries implications for a fragile ecosystem. *ICES Journal of Marine Science*, **57**, 548-557.
Koslow, J.A., K. Gowlett-Holmes, J.K. Lowry, T. O'Hara, G.C.B. Poore and A. Williams, (2001). Seamount benthic macrofauna off southern Tasmania: Community structure and impacts of trawling. *Marine Ecology Progress Series*. **213**, 111-125.

244   Marine Conservation and Biology institute website:
http://www.mcbi.org/DSC_statement/sign.htm

245   Pauly, D. and Christensen, V. (1995). Primary production required to sustain global fisheries. *Nature*, **374**, 255-257.

246   Vitousek, P.M., Ehrlich, P.R., Ehrlich, A.H. and Matson, P.A. (1986). Human appropriation of the products of photosynthesis. *BioScience* **36**(6), 368-373.

247   Jennings and Kaiser (1998).
Collie *et al*. (2000).
Kaiser *et al*. (2002).
Kaiser and de Groot (2000).
Schratzberger *et al*. (2002).

248   Frank, K.T. and Leggett, W.C. (1994). Fisheries ecology in the context of ecological theory and evolutionary theory. *Annual Review of Ecology and Systematics*, **25**, 401-422.

249   Smith, T.D. (1994).

250   Sherman K. and Alexander, L.M. (1986). *Variability and management of large ecosystems*. Westview Press. Boulder.

251   Cushing, D.H. 1988. *The Provident Sea*. Cambridge University Press, Cambridge UK.

252   Pitcher, T. (2001).

253   Pauly, D. (1995). "Anecdotes and the shifting baseline syndrome of fisheries." *Trends in Ecology and Evolution*, **10** (10),430.
Pauly, D., V. Christensen, J. Dalsgaard, R. Froese, and F. Torres, Jr. (1998). "Fishing down marine food webs." *Science*, **279**, 860-863

254   Jennings and Kaiser (1998).
        Collie *et al.* (2000).
        Kaiser *et al.* (2002).
        Kaise and de Groot (2000).
        Schratzberger *et al.* (2002).
255   Pitcher, T. (2001).
256   Pauly, D. (1995).

CHAPTER SIX

1   Bellona Foundation (2003). *The Environmental Status of Norwegian Aquaculture.* Bellona Report No. 7. Bellona Foundation, Norway.

2   IFPRI (2003).

3   University of Stirling, Institute of Aquaculture and Department of Marketing (2003). *The potential impact of technological innovation on the aquaculture industry.* Report to the Royal Commission on Environmental Pollution.

4   Scottish Executive (2003a). *A Strategic Framework for Scottish Aquaculture.* Scottish Executive, Edinburgh.

5   University of Stirling (2003).

6   IFPRI (2003).

7   University of Stirling (2003).

8   Naylor, R.L., Goldburg, R.J., Primavera, J.H., Katusky, N., Beveridge, M.C.M., Clay, J., Folke, C., Lubchenco, J., Mooney, H. and Troell, M. (2000). Effects of aquaculture on world fish supplies. *Nature*, **405**, 1017–1024.

9   University of Stirling (2003).

10   Bellona Foundation (2003).

11   University of Stirling (2003).

12   Barlow, S.M. (2002). *The world market overview of fishmeal and fish oil.* Paper presented to the Second Seafood By-products Conference, Alaska, November 2002. Available at: www.iffo.org.uk/tech/alaska.htm.

13   *Ibid.*

14   *Ibid.*

15   Tuominen, T.R. and Esmark, M. (2003). Food for Thought: The use of marine resources in fish feed. Report 2/03. WWF-Norway.

16   Barlow (2002).

17   IFPRI (2003).

18   University of Stirling (2003).

19   Hardy, R.W., Higgs, D.A., Lall, S.P. and Tacon A.G.J. (2001). *Alternative Dietary Protein and Lipid Sources for Sustainable Production of Salmonids.* Norwegian Institute of Marine Research.

20   University of Stirling (2003).

21   Fish oil and meal replacement (FORM) network website at: http://www.formnetwork.net/

22   Australian Government Fisheries Research and Development Corporation. Project 93/120-05. *Fishmeal Replacement in Aquaculture Feeds for Atlantic Salmon.* See websites at: http://www.agriculture.gov.au/product2.cfm?display2=Fisheries%20and%20Aquaculture http://www.frdc.com.au/pub/reports/files/93-120-05.htm

23   Tuominen and Esmark (2003).

24   *Ibid.*

25   See Norferm DA website at: http://www.norferm.com/norferm/svg02630.nsf

26   Hardy *et al.* (2001).

27   *Ibid.*

28   IFPRI (2003).

29   Figures available at: http://www.lantra.co.uk/scotland/

30   Sabaut, J.J. (2002). *Feeding Farmed Fish*. Paper presented to the Fisheries Committee of the European Parliament at its hearing on European Aquaculture, Brussels, 1 October 2002. Available at: http://www.feap.info/production/feeds/sabautcipa_en.asp

31   Stewart, J.A. (1995). *Assessing sustainability of aquaculture development*. Ph.D. Thesis, University of Stirling.

32   University of Stirling (2003).

33   Kapuscinski, A.R. (2002). *Genetic Impacts of Aquaculture*. Aquaculture and Environment Symposium, September 18 2002, Bordeaux.

34   Convention on Biological Diversity *ad hoc* Technical Expert Group on Mariculture (2002). *The Effects of Mariculture on Biodiversity*. Rome, 1-5 July 2002. UNEP/CBD/AHTEG-MAR/1/2.

35   Porter, G. (2003). *Protecting Wild Atlantic Salmon from Impacts of Salmon Farming. A country-by-country progress report*. World Wildlife Fund for Nature and Atlantic Salmon Federation. WWF, Washington DC

36   McGinnity, P., Prodöhl, P., Ferguson, A., Hynes, R., Ó Maoiléidigh, N., Baker, N., Cotter, D., O'Hea, B., Cooke, D., Rogan, G., Taggart, J. and Cross, T. (2003). Fitness reduction and potential extinction of wild populations of Atlantic salmon, *Salmo salar*, as a result of interactions with escaped farm salmon. Agriculture and Catchment Management Services Marine Institute, Newport, Ireland.

37   Evidence from the Association of Salmon Fishery Boards, May 2003.

38   *Scottish Salmon and Sea Trout Catches 2002*. Scottish Statistical Bulletin – Fisheries Series.

39   Scottish Executive (2003a).

40   Evidence from the Environment Agency, May 2003.

41   *Resolution by the Parties to the Convention for the Conservation of Salmon in the North Atlantic Ocean To Minimise Impacts from Aquaculture, Introductions and Transfers, and Transgenics on the Wild Salmon Stocks*. Council Paper CNL(03)57.

42   Porter (2003).

43   Personal communication from Professor Andrew Fergusson, April 2004.

44   Porter (2003).

45   Pew Initiative on Food and Biotechnology (2003). *Future Fish: Issues in science and regulation of transgenic fish*. Washington DC, cited as Pew Initiative on Food and Biotechnology (2003).

46   MacLean, N. and Laight, R.J. (2000). Transgenic fish: an evaluation of benefits and risks. *Fish and Fisheries*, **1,**146–172.

47   Goldburg, R. and Triplett, T. (1997). *Murky Waters: Environmental effects of aquaculture in the US*. The Environmental Defense Fund.

48   Compassion in World Farming (2002). *In Too Deep – The Welfare of Intensively Farmed Fish*. Lymbery, UK.

49   Pew Initiative on Food and Biotechnology (2003).

50   *Ibid.*

51   Pew Oceans Commission (2003). *America's Living Oceans – Charting a Course for Sea Change*. A report to the nation. Arlington, Virginia.

52   Department for Environment, Food and Rural Affairs (Defra) website: http://www.defra.gov.uk/environment/gm/background/animals.htm

53   Scottish Executive (2003a).

54   Agriculture and Environment Biotechnology Commission (2002). *Animals and Biotechnology.* DTI, London. Available at: http://www.aebc.gov.uk/aebc/pdf/animals_and_biotechnology_report.pdf

55   Joint Group of Experts on the Scientific Aspects of Marine Environmental Protection (1986). *Environmental Capacity. An approach to marine pollution prevention.* GESAMP, Reports and Studies (30):49p.

56   Scottish Parliament Environment and Transport Committee (2002). 5th Report. *Report on Phase 1 of the Inquiry into Aquaculture,* cited as Scottish Parliament Environment and Transport Committee.

57   Evidence from the Scottish Environment Protection Agency (SEPA), May 2003.

58   Scottish Parliament Environment and Transport Committee (2002).

59   Figure adapted from Nautilus Consultants (2002). *Environmental risk assessment and communication in coastal aquaculture.* Available at: http://www.nautilus-consultants.co.uk/pdfs/Aqua%20Env%20Risk%20Assessment.pdf

60   Figure supplied by University of Aberdeen Department of Zoology.

61   EC Council Directive on 28 January 1991 concerning the animal health conditions governing the placing on the market of aquaculture animals and products (91/67/EEC). *Official Journal of the European Communities,* **L046.**

62   Information from Centre for Environment, Fisheries and Aquaculture Science (CEFAS) website at: http://www.cefas.co.uk/fhi/background.htm

63   Farm Animal Welfare Council (1996). *Report on the Welfare of Farmed Fish.* Report Number 2765. Farm Animal Welfare Council, UK

64   *Ibid.*

65   Personal communication from A. Rosie, SEPA, June 2004.

66   Veterinary Medicines Directorate (2003). *Sales of antimicrobial products authorised for use as veterinary medicines, antiprotozoals, antifungals, growth promoters and cocciostats in the UK in 2002.* Veterinary Medicines Directorate, UK.

67   The Scottish Association for Marine Science (SAMS) and Napier University (2002). *Review and Synthesis of the Environmental Impacts of Aquaculture.* Scottish Executive Central Research Unit.

68   Bellona Foundation (2003).

69   SAMS/Napier University (2002).

70   Bellona Foundation (2003).

71   *Ibid.*

72   Davenport, J., Black, K., Burnell, G., Cross, T., Culloty, S., Ekaratne, S., Furness, B., Mulcahy, M. and Thetmeyer, H. (2003). *Aquaculture – The Ecological Issues.* British Ecological Society.

73   SAMS/Napier University (2002).

74   Bellona Foundation (2003).

75   SAMS/Napier University (2002).

76   Bellona Foundation (2003).

77   University of Stirling (2003).

78   University of Stirling (2003).

79   SAMS/Napier University (2002).

80   *Ibid.*

81   Evidence from the British Geological Survey, May 2003.

82   Dolapsakis, N.P. (1996). *Primary resources and aquaculture development beyond the year 2000*. M.Sc. Thesis, University of Stirling.

83   Bellona Foundation (2003).

84   Source for figure 6-VII: United Nations Economic Commission for Europe (UNECE) Kiev Report. Sources: FAO Fishstat Plus; Jonsson and Alanara, 1998; Ospar Commission, 2000; Haugen and Englestad, 2001; Beveridge, pers. comm.; HELCOM, 1998
Notes: The data on 'other coastal nutrient discharges' comprise riverine inputs and direct discharges as reported for 1999 in the OSPAR Study on Riverine Inputs and Direct Discharges (RID). Nutrient discharge from mariculture is estimated from production using the mid-range of values stated in the OSPAR report (Ospar Commission, 2000) (55g N/kg production and 7.5g P/kg production). The figures for Finland are based upon the HELCOM 1998 data. Nitrogen limited to riverine discharge only (no data on direct inputs). Phosphorus discharge: average of lower and upper estimates. Total N for riverine discharge estimated as NH3- N+NO3-N. This will overestimate the relative N discharge from aquaculture. Nutrient discharge applicable to sea areas in which the bulk of marine and/or brackish water finfish aquaculture takes place have been used. These figures do not include N and P discharges from inland aquaculture production. Production figures relate to marine species only, except Finland, which refer to brackish water production.

85   Tett, P. and Edwards V. (2002). *Review of Harmful Algal Blooms in Scottish Coastal Waters*. Report to SEPA, June 2002.

86   See IFREMER website, www.ifremer.fr/envlit/documentation/dossiers/ciem/aciem-c1.htm. Copyright Ifremer.

87   Rydberg, L., Sjöberg, B. and Stigebrandt, A. (2003). *The Interaction between Fish Farming and Algal Communities of the Scottish Waters – A Review*. Final Report to the Scottish Executive; Environment Group Research Report 2003/04.

88   Kennedy, C.M. (1994). An integrated approach to control systems and risk management practical observations. In: *Measures for Success*. Ed. P. Kestemont, J.F. Muir, F. Sevila and P. Williot. CEMAGREF, Paris. Pages 37-41.

89   Farm Animal Welfare Council (1996).

90   Adapted from *Recommendations on Salmon and Trout* in Farm Animal Welfare Council (1996); available at: http://www.fawc.org.uk/reports/fish/fishrtoc.htm

91   Compassion in World Farming (2002).

92   Defra Science Directorate (2002). *Report of the Workshop on Farmed Fish Welfare*. 28 October 2002.

93   University of Stirling (2003).

94   *Ibid.*

95   SAMS/Napier University (2002).

96   Evidence from the Environment Agency, May 2003.

97   Oral evidence from Defra, May 2004.

98   Defra and National Statistics (2002). *UK Sea Fisheries Statistics 2001*. HMSO, London.

99   Dankers and Zuidema (1995). The role of the mussel and mussel culture in the Dutch Wadden sea. *Estuaries*, **18**,71-80.

100   Evidence from the Joint Nature Conservation Committee (JNCC), June 2003.

101   The Food Safety (Live Bivalve Molluscs and Other Shellfish) Regulations 1992.

102   Evidence from the Environment Agency, May 2003.

103   Evidence from SEAFish, July 2003.

104   Oral evidence from Defra, May 2004.

105   Evidence from JNCC, June 2003.

106   Davies, I.M. and McLeod, D. (2003). *Scoping study for research into aquaculture (shellfish) carrying capacity of GB coastal waters.* Aberdeen Collaborative Report No. 04/03. FRS Marine Laboratory, Aberdeen; cited as Davis and McLeod (2003).

107   SEERAD (2003).

108   Gardiner and Egglishaw (1986). *Map of distribution in Scottish Rivers of the Atlantic Salmon.* Scottish Fisheries Publication.

109   Davies and McLeod (2003).

110   Scottish Executive (2003a).

111   Evidence from the Scottish Executive, May 2003.

112   Evidence from the Association of Salmon Fishery Boards, May 2003.

113   *A strategy for the sustainable development of European aquaculture.* Communication from the European Commission to the European Council and the European Parliament. COM(02) 511.

CHAPTER SEVEN

1   Sissenwine, M.P. and Mace, P.M. (2003). Governance for responsible fisheries: An ecosystem approach. In: *Responsible Fisheries in the Marine Ecosystem.* Eds. M. Sinclair and G. Valdimarsson. FAO, Rome.

2   Oral evidence from Dr Joe Horwood, Centre for Environment, Fisheries and Aquaculture Science (CEFAS) (November 2003).

3   Roberts, C.M. (1997). Ecological advice for the global fisheries crisis. *Trends in Ecology and Evolution,* **12**, 35–38.

4   Royal Society of Edinburgh (RSE). (2004). Inquiry into the Scottish Fishing Industry. RSE, Edinburgh.

5   Christensen, V. and Walters, C. (2003). *ECOPATH with ECOSIM methods capabilities and limitations.* Sea Around Us Project Methodology Review. University of British Columbia.

6   De Young, B., Heath, M., Werner, F., Chai, F., Megrey, B. and Monfray, P. (2004). Challenges of modelling ocean basin ecosystems. *Science,* **304**, 463–1466.
     Patterson, K.R., Cook, R.M., Darby C.D., Gavarais, S., Mesnil, B., Punt, A.E., Restrepo, V.R., Skagen, D.W., Stefansson, G. and Smith, M. (2000). *Validating three methods for making probability statements in fisheries forecasting.* ICES CM 2000. International Council for the Exploration of the Sea (ICES), Bruges, Belgium. 27 and 30 September 2000.

7   De Young *et al.* (2004).

8   Patterson *et al.* (2000).

9   Yodzis, P. (2001). Must top predators be culled for the sake of fisheries? *Trends in Ecology and Evolution,* **16**, 78–84;
     Robinson, L.A. and Frid, C.L.J. (2003). Dynamic ecosystem models and the evaluation of ecosystem effects of fishing: can we make meaningful predictions? *Aquatic Conservation Marine Freshwater Ecosystems,* **13**, 5–20.

10   Degnbhol, P. (2002). *The ecosystem approach and fisheries management institutions: The noble art of addressing complexity and uncertainty with all onboard and on a budget.* International Institute of Fisheries Economics and Trade (IIFET) Paper No. 171.

11   ICES (2003).

12   Pontecorvo, G. (2003). Insularity of scientific disciplines and uncertainty about supply: The two keys to the failure of fisheries management. *Marine Policy,* **27,** 69–73.

13   Sharp, G.D. (2000). The past, present and future of fisheries oceanography: Refashioning a responsible fisheries science. In: *Fisheries Oceanography: An integrative Approach to Fisheries Ecology and Management.* Eds. P.J. Harrison and T.R. Parsons. Blackwell Sciences, Oxford.

14   Pitcher, T. (2001).

15   Pontecorvo (2003).

16   Hutchings, J.A., Walters, C. and Haedrich, R.L. (1997). Is scientific inquiry incompatible with government information control? *Canadian Journal of Fisheries and Aquatic Science,* **54**, 1198–1210;
     European Environment Agency (EEA) (2001). *Late lessons from early warnings: The precautionary principle 1896-2000.* Environmental Issue Report No. 22.

17   Pikitch, K., Santora, C., Babcock, E.A., Bakun, A., Bonfil, R., Conover, D.O., Dayton, P., Doukakis, P., Fluharty, D., Heneman, B., Houde, E.D., Link, J., Livingston, P.A., Mangel, M., McAllister, M.K., Pope, J. and Sainsbury, K.J. (2004). Ecosystem based fishery management. *Science*, **305**, 346–347.

18   Jennings and Blanchard (2004).

19   Van Winkle, W., Rose, K.A. and Chambers, R.C. (1993). Individual based approach to fish population dynamics: An overview. *Transactions of the American Fisheries Society*, **122**, 397–403;
DeYoung *et al.* (2004).

20   Woods, J. (2004). Predicting fisheries in the context of the ecosystem. *Philosophical Transactions of the Royal Society of London – Biology* (*in Press*)

21   Fulton, E.A. (2001). *The effects of model structure and complexity on the behaviour and performance of marine ecosystem models*. Ph.D. Thesis. School of Zoology, University of Tasmania;
Fulton, E.A., Smith, A.D.M. and Johnson, C.R. (2003). Effect of complexity on marine ecosystem models. *Marine Ecology Progress Series*, **253,** 1–16.

22   Walters, C., Pauly, D. and Christensen, V. (1999). Ecospace: Prediction of mesoscale spatial patterns in trophic relationships of exploited ecosystems, with emphasis on the impacts of marine protected areas. *Ecosystems*, **2**, 539–554.

23   Cox, P.M., Betts, R.A., Jones, C.D., Spall, S.A. and Totterdell, I.J. (2000). Acceleration of global warming due to carbon-cycle feedbacks in a coupled climate model. *Nature*, **408**, 184–187;
Palmer, J.R. and Totterdell, J. (2001). Production and export in a global ecosystem model. *Deep-Sea Research*, **48**, 1169–1198.

24   Allen, J.I., Blackford, J., Holt, J., Proctor, R., Ashworth, M. and Siddorn, J. (2001). A highly spatially resolved ecosystem model for the North West European Continental Shelf. *Sarsia,* **86**,423–440.

25   Borges, M.F., Moustakas, A., Salgado, M., Silvert, W., Taylor, L., Sigurdardottir, A.J., Olafsdottir, E.I. and Begley, J. (2004). *European Fisheries Ecosystem Plan: Managing uncertainty in ecosystem model dynamics and the implications and feasibility of specific management scenarios.* European Project No. Q5RS-2001-01685.

26   Halliday, R.G., Fanning, L.P. and Mohn, R.K. (2001). Use of traffic light method in fisheries management planning. S.B. Marine Fish Division, Scotia-Fundy Region, Department of Fisheries and Oceans, Bedford Institute of Oceanography, Dartmouth, Canadian Science Advisory Secretariat;
Hall, S.J. and Mainprize, B. (2004). Towards ecosystem-based fisheries management. *Fish and Fisheries,* **5**, 1–20.

27   Caddy, J.F. and Mahon, R. (1995). *Reference points for fisheries management.* Food and Agriculture Organizations (FAO) Fisheries Technical Paper 347. FAO, Rome;
Hall and Mainprize (2004).

28   ICES (2001a). *Report of the Working Group on Ecosystem Effects of Fishing*. ICES CM/ACME/E:09.

29   Figure supplied by Dr Simon Jennings, CEFAS.

30   Cochrane, K.J. (2002). The use of scientific information in the design of management strategies. Chapter 5 in: *A Fishery Manager's Guidebook. Management measures and their application.* Ed. K.J. Cochrane. FAO Fisheries Technical Paper 424. FAO, Rome; Hall and Mainprize (2004).

31   Garcia, S.M. (1995). *Precautionary approach to fisheries. Part II: Scientific papers.* FAO Fisheries Technical Paper 350/2. FAO, Rome; Hall and Mainprize (2004).

32   Caddy, J.F. (1998). *A short review of precautionary reference points and some proposals for their use in data-poor situation.* FAO Fisheries Technical Paper 379. Pages 1-29. FAO, Rome.

33   Jennings and Kaiser (1998). Rice, J. (2003). Environmental health indicators. *Ocean and Coastal Management,* **46,** 235–259.

34   Hall and Mainprize (2004).

35   Hall and Mainprize (2004).

36   ICES (2003b). *Report of the Working Group on Ecosystem Effects of Fishing Activities.* ICES CM 2003/ACE:05. Copenhagen.

37   Hall and Mainprize (2004). ICES (1997). *Report of the Comprehensive Fishery Evaluation Group.* Advisory Committtee on Fishery Management. ICES CM 1997/Assess:15.

38   Hall, S.J. (1999). *The Effects of Fishing on Marine Ecosystems and Communities.* Blackwell, Oxford; University of Alsaka (1999). *Ecosystem approaches for fisheries management.* University of Alaska Sea Grand, Fairbanks. AK-SG-99-01. ICES (2000). Ecosystem effect of fishing. Proceedings of an ICES/SCOR Symposium held in Montpellier, France, 16–19 March 1999. *ICES Journal of Marine Science,* **57,** 465–792; Trenkel, V.M. and Rochet, M.J. (2003) Performance of indicators derived from abundance estimates for detecting the impact of fishing on a fish community. *Canadian Journal of Fisheries and Aquatic Science,* **60**(1), 67–85.

39   ICES (2001); Hall and Mainprize (2004).

40   Die, D.J. and Caddy, J.F. (1997). Sustainable yield indicators from biomass: Are there appropriate reference points for use in tropical fisheries? *Fisheries Research,* **32,** 69–79; FAO (1999). *Indicators for sustainable development of marine capture fisheries.* FAO Technical Guidelines for Responsible Fisheries 8; Rice, J. (2000). Evaluating fishery impacts using metrics of community structure. *ICES Journal of Marine Science,* **57,** 682–688; Frid, C. and Robinson, L. (2000). *Ecological reference points for North Sea benthos: can we manage benthic biodiversity?* ICES CM 2000/Mini:01; Frid, C., Rogers, S., Nicholson, M., Ellis, J. and Freeman, S. (2000). *Using biological characteristics to develop new indices of ecosystem health.* ICES CM 2000/Mini:02; Trenkel and Rochet (2003); Rochet, M.J. and Trenkel, V.M.(submitted). Which community indicators can measure the impact of fishing? A review and proposals. *Canadian Journal of Fisheries and Aquatic Science*

41   Rice, J. (2003).

42   ICES (2001);
      Hall and Mainprize (2004).

43   Degnbhol, P. (2002);
      Hall and Mainprize (2004).

44   Sissenwine and Mace (2003);
      Defra (2004).

45   FAO (2001). Reykjavik Conference on Responsible Fisheries in the Marine Ecosystem,
      1-4 October 2001. See website at: http://www.refisheries2001.org

46   Garcia, S.M. and de Leiva Moreno, I. (2003). Global overview of marine fisheries. In:
      Responsible fisheries in the marine ecosystem. Eds. M. Sinclair and G. Valdimarsson. FAO,
      Rome; CABI Publishing, Wallingford. Pages 1–24.

47   Hall, M.A. (1998). An ecological view of the tuna–dolphin problem – Impacts and trade-
      offs. *Review of Fish Biology and Fisheries,* **8,** 1–34:
      Casey, J.M. and Myers, R.A. (1998). Near extinction of a large widely distributed fish.
      Science, **281,** 690–691;
      Lindeboom and de Groot (1998).
      Collie *et al.* (2000).
      Law, R. (2000).
      Tasker, M.L., Camphuysen, C.J., Cooper, J., Garthe, S., Montevecchi, W.A. and Blaber,
      S.J.M. (2000). The impacts of fishing on marine birds. *ICES Journal of Marine Science,*
      57, 531–547;
      Kaiser *et al.* (2002).
      Schindler, D.E., Essington, T.E., Kitchell, J.F., Boggs, C. and Hilborn, R. (2002). Sharks
      and tunas: Fisheries impacts on predators with contrasting life histories. *Ecological
      Applications,* **12,** 735–748;
      Schratzberger *et al.* (2002).
      Dulvy *et al.* (2003).
      Kenchington *et al.* (2003).
      ICES (2004). *Report of the Working Group on the Applications of Genetics in Fisheries and
      Mariculture (WGAGFM).* ICES CM 2004/F: 04, Ref. ACFM, ACME, I.

48   Laffoley, D.d'A., Maltby, E. Vincent, M.A., Dunn, E., Gilland, P., Hamer, J. and Pound, D.
      (2004) *Sustaining benefits, for all, forever. Realising the benefits of the Ecosystem
      Approach: priorities for action in maritime environments.* A report to the UK
      Government and the European Commission. English Nature, Peterborough.

49   CBD (2000). Conference of Parties, Fifth meeting, decision V6. Available on CBD
      website at http://www.biodiversity.org/decisions/default.aspx?lg=O&m=cop-05&d=06

50   CBD (2003) Report of the Expert Meeting on the Ecosystem Approach
      (UNEP/CBD/SBSTTA/9/INF/4), available at:
      http://www.biodiv.org/doc/meetings/sbstta/sbstta-09/information/sbstta-09-inf-04-en.doc

51   Personal communication, David Griffith, ICES general secretary, 2004.

52   European Marine Strategy (2004). *Guidance to the application of the Ecosystem Approach
      to Management of human activities.* EAM(3) 04/2/1.

53   Pitcher, P. (2001).

54   Defra (2002).

55   Laffoley *et al.* (2004).

56   Oral evidence from CEFAS, date 2003.

57   Degnbhol, P. (2002).

58   WWF (2002). *Policy Proposals and Operational Guidance for Eosystem-Based Management of Marine Capture Fisheries.* World Wide Fund For Nature, Australia; European Marine Strategy (2004).

59   Jennings, S. (2004). The ecosystem approach to fishery management: A significant step towards sustainable use of the marine environment? *Marine Ecology Progress Series*, **274,** 269–303.

60   Degnbhol, P. (2002).

61   Walters, C. (1997). Challenges in adaptive riparian and coastal ecosystems. Conservation Ecology, 1(1), available online at http://www.consecol.org/voll/iss2/artl.htm
      Lee, K.N. (1993). *Compass and gyroscope: integrating science and politics for the environment.* Island Press, Washington, D.C.

62   Walker, B.S., Carpenter, J., Andries, N., Abel, G.S., Cumming, M., Jassen, L., Lebel, G., Norberg, D., Petersen, R. and Richard, R. (2002). Resilience management in social-ecologcial systems: a working hypothesis for the participatory approach. Conservation Ecology (on-line) 6(1): 14.
      Laffoley *et al.* (2004).

63   Mee, L.D. (2004 in press). Assessment and monitoring requirements for the adaptive management of Europe's regional seas. In: Turner and Salomons (Eds.) *Managing European Coasts: Past, present and future.* Elsevier. London.

64   Website of WWF: www.wwf.org.uk/investinfish/investinfish.pdf

65   *Ibid.*

66   European Marine Strategy (2004).

67   Degnbhol, P. (2002).

68   ICES (1998). *Report of the ICES Advisory Committee on Fishery Management, 1997. Part 1.* ICES Cooperative Research Report No. 223. ICES, Copenhagen.

69   Dunn, E. (1998). The case for greater environmental safeguards in the revision of the Common Fisheries Policy. *Marine Environmental Management Review of 1997 and Future Trends*, **5**, Paper 20, 144146.

70   Bailey, P.D. (2000). *Discourse and the Regulation of the Environment and Technology: Overfishing and Vessel Monitoring in European Fisheries.* Institute for Fisheries Management Research Report No. 54.

71   Hutchings *et al.* (1997);
      EEA (2001).

72   *Ibid.*

73   Available at: http://europa.eu.int/comm/fisheries/doc_et_publ/factsheets/legal_texts/docscom/en/com_04_58_en.pdf

74   Personal communication from Professor Callum Roberts, 2004.

75   Laffoley *et al.* (2004).

76   *Ibid.*

77   *Ibid.*

78   ICES (2001);
      Cochrane, K.J. (2002);
      Hall and Mainprize (2004).

79   Degnbhol, P. (2002).

311

80   Pitcher, T.J. and Preikshot, D.B. (2001). RAPFISH: A rapid appraisal technique to evaluate the sustainable status of fisheries. *Fisheries Research*, **49,** 255–270;
      Link, J.S., Brodziak, J.K.T. and Edwards, S.F. (2002). Marine ecosystem assessment in a fisheries management context. *Canadian Journal of Fisheries and Aquatic Science*, **59**, 1429–1440;
      Hall and Mainprize (2004).

81   Rice, J. (2000);
      ICES (2001);
      Rice, J.C. (2001). *From science to advice – how to find ecosystem metrics to support management*. ICES CM/T:12;
      Hall and Mainprize (2004).

82   Garcia, S.M. (1996). *The Precautionary approach to fisheries, its implications for fishery research, technology and management: an updated review*. FAO Fisheries Technical Paper No. 350, Part 2. Pages 1–76;
      Rice, J. (2003).

83   Laffoley, *et al.* (2004).

84   Jennings, S. (2004).

85   Symes, D. (1998). *The Integration of Fisheries Management and Marine Wildlife Conservation*. JNCC Report No. 287.

86   Witherell, D., Pautzke, C. and Fluharty, D. (2000). An ecosystem-based approach for Alaska groundfish fisheries. *ICES Journal of Marine Science*, **57**, 771–777;
      Pikitch *et al.* (2004).

87   Degnbhol, P. (2002).

CHAPTER EIGHT

1   IUCN (1994).

2   Palumbi, S. (2002). *Marine Reserves. A tool for ecosystem management and conservation.* Pew Oceans Commission, Arlington, Virginia.

3   Following a recommendation of the Bahamas National Trust, the Bahamas Government established the Exuma Cays Land and Sea Park in 1958, one of the first of its kind anywhere in the world.

4   Lubchenco, J., Palumbi, S.R., Gaines, S.D. and Andelman, S. (2003). Plugging a hole in the Ocean: The emerging science of marine reserves. *Ecological Applications* **13**(1) Supplement S3-S7.

5   Hastings, A. and Botsford, L. (1999). Equivalence in yield from marine reserves and traditional fisheries management. *Science* **284**, 1-2.

6   Roberts, C.M. and Hawkins, J.P. (2000). *Fully protected marine reserves; a guide.* WWF Endangered Seas Campaign, Washington DC, and Environment Department, University of York, UK.

7   United Nations (2003). List of Protected Areas 2003. IUCN, Gland, Switzerland and UNEP World Conservation Monitoring Centre, Cambridge, UK.

8   Evidence from the Centre for Environment, Fisheries & Aquaculture Science, May 2003.

9   Day, J.C. (2002). Zoning – lessons from the Great Barrier Reef Marine Park. *Ocean & Coastal Management* **45** 139-156. Original figures were A$1.2billion, A$700 million, A$250 million, and A$270 million respectively. At October 2004 conversion rates 1A$ = approx. 0.39 UK£; Figure 8-I from GBRMPA website, RAP poster available at www.gbrmpa.gov.au/corp_site/management/zoning/documents/Rap_poster.pdf. Copyright GBRMPA.

10   *Ibid.*

11   Gell, F.R. and Roberts, C.M. (2003a) Benefits beyond boundaries: the fishery effects of marine reserves. *Trends in Ecology and Evolution.* **18**, 148-155;
     Roberts, C.M., Gell, F.R. and Hawkins, J.P. (2003) *Protecting nationally important marine areas in the Irish Sea Pilot Project Region.* Report for the JNCC. Available at http://www.jncc.gov.uk/marine/irishsea_pilot/pdfs/finalreports_2004/york/default.htm

12   Executive Order 13158. (2000). Marine Protected Areas. *Federal Register,* **65** (105), 34909-11. Washington, DC.

13   Website for the US National Marine Protected Areas Center, www.mpa.gov

14   Halpern, B.S. (2003). The impact of Marine Reserves: Do reserves work and does reserve size matter? *Ecological Applications* **13**(1) Supplement: S113-S137.

15   National Research Council (NRC) (1999). *Sustaining Marine Fisheries.* National Academy Press, Washington DC.
     NRC (2001). *Marine Protected Areas: tools for sustaining ecosystems.* National Academy Press, Washington DC.
     NRC. (2002). *Effects of trawling and dredging on seafloor habitat.* National Academy Press, Washington DC.

16   Gell, F.R. and Roberts, C.M. (2003b). *The fishery effects of marine reserves and fishery closures,* WWF-US (available at http://www.worldwildlife.org/oceans/fishery_effects.pdf)

17   Laffoley *et al.* (2004).

18   Halpern, B.S. and Warner, R.R. (2002). Marine Reserves have rapid and lasting effects. *Ecological Letters* **5**: 361-366;
     Gell and Roberts (2003a).

313

19   Yamaski, A. and Kuwarhara, A. (1990). Preserved area to effect recovery of over-fished Zuwai crab stocks off Kyoto Prefecture. In: *Proceedings of the International Symposium on King and Tanner Crabs*, November 1989, Anchorage, Alaska pp. 575-585 University of Alaska Fairbanks, Alaska Sea Grant Program Report **90-04**.

20   Rowe, S. (2001). Movement and harvesting mortality of American lobsters (*Homarus americanus*) tagged inside and outside no-take reserves in Bonavista Bay, Newfoundland. *Canadian Journal of Fisheries and Aquatic Science.* **58**, 1336–1346

21   Kelly, S. (2001). Temporal variation in the movement of the spiny lobster, *Jasus edwardsii. Marine Freshwater Research.* **52**, 323–331.

22   Willis, T.J., Parsons, D.M. Babcock, R.C. et al. (2001). Evidence of long-term site fidelity of snapper (*Pagrus auratus*) within a marine reserve. *New Zealand Journal of Marine Freshwater Research* **35**, 581–590.

23   McClanahan, T.R. and Mangi, S. (2000). Spillover of exploitable fishes from a marine park and its effect on the adjacent fishery. *Ecological Applications* **10**, 1792–1805.

24   Francour, P. (1991). The effect of protection level on a coastal fish community at Scandola. Corsica. *Revue d'Ecologie (la Terre et la Vie)* **46**, 65–81.

25   Goni, R., Renones, O., Quetglas, A. and Mas, J. (2001). Effects of protection on the abundance and distribution of red lobster (*Palinurus elephas, Fabricius*, 1787) in the marine reserve of Columbretes Islands (Western Mediterraneo) and surrounding areas. In: *First International Workshop in Marine Reserves, Murcia, Spain 1999*, pp. 117–133, Ministerio de Agricultura, Pesca y Alimentacio´n, Spain.

26   Cowley, P.D., Browser, S.L. and Tilney, R. T. (2002). The role of the Tsitsikamma National Park in the management of four shore-angling fish along the south-eastern cape coast of South Africa. *South African Journal of Marine Science* **24**, 27–36.

27   Russ, G.R. and Alcala, A.C. (1996). Marine reserves: rates and patterns of recovery and decline of large predatory fish. *Ecological Applications.* **6**, 947–961.

28   Davidson, R.J. (2001). Changes in population parameters and behaviour of blue cod (*Parapercis colias; Pinguipedidae*) in Long Island – Lokomohua Marine Reserve, Marlborough Sounds, New Zealand. *Aquatic Conservation Marine & Freshwater Ecosystems.* **11**, 417–435.

29   Edgar, G.J. and Barrett, N.S. (1999). Effects of the declaration of marine reserves on Tasmanian reef fishes, invertebrates and plants. *Journal of Experimental Marine Biology and Ecology.* **242**, 107–144.

30   Faunce, C.H., Lorenz, J.J., Ley, J.A. and Serafy, J.E. (2002). Size structure of gray snapper (*Lutjanus griseus*) within a mangrove 'no-take' sanctuary. *Bulletin of Marine Science* **70**, 211–216.

31   Kelly, S., Scott, D., MacDiarmid, A.B., Babcock, R.C. (2000). Spiny Lobster, *Jasus edward sii*, recovery in New Zealand marine reserves. *Biological Conservation* 92, 359-369.

32   Willis, T.J., Miller, R.B., and Babcock, R.C. (2003). Protection of exploited fishes in temperate regions: high density and biomass of snapper *Pagrus auratus (Sparidae)* in northern New Zealand marine reserves. *Journal of Applied Ecology.* **40**, 214–227.

33   Palsson, W.A. and Pacunski, R.E. (1995). The Response of Rocky Reef Fishes to Harvest Refugia in Puget Sound. *Proceedings Puget Sound Research*, **1** Puget Sound Water Quality Authority.

34   Gell and Roberts (2003b).

35   Roberts and Hawkins (2000).

36   *Ibid.*

37   ICES WEGCO (2004). *Report of the Working Group on Ecosystem Effects of Fishing Activities*, 14-21 April 2004. ICES, Copenhagen.

38   Murawski, S.A, Rago, P. and Fogarty, M. (2004). Spillover effects from temperate marine protected areas. In: Shipley, J.B. Ed. *Aquatic Protected areas as fishery management tools.* American Fisheries Society, Symposium 42, Bethesda MD.

39   Murawski, S.A., Brown, R. Lai, H-L., Rago, P.J. and Hendrickson, L. (2000). Large-scale closed areas as a fisheries management tool in temperate marine systems: the Georges Bank experience. *Bulletin of Marine Science*, **66**, 775-798.

40   Gell and Roberts (2003a).

41   Website of the Wadden Sea Secretariat, http://www.waddensea-secretariat.org/tgc/pssa/pssa-designation.html.

42   Greenpeace (2004).

43   Rogers, S. I. (1997). A review of closed areas in the United Kingdom Exclusive Economic Zone. *Science Series Technical Report* No. **106**. Centre for Environment Fisheries and Aquaculture Science, Lowestoft, UK.

44   Gibson, J. and Warren, L. (1995). Legislative requirements. In: *Marine protected areas: principles and techniques for management.* Ed S. Gubbay, Chapman and Hall, London, UK, p32-60.

45   Website of UK Marine Special Areas of Conservation Project, www.ukmarinesac.org.uk/uk-sites.htm. Website of JNCC, www.jncc.gov.uk/ProtectedSites/marine

46   Personal communication, Steve Atkins, JNCC, March 2004.

47   D. Laffolley personal communication to RCEP, 2004.

48   Coffey and Richartz (2004).

49   *Ibid.*

50   *Ibid.*

51   Bradshaw, C., Veale, L.O., Hill, A.S. and Brand, A.R. (2000). The effects of scallop dredging on gravely sea-bed communities. In: *Effect of Fishing on Non-target Species and Habitats.* Eds M.J.Kaiser and S.J. De Groot. Blackwell Science, Oxford, pp. 83-104.
      Bradshaw, C., Veale, L.O., Hill, A.S. and Brand, A.R. (2001). The effect of scallop dredging on the Irish Sea benthos: experiments using a closed area. *Hydrobiologia* **465**, 129-138.
      Bradshaw, C., Veale, L.O., Hill, A.S. and Brand, A.R. (2002). The role of scallop dredge disturbance in long-term changes in Irish Sea benthic communities: a reanalysis of an historical dataset. *Journal of Sea Research* **47**: 161-184.

52   Blyth, R.E., Kaiser, M.J., Edwards-Jones, G. and Hart, P.J.B. (2002). Voluntary management in an inshore fishery has conservation benefits. *Environmental Conservation* **29**, 493-508.

53   Lewis, M. (2003). St Abbs-Eyemouth area fixed gear reserve. *Marine Conservation* **6**-11.

54   Stanford, R. (2003). *Lyme Bay reefs grow back! Marine Conservation* **6**, 11.

55   Lock, K., Newman, P. (2001). *Skomer Marine Nature Nature Reserve scallop survey 2000.* Countryside Council for Wales Report.

56   Fisheries Society of the British Isles (2001). *Marine protected areas in the North Sea.* Briefing Paper 1, Fisheries Society of the British Isles, Granta Information Systems, Cambridge.

57   ICES WEGCO (2004).

58   *Ibid.*

59   Ballantine, W.J. (1999). *Marine reserves in New Zealand. The Development of concept and principles.* Paper for workshop on MPAs: KORDI, Korea, November 1999.
       Roberts, Gell and Hawkins (2004).

60   Roberts, C.M. (2000). Selecting marine reserve locations: optimality versus opportunism. *Bulletin of Marine Science,* **66**(3): 581-592.

61   Great Barrier Reef Marine Park Authority (2003). *Biophysical Operational Principles as recommended by the Scientific Steering Committee for the Representative Areas Program.* Available on Great Barrier Reef Marine Park Authority website - http://www.gbrmpa.gov.au/corp_site/key_issues/conservation/rep_areas/info_sheets.html #Biophysical%20operational%20principles

62   Ocean Studies Board Commission on Geosciences, Environment and Resources National Research Council. *Marine protected areas: tools for sustaining ocean ecosystems.Committee on the Evaluation, Design, and Monitoring of Marine Reserves and Protected Areas in the United States.* Relevant except available from secretariat on request.

63   Marine Conservation Biological Institute (1997). *Troubled Waters: A call for Action.*

64   Great Barrier Reef Marine Park Authority (2003).

65   Pauly and Maclean (2003).

66   Gell and Roberts (2003b).

67   Ball, I. and Possingham, H. (2000). *Marine Reserve Design using Spatially Explicit Annealing. A Manual prepared for the Great Barrier Reef Marine Park Authority.* MARXAN (v 1.8).

68   Website of the Great Barrier Reef Marine Park Authority, http://www.gbrmpa.gov.au/corp_site/management/zoning/rap/rap/index.html)

69   Airamé, S., Dugan, J.E., Lafferty, K.D., Leslie, H., McArdle, D.A. and Warner, R.R. (2003). Applying ecological criteria to marine reserve design: a case study from the California Channel Islands. *Ecological Applications* **13**, S170-S184.

70   Vincent, C.A., Atkins, S.M., Lumb, C.M, Golding, N., Lieberknecht, L.M. and Webster, M. (2004). *Marine Nature Conservation and sustainable development – the Irish Sea Pilot.* Report to Defra by the Joint Nature Conservation Committee, Peterborough.

71   Roberts, C.M., Andelman, S., Branch, G., Bustamante, R., Castilla, J.C., Dugan, J., Halpern, B., Lafferty, K.D., Leslie, H., Lubchenco, J., McArdle, D., Possingham, H., Ruckelhaus, M., and Warner, R.R. (2003a). Ecological criteria for evaluating candidate sites for marine reserves. *Ecological Applications* **13**, S199-S214.
       Roberts, C.M., Branch, G., Bustamante, R., Castilla, J.C., Dugan, J., Halpern, B., Lafferty, K.D., Leslie, H., Lubchenco, J., McArdle, H., Ruckelhaus, M., and Warner, R.R. (2003b) Application of ecological criteria in selecting marine reserves and developing reserve networks. *Ecological Applications* **13**, S215-S228.

72   Roberts, C.R. and Mason, L. (2004). *Design of marine protected area networks in the North Sea and Irish Sea. In press.*

73   *Ibid.*

74   Vincent, *et al* (2004).

75   Balmford, A., Gravestock, P., Hockley, N., McClean, C.J. and Roberts, C.M. (2004). The worldwide costs of marine protected areas. *Proceedings of the National Academy of Sciences* **101**: 9694-9697.

76   Gravestock, P. (2003) *Towards a better understanding of the income requirements of marine protected areas.* MSc Thesis, Cranfield University, Silsoe.

77   Costanza, R., d'Arge, R., deGroot, R., Farber, S., Grasso, M., Hannon, B., Limburg, K., Naeem, S., O'Neill, R.V., Paruelo, J., Raskin, R.G. and Sutton, P. (1997). The value of the world's ecosystem services and natural capital. *Nature* **387**: 253-260. Original figures were US$ 12 billion, US$14 billion, US$15 billion and US$30 billion respectively. At October 2004 conversion rates, 1US$ = approx. 0.54 UK£.

78   Prime Minister's Strategy Unit. (2004).

79   Defra (2003). Sea Fisheries Statistics. Available on the Defra website: http://statistics.defra.gov.uk/esg/publications/fishstat/uksfs03.pdf.

80   Roberts and Mason (2004).

81   *Ibid.*
      Using a deflationary index of 2.6% per year based on official HM Treasury figures.

82   Murawski *et al* (2000).

83   Roberts, C.M., Bohnsack, J.A., Gell, F.R., Hawkins, J.P. and Goodridge, R. (2001). Effects of marine reserves on adjacent fisheries. *Science* **294**, 1920-1923.

84   Website of the Queensland Rural Adjustment Authority, www.qraa.qld.gov.au

85   Horwood, J. W., Nichols, J. H., Milligan, S. (1998). Evaluation of closed areas for fish stock conservation. *Journal of Applied Ecology* **35**, 893-903.

86   Kaiser, M.J. (2003). Are closed areas the solution? *Marine Conservation* **5**, 11.

87   Dinmore, T.A., Duplisea, D.E., Rackham, B.D., Maxwell, D.L. and Jennings, S. (2003). Impact of large-scale area closure patterns of fishing disturbance and consequences or benthic communities. *ICES Journal of marine Science*, **60**: 371-380.

88   Murawski *et al* (2000).

89   Vincent *et al* (2004).

90   Coleman, F.C., Baker P.B. and Koenig C.C. (2004). A review of Gulf of Mexico Protected Areas: Success, Failures, and Lessons Learned. *Fisheries Management* **29** (2).

91   John, D. (1996). *Protecting a profitable paradise: the national ocean service leads multi-agency planning in the Florida Keys.* National Academy of Public Administration, Washington DC

92   Thompson, L., Jago, B., Fernandes, L. and Day, J. (2004). *Barriers to communication – how these critical aspects were addressed during the public participation for the rezoning of the Great Barrier Reef Marine Park.* Available on the Great barrier reef marine park authority website: http://www.gbrmpa.gov.au/corp_site/management/zoning/documents/Breaking_through_the_barriers_15April04%20FINAL.pdf.

93   Coffey and Richartz (2004).

94   Murawski *et al* (2000).

CHAPTER NINE

1   Website of the European Commission
     http://europa.eu.int/comm/fisheries/pcp/faq2_en.htm.

2   Website of Danish Research Institute of Food economics
     http://www.foi.dk/Publikationer/wp/1998-99wp/wp199919.pdf

3   Goodlad, J. (1998). Sectoral quota management: fisheries management by fish producer
     organisation. In: *The Politics of Fishing*, Ed. T.S. Gray. Macmillian Press, London

4   Prime Minister's Strategy Unit (2003a). *Fisherman's incentives and policy. Analytical
     paper produced to support the Net.* Benefits Report. Available on cabinet office website
     http://www.strategy-unit.gov.uk
     Project No96/96C. *Compliance with fish regulation.* Available on E.C. website
     http://europa.eu.int/comm/fisheries/doc_et_publ/liste_publi/studies/biological/1309RO3B
     96090.pdf

5   Clover, C. (2004).

6   House of Lords Select Committee on the European Union (2003). *Progress of reform of
     the common fisheries policy session 2002–03 25th report.* HL Paper 109.

7   Prime Minister's Strategy Unit (2003a).

8   Prime Minister's Strategy Unit (2003). *UK Fisheries Industry – Current Situation Analysis.*

9   Figures supplied by Defra

10   Prime Minister's Strategy Unit (2004).

11   *Ibid.*

12   *Ibid.*

13   The Royal Society of Edinburgh (2004).

14   Website of DG Fisheries.
     http://europa.eu.int/comm/fisheries/reform/conservation_en.htm.

15   Charles, A. (2001). *Sustainable Fishery Systems. Fish and Aquatic Resources Series 5.*
     Blackwell Sciences. Oxford.

16   The Royal Society of Edinburgh (2004).

17   Council Regulation (EC) No 1954/2003 of 4 November 2003 on the management of the
     fishing effort relating to certain Community fishing areas and resources and modifying
     Regulation (EC) No 2847/93 and repealing Regulations (EC) No 685/95 and (EC) No
     2027/95. *Official Journal of the European Union* **L289/1**, 7.11.2003.

18   Diagram from DG Fisheries website, available at:
     www.europa.eu.int/comm/fisheries/news_corner/pictures/westwaters_10_03.jpg

19   DG Fisheries (2003) Press Release – Outcome of Agriculture/Fisheries Council of
     October 2003. On DG fisheries website at
     http://europa.eu.int/comm/fisheries/news_corner/press/inf03_44_en.htm.

20   The Royal Society of Edinburgh (2004).

21   Pascoe, S., Tingley, D. and Mardle, S. (2002). *Appraisal of Alternative Policy Instruments
     to Regulate Fishing Capacity.* Centre for the Economics and Management of Aquatic
     Resources (CEMARE) University of Portsmouth. Final Report ER0102/6.

22   Chuenpagdee, R., Morgan, L.E., Maxwell, S.M., Norse, E.A. Pauly, D. (2003). Shifting
     gears: assessing collateral impacts of fishing methods in US waters. *Frontiers in Ecology
     and Environment,* **1** (10), 517-524.

23   CEFAS (2003). *A study on the consequences of technological innovation in the capture fishing industry and the likely effects upon environmental impacts.* CEFAS Contract Report C1823. Available at; http://www.rcep.org.uk/fisheries.htm.

24   Available on Defra website at http://www.defra.gov.uk/environment/consult/copcar/chap6.htm

25   Symes, D. and Ridgway, S. (2003). *Inshore fisheries regulation and management in Scotland; Meeting the challenges of Environmental Integration.* Scottish Natural Heritage Commissioned Report F02AA405.

26   English Nature's website http://www.english-nature.org.uk/about/teams/NewsDetails.asp?Id=9&NewsId=51

27   National Audit Office (2003). *Fisheries enforcement in England.* Report HC563 Session 2002-2003: 3 April 2003

28   Personal communication, Professor Chris Frid, 2004.

29   Symes and Ridgway (2003).

30   Evidence from CEFAS, November 2003.

31   Lack, M., Short, K. and Willock, A. (2003). *Managing risk and uncertainty in deep-sea fisheries: lessons from Orange Roughy.* TRAFFIC Oceania and WWF Australia.

32   IUCN (2004). *High seas bottom fisheries and their impact on the biodiversity of vulnerable deep-sea ecosystems: summary findings.*

33   *Ibid.*

34   Roberts, C.M. (2002). Deep impact: the rising toll of fishing in the deep sea. *Trends in Ecology and Evolution.*

35   IUCN website: http://www.iucn.org/themes/marine/pdf/DeepSeaOptions.pdf Convention on Biological Diversity website: http://www.biodiv.org/doc/meetings/cop/cop-07/information/cop-07-inf-25-en.pdf

36   CEFAS (2003).

37   ASCOBANS (2000). *Resolution on incidental take of small cetaceans.* Annex 9c of Proceedings of Third Meeting of the Parties to ASCOBANS. ASCOBANS, Bonn, Germany.

38   COM (2003) 451 final. *Proposal for a regulation laying down measures concerning incidental catches of cetaceans in fisheries and amending regulation* (EC) 88/98.

39   House of Commons Environment, Food and Rural Affairs Committee (2004). *Caught in the net: by-catch of dolphins and porpoises off the UK coast Third Report of Session 2003–04.* HC 88 (including HC 1244 i and ii Session 2002-03).

40   Prime Minister's Strategy Unit (2004).

41   House of Commons Committee of Public Accounts (2003). *Fisheries enforcement in England Forty-third Report of Session 2002–03* HC725

42   Science and Technology Options Assessment of the European Parliament (1999). *The problem of discards in fisheries.* STOA report No. EP/IV/B/STOA/98/17/01.

43   Evidence from the Joint Nature Conservation Committee, May 2003.

44   Reeves, S.A, Furness, R.W. (2002). *Net loss – seabirds gain? Implications for seabirds scavenging discards in the North Sea.* The RSPB, Sandy, UK.

45   Votier, S.C., Furness, R.W., Bearhop, S., Crane, J.E., Caldow, R.W.G., Catry, P., Ensor, K., Hamer, K.C., Hudon, A.V., Kalmbach, E., Klomp, N.I., Pfeiffer, S., Phillips, R.A., Prieto, I. and Thompson, D.R. (2004). Changes in fisheries discard rates and seabird communities. *Nature,* **427** (6976), 727-731.

46   COM(2002)656 Final.

47   UN FAO Fisheries Circular No. 928 FIIU/C928 *A Study of the Options for Utilisation of Bycatch and Discards from Marine Capture Fisheries.*

48   House of Commons Committee of Public Accounts (2003).

49   *Ibid.*

50   Evidence to the Commission from SeaFish, December 2003.

51   Defra (2004). *Letter advising of Action to strengthen fisheries monitoring, control and surveillance in the United Kingdom.* Available on the Defra website at; http://www.defra.gov.uk/fish/fish-controls/letter.htm.

52   COM (2002). 181 final. *Communication from the commission on the reform of the of the Common Fisheries policy.*

53   Evidence to the Commission from WWF, November 2003.

54   Evidence to the Commission from the Joint Nature Conservation Committee, October 2003.

55   Watson, J. and Martin, A. (2001). *Economic survey of the UK fishing fleet.* Published by Seafish. Edinburgh, UK. ISBN 0 0903941 75 9.

56   Pauly and MacLean (2003) *In a perfect ocean. Island Press.*

57   Personal communication from Dr John Watterson, Manager of the National Atmospheric Emissions Inventory, AEA Technology Environment, November 2004.

58   European Union website, http://europa.eu.int/rapid/pressReleasesAction.dc?reference=IP/02/ 1719&format=HTML&aged=0&language=EN&guiLanguage=en

59   *Ibid.*

CHAPTER TEN

1 House of Commons Environment, Food and Rural Affairs Committee (EFRA) (2004). *Marine Environment.* House of Commons, London.

2 WS Atkins (2004). *ICZM in the UK: A Stocktake.* Final Report. HMSO, London. Based on figure supplied by MECU, Defra.

3 Defra (2004).

4 European Council (2002). *Recommendation of the European Parliament and of the Council of 30 May 2002 concerning the in implementation of Integrated Coastal Zone Management.* (2002/413/EC). European Council, Brussels.

5 English Nature (2004). *Our coasts and seas.* English Nature, Peterborough, UK.

6 OSPAR (2002). *Ministerial Declaration of the Fifth International Conference on the Protection of the North Sea.* Bergen, Norway 20-21 March 2002.

7 European Commission (2004a). *The European Marine Strategy – policy document.* SGO(2)04/4/1. European Commission, Brussels.

8 *Ibid.*

9 National Oceans Office (2003). *Oceans Policy: Principles and Processes.* National Oceans Office, Hobart, Australia.

10 Pew Oceans Commission (2003).

11 National Oceans Office (2003).

12 National Oceans Office (2003a). *Draft South-east regional marine plan.* National Oceans Office, Hobart, Australia.

13 EFRA (2004).

14 Vincent *et al.* (2004).

15 Defra (2004).

16 Twenty-third Report, paragraph 9.62.

17 Defra (2004a).

18 WS Atkins (2004).

19 Vincent *et al.* (2004).

20 European Commission (2004a).

21 Scottish Executive Environment Group (2004). *Developing a strategic framework for Scotland's marine environment. A consultation.* Scottish Executive, Edinburgh.

22 Defra (2004).

23 Defra (2004b). Specification – programme of work. Research into Marine Spatial Planning, Defra, London.

24 Twenty-third Report, paragraph 9.62.

25 Royal Commission visit to Washington DC, May 2004.

26 Prime Minister's Strategy Unit (2004).

27 Royal Society of Edinburgh. (2004).

28 Vincent *et al.* (2004).

29 *Ibid.*

30 Osmundsen, P. and Tveterås, R. (2003). Decommissioning of Petroleum Installations – major policy issues. *Energy Policy* **31**, 1579-1588.

31 Watson, T. (2001). *The Environment and the Decommissioning of Offshore Installations.* MSc Dissertation, University of Greenwich, London.

32 Defra (2002).

33 Scottish Executive Environment Group (2004).

34   Defra (2002).

35   Pew Oceans Commission (2003).

36   Coffey and Richartz (2004).

37   Scottish Executive Environment Group (2004).

38   Pew Oceans Commission (2003).

39   Magnuson-Stevens Fishery Conservation and Management Act Public Law 94-265 As amended through October 11, 1996. Article 104-297.

40   Pew Oceans Commission (2002). *Managing marine fisheries in the United States. In: Proceedings of the Pew Oceans Commission workshop on marine fishery management, Seattle, Washington 18-19 July 2001.* Pew Oceans Commission, Arlington, Virginia

41   US Commission on Ocean Policy (2004). *Preliminary Report. Governors' Draft.* Washington, DC.

42   National Academy of Public Administration (NAPA) (2002). *Courts, Congress and Constituencies: Managing fisheries by default.* NAPA, Washington DC.

43   US Commission on Ocean Policy (2004).

44   European Commission website, press release 20.10.03 available at www.europa.eu.int/comm/fisheries/new_corner/press/inf03_45_en.htm

45   European Commission (2003). COM (2003) 607 final. *Proposal for a Council decision establishing Regional Advisory Councils under the Common Fisheries Policy.* Article 2 of Decision. European Commission, Brussels.

46   Prime Minister's Strategy Unit (2004).

47   The Royal Society of Edinburgh (2004).

48   Evidence submitted by the Local Government Association Coastal Special Interest group, May 2003.

49   The Royal Society of Edinburgh (2004).

50   Prime Minister's Strategy Unit (2004).

51   US Commission on Ocean Policy (2004).

52   Defra (2002).

53   Defra (2004a).

54   Twenty-first Report, paragraph 7.20.

55   Defra (2004c). *The Government's response to its Seas of Change consultation to help deliver our vision for the marine environment.* Defra, London.

56   GM nation website, www.gmnation.org.uk

57   Hunt, J. and Wynne, B. (2000). *Forms for Dialogue: Developing legitimate authority through communication and consultation. A contract report for Nirex.* Centre for the study of Environmental Change, Lancaster University.

58   Prime Minister's Strategy Unit (2004).

59   Thompson, L. Jago, B., Fernandes, D. and Day, J. (2004). *Barriers to communication – how these critical aspects were addressed during the public participation for the rezoning of the Great Barrier Reef Marine Park.* http://www.gbrmpa.gov.au/corp_site/ management/zoning/documents/Breaking_through_the_barriers_15April04%20FINAL.pdf

60   Kelleher, G. (2000). The development and establishment of coral reef marine protected areas. *Proceedings 9th International Coral Reef Symposium,* **2**, 609-613.

61   GM nation website, www.gmnation.org.uk

62   US Ocean Policy Commission (2004).

63   Clark, B. (2002). *Good Fish Guide.* Marine Conservation Society, Ross on Wye.

64   Clover, C. (2004).

65   Marine Stewardship Council website, www.msc.org

66   Evidence submitted by SeaFish, July 2003.

67   Clark, B. (2002).

68   Personal communication from D. Gregory and A. Mallison. Marks & Spencer, December 2003.

69   European Commission (2001). *Green Paper on the future of the Common Fisheries Policy.* European Commission, Brussels.

70   Twenty-first Report, paragraphs 7.19-7.23, box 7A.

CHAPTER ELEVEN

1   Muraswki *et al* (2000).

2   Pew Oceans Commission (2003).

3   *Ibid.*

4   Prime Minister's Strategy Unit (2004).

5   House of Commons Committee of Public Accounts (2003). *Fisheries enforcement in England*. Forty-third Report of Session 2002–03 HC725. House of Commons, London.

6   IUCN (2004).

7   US Commission on Ocean Policy (2004).

Subject Index

INDEX

Note: Index entries under UK and EU have been kept to a minimum and readers are advised to seek more specific topics. Page numbers in **bold italics** refer to definitions.

Printed in the UK for The Stationery Office Limited
on behalf of the Controller of Her Majesty's Stationery Office
173179 11/04